S0-DCJ-485

Liquefied Petroleum Gases Handbook

Fourth Edition

Based on the 1995 edition of NFPA 58,
Standard for the Storage and Handling of Liquefied Petroleum Gases

Edited by

Theodore C. Lemoff, P.E.

Senior Gases Engineer

NFPA®

National Fire Protection Association
Quincy, Massachusetts

This fourth edition of the *Liquefied Petroleum Gases Handbook* builds upon the highly successful third edition to provide an essential reference for everyone associated with the production, distribution, or use of liquefied petroleum gases. This edition includes expanded commentary on the newly required two-stage pressure regulation, including a recent photograph of the new integral two-stage regulator, as well as a useful supplementary guide for conducting fire safety analysis in large capacity installations. In addition, the handbook explains the application of the standard to marine terminals, pipeline terminals, natural gas processing plants, refineries, and petrochemical plants. The handbook explains the requirements found in NFPA 58, *Standard for the Storage and Handling of Liquefied Petroleum Gases*, especially changes to the standard from previous editions. It also serves as a helpful guide in applying the standard's provisions.

All NFPA codes and standards are processed in accordance with NFPA's *Regulations Governing Committee Projects*. The content of this handbook is the opinion of the author(s), recognized experts in the field of LP-Gases. It is not, however, processed in accordance with the NFPA *Regulations Governing Committee Projects*, and therefore shall not be considered to be, nor relied upon as, a formal interpretation of the meaning or intent of any specific provision or provisions of NFPA codes or standards. The language contained in NFPA codes and standards, rather than the content of this handbook, represents the official position of the NFPA.

The handbook contains the complete text of the standard, along with explanatory commentary, which is integrated between requirements. Commentary text answers questions frequently asked about the standard and includes: intent and interpretations; historical perspective on provisions; application of requirements; illustrations of actual LP-Gas equipment installations; many new photographs of equipment addressed in the standard; new requirements to permit both factory-assembled and field-assembled risers for polyethylene pipe and tubing; and a step-by-step guide for conducting fire safety analysis in installations over 4,000 gallon capacity.

This handbook is a must for all who design, install, inspect, approve, or operate LP-Gas facilities.

Project Manager: Jennifer Evans
Project Editor: Debbie Liehs
Composition: Kathleen Barber, Claire Ross
Illustrations: George Nichols
Art Coordinator: Debbie Liehs
Cover Design: Boston Marketing Services Group, Inc.
Interior Design: Joyce Weston

NFPA 58HB95
ISBN: 0-87765-403-4
Library of Congress No.: 92-60521
Printed in the United States of America

98 97 96 95 4 3 2 1

Dedication

This handbook is dedicated to my son, Brian E. Lemoff, Ph.D., in honor of completing the academic portion of his life by graduating from the physics department at Stanford University. His accomplishments are significant, and his father is sure that they will be eclipsed by his work in the future.

Contents

Standard for the Storage and Handling of Liquefied Petroleum Gases, NFPA 58

Supplement

Preface

The fourth edition of the *Liquefied Petroleum Gases Handbook* continues the tradition begun by Wilbur L. Walls, editor of the first edition, to provide commentary that brings the cold, legalistic standard to life. The standard must be written in a style that is technically clear, useable, enforceable, and adoptable, without the nonmandatory guidance that the Technical Committee would like to include.

The purpose of this handbook is to assist you, the user or enforcer of NFPA 58, *Standard for the Storage and Handling of Liquefied Petroleum Gases.* The commentary provides guidance, recommendations, and common practices of the propane distribution industry.

The 1995 edition contains several important changes, including:

- Creation of a foreword to which the general advisory material from Chapter 1 has been relocated
- Adding LP-Gas used as a closed cycle refrigerant to the nonapplication list of NFPA 58
- Marking requirements for unodorized LP-Gas
- Marking requirements to assist emergency responders
- New requirements permitting risers for terminating polyethylene piping aboveground
- Protection requirements for areas of very heavy snowfall
- Overpressure protection of building piping systems, accomplished by mandating two-stage pressure regulation in most buildings
- Revisions to the dispensing section
- Revisions to Chapter 10, Marine Shipping and Receiving

A handbook of this type is never complete. In order to stay current, it must change as technology changes. Users may find that problems they encounter may not be addressed in the commentary, or that the commentary appears incomplete when applied to a real problem. The editor invites suggestions from any reader for improvements to the next edition.

About the Editor

After graduating from the City College of New York with a Bachelor of Engineering (Chemical) degree, Theodore C. Lemoff was employed in the chemical and petrochemical industries by the Proctor and Gamble Company, Sun Chemical Corporation, and Badger Engineers, Inc. Since 1985, when he joined NFPA, he has served as the secretary to the Technical Committee on Liquefied Petroleum Gases and as staff liaison to the other NFPA technical committees that are responsible for standards on flammable gases.

He represents NFPA on the ANSI *Code for Pressure Piping* (B31), *Safety in Cutting and Welding* (Z49), and *Gas Utilization Equipment* (Z21 and Z83) projects and on the National Propane Gas Association Technology and Standards and Safety Committees.

Mr. Lemoff is a registered Professional Engineer in the Commonwealth of Massachusetts and is a member of the American Institute of Chemical Engineers and the Society of Fire Protection Engineers.

Acknowledgments

A book that compiles the history, intent, and application of an NFPA standard, as the *Liquefied Petroleum Gases Handbook* does, is the work of many contributors, many of whom are members of the Technical Committee on Liquefied Petroleum Gases, and all of whom work with LP-Gas daily. The foundation for this fourth edition was laid by the three authors who contributed to the first edition: Wilbur L. Walls, who represented, at the time, the expertise of NFPA on flammable gases; Walter H. Johnson, who possessed a broad knowledge of the LP-Gas industry from the industry association viewpoint; and H. Emmerson Thomas, who offered his perspective as one of the founders of the propane industry in the United States.

This fourth edition benefits from the contributions of others who merit recognition. Readers will note that references to specific fires and accidents have been added, and that the statistical references have been revised. This was the result of a review by Dr. John Hall of the NFPA Fire Analysis Department, and the editor is in his debt for the improvements resulting from his comments. Readers interested in specific incidents will appreciate the references.

New artwork that explains the materials, equipment, and installations covered in the standard was contributed by Brian Clayton of Ace Gas, William Young of Plant Systems, Inc., Perfection Corporation, R. W. Lyle Company, and again by Fisher Corporation, Engineered Controls Inc., and Sherwood Selpac. Their cooperation is very much appreciated. The editor invites anyone who believes that the commentary could be improved by photos or drawings, to submit them for consideration in future editions.

The handbook could not exist without the work done by the Liquefied Petroleum Gases Committee that writes NFPA 58. They worked long and hard to thoroughly address the 188 proposals and 84 comments received on

NFPA 58. I thank each of them for the long hours spent preparing for meetings, attending meetings, and carefully considering the follow-up ballots. I also must recognize the good work of Mr. Sam McTier, who personally submitted 45 proposals and 32 comments to provide clarifications, improve references, and correct the minor errors that inevitably creep into a document as complex as NFPA 58. I know that Sam spent many long hours checking references and improving wording, and the users of NFPA 58 are in his debt for his selfless work.

In addition to the LP-Gas Committee of NFPA, a number of other individuals provided expertise in developing the fourth edition. I especially appreciate the efforts of the members of the Technology and Standards Committee of the National Propane Gas Association, who provided input on the real world. Without their ongoing assistance, this handbook would not be as informative and practical as it is. I would also like to thank Wilbur Walls, NFPA Gases Engineer emeritus, who went out of his way to assist me when I joined NFPA, and took over responsibility for NFPA 58 and the other gases standards. While Bill was already retired when I began at NFPA, he spent several days with me, and was always available over the phone for any questions.

I also appreciate the efforts of the NFPA staff members who attended to the countless details that went into preparing this handbook. In particular, I'd like to thank Debbie Liehs, Project Editor and Art Coordinator; and Kathleen Barber and Claire Ross, Composition.

Finally, I wish to thank my wife, Sharon, for her patience and understanding during the many evenings I worked on this handbook and other projects at home.

Theodore C. Lemoff, P.E.
Editor

History of NFPA 58

NFPA standards concerned with gases date from 1900, only four years after the establishment of NFPA itself. The first standards were concerned with acetylene, which was actually used in those days as a household cooking, lighting, and heating fuel, and with manufactured, or "city," gas derived from coal and oil.

These standards were developed by the NFPA Technical Committee on Gases. By 1924, the use of liquefied petroleum gas (LP-Gas), primarily as a cooking and heating fuel in rural areas, had become common, and the need for a national fire safety standard was recognized. The LP-Gas for these systems was stored in compressed gas cylinders. Because of their cylindrical shape, the LP-Gas they contained was widely referred to simply as "bottled gas," an identification rather lacking in specificity (acetylene was also a "bottled gas") but one still used today.

In 1927, the Committee on Gases secured NFPA approval of the first NFPA standard on LP-Gas, a four-page document whose title, "Regulations for the Installation and Operation of Compressed Gas Systems Other than Acetylene for Lighting and Heating," comprised a fair proportion of the text of the standard itself. The standard covered only systems in which the LP-Gas was stored in cylinders fabricated to regulations of the U.S. Interstate Commerce Commission (ICC), today known as the U.S. Department of Transportation.

In those days, NFPA published its standards only in the Proceedings of the Annual Meeting at which the standards were adopted. To make them more available to users, the National Board of Fire Underwriters (NBFU), now known as the American Insurance Association, obtained NFPA permission to publish many of them in pamphlet form. These were identified as standards of the NBFU "as recommended by the NFPA." In those days, NFPA

An early magazine advertisement for acetylene lighting.

did not identify many of its standards by number. However, the NBFU did, and the standard was thus designated NBFU 52. Amended editions of the standard were adopted in 1928, 1933, and 1937.

Because the LP-Gas "bottles" had to be refilled—in those days, only at a plant built for that purpose—which required the use of a much larger container for plant storage, new specifications had to be developed. In addition, major consumers found a container larger than an ICC cylinder to be advantageous. These containers required different design and siting criteria.

In 1931, NFPA tentatively adopted "Regulations for the Design, Installation and Construction of Containers and Pertinent Equipment for the Storage and Handling of Liquefied Petroleum Gases," which was officially adopted at the next NFPA meeting in 1932. This standard covered the larger containers, which are the ASME containers of today. Also published by the NBFU, under the designation NBFU 58, this standard had 14 pages, 10 more than NBFU 52.

At the 1931 Annual Meeting, Committee Chairman Harry E. Newell of the NBFU noted in his report that the proposed standard had been prepared by a joint subcommittee of the Committee on Gases and the Committee on Flammable Liquids. He attributed this to the fact that "it was difficult to decide whether liquefied petroleum gases were flammable liquids or gases." Even today, confusion exists on this score. Many municipal ordinances, undoubtedly of considerable vintage, equate them, and many communities have failed to adopt LP-Gas regulations under the erroneous impression that their flammable liquid regulations apply to LP-Gas.

For many years, the Committee on Gases was composed of 11 to 20 members. In 1932, there were 15 members, representing the Associated Factory Mutual Fire Insurance Companies, American Gas Association, Board of Fire Underwriters of Allegheny County, Western Factory Insurance Association, Conference of Special Risk Underwriters, Manufacturing Chemists Association, Boston Board of Fire Underwriters, International Acetylene Association, Underwriters Laboratories Inc., Compressed Gas Manufacturers Association, American Petroleum Institute, Railway Fire Protection Association, and the U.S. Bureau of Standards, as well as the NBFU. With 6 of the 15 members, including the chairman, representing the insurance industry, the standard could be presumed to be rather conservative, and this lack of balance of interests could not exist under later NFPA procedures.

The extreme versatility of "gas in a bottle" soon led to more and more complex uses, and the Committee on Gases was hard-pressed to keep up.

Amended editions of the 1931 standard were adopted in 1934, 1937, 1938, and 1939.

In 1935, a standard on LP-Gas cargo vehicles, whose use had become widespread in the early 1930s, was adopted and published by NBFU as NBFU 59. In 1937, a standard was adopted to regulate the use of LP-Gas as a fuel to power vehicles, a practice that had become prevalent by the mid-1930s. In 1938, however, this became a part of the ASME Code container standard (NBFU 58).

By now, it was apparent that the various standards contained considerable duplicate material and could be combined into a single standard. This was done in 1940. The resulting 47-page standard (in the NBFU 58 pamphlet version) combined the 1937 edition of NBFU 52 and the 1939 editions of NBFU 58 and 59. This single standard also replaced the *Liquefied Petroleum Gas Code*, which was adopted in 1937 in response to a request by regulatory and insurance interests to provide container siting and basic fabrication criteria that were not as detailed as those found in the existing standards. This code ultimately proved inadequate from the industry's point of view, and it was withdrawn when the combined standard was introduced. The 1943 edition of the standard was the first edition to be designated NFPA 58. The first pamphlet edition of NFPA 58 was published in 1950.

The dates of each edition of NFPA 58 are given in the "Origin and Development" section of each edition. To date, there have been 25 editions since the 1940 edition, an average of about one every two years. From 1950 to 1961, in fact, a new edition was adopted each year. This is a very high rate, and it placed an impossible burden upon its use as a public safety regulatory instrument. Since 1961, new editions have been adopted about every three years.

The Committee on Gases itself developed NFPA 58 until 1956. By that time, the number and variety of NFPA standards covering various gases had grown so large that the size of the committee was becoming difficult to manage. The solution was to establish a number of smaller committees, known as sectional committees, composed of experts on the different gas applications. The developmental and interpretative responsibilities for NFPA 58 were thus assigned to the Sectional Committee on Liquefied Petroleum Gases. However, the sectional committee could only recommend amendments, formal interpretations, or Tentative Interim Amendments for adoption by the Committee on Gases.

Because the sectional committee lacked authority, its membership essentially duplicated that of the Committee on Gases, and the purpose of the sectional committee was defeated. It was also becoming evident that the Committee on Gases seldom overturned a sectional committee recommendation.

In 1966, the sectional committees—there were four at the time—were organized into three full-fledged technical committees, and the Committee on Gases ceased to exist. NFPA 58 thus became the complete responsibility of the Technical Committee on Liquefied Petroleum Gas.

From 1940 to 1969, NFPA 58 consisted of a chapter entitled "Basic Rules," followed by a number of chapters entitled "Divisions," and by appendices. For example, the 1969 edition was as follows:

Introduction
Basic Rules

Division I	Cylinder Systems
Division II	Systems Utilizing Containers Other than ICC
Division III	Truck Transportation of Liquefied Petroleum Gas
Division IV	Liquefied Petroleum Gas as a Motor Fuel
Division V	Storage of Containers Awaiting Use or Resale
Division VI	LP-Gas Exhangeable (Readily Portable Container) System Installations on Travel Trailers, Camper Trailers, Self-Propelled Campers, or Mobile Homes
Division VII	LP-Gas System Installations on Commercial Vehicles and Certain Self-Propelled or Trailer Type Mobile Living Units
Division VIII	Liquefied Petroleum Gas Service Stations
Appendices A-G	Covering Design of Relief Valves, Container Filling Volumes, a Pictorial Presentation of Container Siting, and Tubing Wall Thicknesses

The "Introduction" and "Basic Rules" included all material common to all applications, such as scope statements, retroactivity, definitions, odorization, approval requirements, container fabrication, siting, design criteria for piping, valves, hoses and fittings, filling levels, and ignition source controls. The eight divisions amplified and sometimes modified the basic provisions for specific applications, some on the basis of kinds of containers and others on the basis of the environment in which the system was used.

As a new application was developed, it was essentially treated as a package and added as a new division. This was easier to do and allowed for the timely inclusion in the standard of new applications. The problem with this approach was that similar hazards were being treated differently, and the basic rules were becoming less and less basic.

In the late 1960s, the Technology and Standards Committee of the National Propane Gas Association (NPGA) undertook the development of a format that would provide consistency in hazard evaluation, would reduce the extent to which particular interests would have to study the entire standard, and would make the format more like that used in other NFPA standards.

In the current format, Chapter 1, General Provisions, includes only material truly relevant to all applications.

Chapters 2 and 3 address the fundamental breakdown of interest that experience has shown exists between manufacturers of equipment and those who assemble this equipment into systems and actually install them. Fire experience clearly reveals that incidents are nearly always due to failure to comply with one or more provisions and that many of these failures are the result of confusion between manufacturers and installers as to who should do what. To help reduce this confusion, Chapter 2 is aimed at the equipment manufacturer and Chapter 3 at the installer.

The same experience has also shown that the great majority of nonappliance-related accidents occur when LP-Gas liquid is being transferred from one container to another. An essential aspect of such operations is the presence and performance of a person or persons conducting the transfer operation: An accident generally involves human behavior, as well as equipment. The operational factors and their relation to equipment factors have been put together in Chapter 4.

Chapter 5, Storage of Portable Containers Awaiting Use, Resale, or Exchange, and Chapter 6, Vehicular Transportation of LP-Gas, retain the old format of Divisions V and III.

Chapter 7 was essentially new in the 1972 edition. Earlier editions had not considered the role of the structural behavior of a room or building in an explosion.

Chapter 8 was created in the 1992 edition by relocating the coverage of engine fuel systems from Chapter 3 to this chapter.

Drawn largely from NFPA 59 and 59A, Chapter 9 was added in the 1989 edition to provide requirements for the new coverage of refrigerated storage containers.

Coverage of marine shipping and receiving was relocated from Chapter 4 to create Chapter 10 in the 1992 edition.

While the 1972 edition went a long way toward bringing the format of NFPA 58 into line with other NFPA standards, it still did not properly segregate mandatory and nonmandatory provisions, and it used a different paragraph numbering system. These differences were corrected in the 1983 edition.

Experience with the format of the 1972 edition and with subsequent editions indicates that it has led to improved comprehension and application. However, it is more difficult to review existing provisions to see if a new application is in fact already covered than it is to start from scratch. Furthermore, there is always pressure to include all provisions applicable to a new and "hot" application in one place. This temptation must be resisted lest the standard evolve into the conflicting and cumbersome document it once was.

The administration of the committees responsible for NFPA 58 has been remarkably stable over its history. Harry E. Newell of the NBFU was chairman from 1932 until 1956, when he retired from NBFU. As an indication of the respect the committee and NFPA had for him, the position of honorary chairman was created for him, the only time this position has existed. He served in this capacity until 1958.

Newell's dedication was even more remarkable in that he also performed the chores of committee secretary, albeit anonymously. It wasn't until 1954 that NFPA was able to assign a staff member, Clark F. Jones, as committee secretary. Jones served in this post until his untimely death in 1962. He was succeeded by Wilbur L. Walls in September 1962. Walls was the committee secretary until his retirement from NFPA in 1984. In May 1985, Theodore C. Lemoff became the NFPA staff secretary.

Newell was succeeded as chairman of the Committee on Gases in 1956 by Franklin R. Fetherston. Initially representing the Liquefied Petroleum Gas Association (now the National Propane Gas Association), he later represented the Compressed Gas Association until he retired in 1966. During Fetherston's tenure, the Sectional Committee on Liquefied Petroleum Gases was chaired by Harold L. DeCamp of the Fire Insurance Rating Organization of New Jersey from 1956 to 1964, and by Myron Snell of the Hartford Accident and Indemnity Company from 1964 to 1966.

Hugh V. Keepers of the Fire Prevention and Engineering Bureau of Texas became the first chairman of the current committee in 1966 and served for 10 years until his retirement. The current chairman, Connor L. Adams, a safety consultant and former building official with the City of Miami, Florida, took over the reins in 1976.

Standard and Commentary for NFPA 58, *Storage and Handling of Liquefied Petroleum Gases*

1995 Edition

The text and illustrations that make up the commentary on the various sections of NFPA 58 are printed in black. The text of the standard itself is printed in blue.

The Formal Interpretations included in this handbook were issued as a result of questions raised on specific issues of the standard. They apply to all previous and subsequent editions in which the text remains substantially unchanged. Formal Interpretations are not part of the standard and therefore are printed in a shaded blue box.

Paragraphs that begin with the letter "A" are extracted from Appendix A of the standard. Appendix A material is not mandatory. It is designed to help users apply the provisions of the standard. In this handbook, material from Appendix A is integrated with the text, so that it follows the paragraph it explains. An asterisk (*) following a paragraph number indicates that explanatory material from Appendix A will follow.

Foreword

General Properties of LP-Gas. LP-Gases, as defined in this standard (*see 1-2.1*), are gases at normal room temperatures and atmospheric pressure. They liquefy under moderate pressure, readily vaporizing upon release of the pressure. It is this property that permits transportation and storage of LP-Gases in concentrated liquid form, although they normally are used in vapor form. The potential fire hazard of LP-Gas vapor is comparable to that of natural or manufactured gas, except that LP-Gas vapors are heavier than air. The ranges of flammability are considerably narrower and lower than those of natural or manufactured gas. For example, the lower flammable limits of the more commonly used LP-Gases are 2.15 percent for propane and 1.55 percent for butane. These figures represent volumetric percentages of gas in gas-air mixtures.

The boiling point of pure butane is 31°F (–0.6°C); and pure propane has a boiling point of –44°F (–42°C). Both products are liquids at atmospheric pressure at temperatures lower than their boiling points. Vaporization is rapid at temperatures above their boiling points; thus, liquid propane normally does not present a flammable liquid hazard. For additional information on these and other properties of the principal LP-Gases, see Appendix B.

Federal Regulations. Regulations of the U.S. Department of Transportation (DOT) are referenced throughout this standard. Prior to April 1, 1967, these regulations were promulgated by the Interstate Commerce Commission (ICC).

The sources of LP-Gases are natural gas and crude oil. In a natural gas processing plant or a refinery, LP-Gases are removed from the base natural gas or crude oil and liquefied. They are then separated from one another for particular uses. Impurities such as sulfur and water are removed.

LP-Gases, by nature, can be easily liquefied at moderate pressure for ease of storage and transportation. For instance, 1.0 cu ft (0.028 m^3) of liquid will expand to 270 cu ft (7.6 m^3) of vapor at atmospheric pressure.

The major uses of propane are as domestic and commercial fuels for cooking, water heating, and building heating. Propane also has numerous uses on farms, e.g., fuel for tractors and power for irrigation pumping, crop drying, and poultry incubating and brooding. It is also used in industry for heat treating, cutting and welding, and building heating, and as fuel for industrial trucks. In addition, propane is widely used by natural gas utility companies as a standby fuel in instances of natural gas shortages or interruptions of natural gas supply. Propane is also used as a vehicle fuel. It was recently estimated that there were about 350,000 vehicles fueled by propane in use in the United States.[1]

Propane is also being discussed as a replacement for freon in closed cycle refrigeration service for small units (residential refrigerators and air conditioners).

In utility gas plants, propane is used for enrichment of natural gas, as a standby fuel, and to supplement natural gas supplies. Propane and butane are also used as feedstocks for chemical processing.

Butane's uses are similar to those of propane, but it is used to a much lesser degree. The use of butane is growing in consumer appliances, such as portable cooking stoves, cigarette lighters, hair-curling irons, and irons, and it has been used to replace freon in aerosol spray cans and as a foaming agent in foaming-plastic products such as meat trays. (This use is being phased out as nonozone depleting propellants are becoming available.) It is also used in refinery gasoline to alter the gasoline's volatility. Some operations use a mixture of propane and butane.

As noted in Appendix B, commercial butane and propane contain other gaseous hydrocarbons (and, therefore, are not "pure"), and their boiling points are usually somewhat lower than those of pure butane and propane. The commercial forms also tend to vaporize more rapidly than do the pure products. These commercial grades of butane and propane are not unique mixtures, but will vary with their source, seasonally, and with the changing market value of their components.

This tendency to vaporize rapidly when they escape from containment is a primary difference in hazards and, consequently, in fire safety standards between LP-Gases and flammable liquids such as gasoline, alcohol, and paint thinners. The lack of appreciation of this vital difference has caused public safety enforcement problems.

A number of provisions of NFPA 58 rely on certain regulations of DOT because their origin and continued development is traced to this agency. At

[1] U.S. Department of Energy. *Liquefied Petroleum Gas Fact Sheet*. Fact sheet on liquefied petroleum gas and propane, developed by the Development and Communications Office, Technical Information Program. NREL SP 220-4066, DE91002114. Rev. Feb. 1992.

the turn of the century, ICC was charged with the duty of promulgating regulations for the safe transportation of hazardous commodities by rail—including shipping container specifications—and utilized the services of the Bureau of Explosives of the American Association of Railroads in this activity. These regulations were subsequently extended to other transportation modes. DOT was established in 1967 and took over this function of ICC. The Hazardous Materials Regulations issued by DOT and published in the *Code of Federal Regulations,* Title 49, Parts 100–199, are the most relevant to NFPA 58.

LP-Gas containers covered by these regulations are cylinders, portable tanks, cargo tanks, tank trucks, and tank cars. The use of LP-Gas cylinders as storage containers or fuel supply containers is recognized in the development of requirements for their use as shipping containers. DOT requirements for containers include specifications for construction, testing, marking, equipping with stipulated pressure relief devices, periodic requalifications, repair, method of filling, and maximum filling limits. In addition to regulations covering cylinders, portable tanks, tank and cylinder trucks, and tank cars, there are also regulations covering pipeline safety (Office of Pipeline Safety Operations) and motor carrier safety (Bureau of Motor Carrier Safety).

1

General Provisions

1-1 Scope.

1-1.1* **Liquefied Petroleum Gas.** LP-Gas stored or used in systems within the scope of this standard shall not contain ammonia. When such a possibility exists (such as resulting from the dual use of transportation or storage equipment), the LP-Gas shall be tested.

A-1-1.1 Allow a moderate vapor stream of the product to be tested to escape from the container. A rotary, slip tube, or fixed level gauge is a convenient vapor source. Wet a piece of red litmus paper by pouring distilled water over it while holding it with clean tweezers. Hold the wetted litmus paper in the vapor stream from the container for 30 sec. The appearance of any blue color on the litmus paper indicates that ammonia is present in the product.

NOTE 1: Since the red litmus paper will turn blue when exposed to any basic (alkaline) solution, care is required in making the test and interpreting the results. Tap water, saliva, perspiration, or hands that have been in contact with water having a pH greater than 7, or with any alkaline solution, will give erroneous results.
NOTE 2: For additional information on the nature of this problem and conducting the test, see *Recommendations for Prevention of Ammonia Contamination of LP-Gas*, published by the National Propane Gas Association.

Since anhydrous ammonia and LP-Gas have about the same vapor pressure characteristics, transportation and storage equipment is sometimes used for ammonia when it is not needed for LP-Gas, and vice versa, for economic reasons. This is common in agricultural areas where ammonia is used as a fertilizer in the warmer months, when the demand for LP-Gas for heating falls. It is very important when switching a tank from ammonia to LP-Gas service that the tank is cleaned properly. If not, serious consequences can result. For example, brass fittings are commonly used on tanks in LP-Gas service, particularly on the tanks of consumers and dealers. Under certain conditions ammonia can cause brass fittings to fail while in service through a process known as stress corrosion cracking. If ammonia gets into an LP-Gas distribution system, it will spread through the system and may cause damage to any brass fitting it contacts.

Maximum concentrations of ammonia that can be tolerated in LP-Gas have been determined through a research project, "A Study of Cleaning Methods for LP-Gas Transports," conducted by Battelle Memorial Institute, Columbus, Ohio. This research indicates that the red litmus paper test for ammonia in LP-Gas, if performed properly during a purging operation, is much more sensitive a test than the absence of ammonia odor and will ensure that tolerable limits are not exceeded. Reference to this test procedure is located in Appendix A as the test is not mandated for testing for ammonia, but rather informative material for the user. Other test methods can be used, and it is the responsibility of the user to verify that any alternate method used is equally effective in identifying ammonia concentrations that will cause corrosion.

Recommended procedures for purging storage and cargo tanks by steam cleaning, water flooding, and a combination of water flooding and steaming methods, when switching from anhydrous ammonia to LP-Gas service, are contained in the National Propane Gas Association publication cited in Note 2.

1-1.2 Application of Standard. This standard shall apply to the operation of all LP-Gas systems including:

(a) Containers, piping, and associated equipment, when delivering LP-Gas to a building for use as a fuel gas,

(b) Highway transportation of LP-Gas,

(c) The design, construction, installation, and operation of marine terminals whose primary purpose is the receipt of LP-Gas for delivery to transporters, distributors, or users.

While the intent has been to make NFPA 58 applicable to all LP-Gas uses, it has not been feasible to do so for a number of reasons. Because of the great versatility of LP-Gas, the number and variety of its applications are large and have steadily increased since its original commercial use as a residential fuel. This is especially true in industry, where use of LP-Gas is not only of great variety but is also quite complex.

This requirement has evolved to its current form. In the 1989 edition, marine and pipeline terminals were added to the scope of NFPA 58 to provide coverage for terminals. This recognized that these facilities are storage and transfer facilities similar to those already covered by the standard and that, in many cases, authorities having jurisdiction and owner-operators were applying NFPA 58 and NFPA 59 to these facilities even though they were specifically excluded in 1-2.3.1 prior to the 1989 edition. Previous editions had referred the reader to American Petroleum Institute (API) Standard 2510, *The Design and Construction of Liquefied Petroleum Gas Installations at Marine and Pipeline Terminals, Natural Gas Processing Plants, Refineries, Petrochemical Plants and Tank Farms.*

This change involved some controversy, as representatives of the API believed it would be difficult to apply NFPA 58 to marine and pipeline terminals, which are frequently an integral part of refineries and chemical plants. A Tentative Interim Amendment was issued to the 1989 edition that addressed the valid concerns of the API and maintained coverage of terminals not associated with refineries and chemical plants, which were previously not covered. This TIA was incorporated into the 1992 edition.

It is vital that the Technical Committee and support staff responsible for NFPA 58 be technically qualified to develop and interpret its provisions. At the same time, the necessary balance of interests on the Committee must be maintained. The size of the Committee also must be kept manageable if the standard is to keep pace with the ever-growing variety of applications. For these reasons, the applications covered by NFPA 58 are those that impact the largest number of common interest groups, e.g., the general public, employees and management of general commerce and industry, and those who must regulate, insure, and handle the emergencies associated with these applications. These represent the bulk of all LP-Gas applications.

The applications not covered by NFPA 58 often are addressed by other NFPA technical committees, other standards-writing bodies, or by technically qualified groups. Where the Technical Committee responsible for NFPA 58 considers these standards adequate, it has been the practice to call attention to them in NFPA 58. In instances where the standards are written in a mandatory format that can be adopted as law, this can take the form of mandatory reference stating that the referenced standard "shall" be used. In this case, if NFPA 58 is adopted as, for example, a regulatory instrument, the referenced document has equal status (unless the adopting agency states otherwise, of course). In cases where the standard is not written in mandatory form, but is a recommended practice or guide, it is referenced in a note or in the appendix to advise the user of its availability.

Whenever a standard or other document covers an LP-Gas application, it is important that its scope be carefully correlated with that of NFPA 58 to avoid conflicts. This is not always easy to do because of the large number of standards-writing bodies, but the effort has been rather successful, as the spirit of cooperation among these bodies has always been high.

Exception No. 1: Marine terminals associated with refineries, petrochemicals, and gas plants.

This exception recognizes that marine terminals supplying refineries and petrochemical and gas plants are frequently designed and associated integrally with the refinery or plant making it difficult, in some cases, to identify the components that fall under NFPA 58. If the terminal has the capability of supplying LP-Gas to other uses by truck or rail loading facilities, those loading facilities would come under NFPA 58, if the facilities are separate. The owner would be responsible for determining the "split" between the loading facilities, which would come under NFPA 58 and the marine terminal.

Exception No. 2: Marine terminals whose purpose is the delivery of LP-Gas to marine vessels.

This exception recognizes that safety standards for these terminals are written by the U.S. Department of Transportation and enforced in the United States by the U.S. Coast Guard, and that the NFPA LP-Gas Committee does not have expertise in the area. If the terminal was used for truck or rail car loading incidental to its main use, those truck or rail loading facilities would come under NFPA 58. See above commentary.

(d) The design, construction, installation, and operation of pipeline terminals that receive LP-Gas from pipelines under the jurisdiction of the U.S. Department of Transportation, whose primary purpose is the receipt of LP-Gas for delivery to transporters, distributors, or users. Coverage shall begin downstream of the last pipeline valve or tank manifold inlet.

This provision applies to pipeline terminals served by pipelines under the jurisdiction of the Department of Transportation. This does not include terminals associated with refineries and petrochemical and gas plants that are served by pipelines built to other standards, usually ANSI/ASME B31.8, *Gas Transmission and Distribution Piping Systems.*

Exception: Those systems designated by 1-1.3.

1-1.3 Nonapplication of Standard.

1-1.3.1 This standard shall not apply to:

(a) Frozen ground containers and underground storage in caverns including associated piping and appurtenances used for the storage of LP-Gas.

This exclusion was changed in the 1989 edition from "LP-Gas refrigerated storage systems." This added LP-Gas refrigerated storage systems to

NFPA 58, and a chapter was added with specific requirements. (Note that LP-Gas refrigerated storage systems at utility gas plants continue to be covered under NFPA 59.) LP-Gas refrigerated storage systems are often located at marine and pipeline terminals, and may also be found at natural gas processing plants, refineries, and petrochemical plants.

Frozen ground containers are a type of refrigerated LP-Gas container made by excavating an open pit, covering it with a gastight cap, and freezing the ground to provide containment. Because of the differences between these and aboveground refrigerated containers, the fact that they are rarely used for LP-Gases, and the lack of standards that can be referenced, they are excluded from the standard. It is not the intent of the Committee to prohibit or discourage the use of such containers. If they are to be used, other sources must be referenced for guidance.

Refrigerated LP-Gas storage systems are those in which the liquid product is stored as a boiling liquid at or near atmospheric pressure. Refrigeration necessary to maintain the LP-Gas as a liquid is often provided by compressing and condensing the vapors that boil off the liquid. The storage containers are characterized by their large size [more than 1 million gallons (4000 m^3) is not unusual]. They are thermally insulated as a matter of operational necessity and to minimize the refrigeration load, and they can usually withstand only very low internal pressures. Such containers are not only quite different from nonrefrigerated LP-Gas containers, they are limited in number and found usually in specialized applications in LP-Gas production facilities, marine and pipeline terminals, and in natural gas peakshaving plants (utility gas plants).

(b) Natural gas processing plants, refineries, and petrochemical plants.

NOTE: For further information on the storage and handling of LP-Gas at natural gas processing plants, refineries, and petrochemical plants see API Standard 2510, *Design and Construction of LP-Gas Installations.*

This excludes refineries and petrochemical and gas plants from the scope of NFPA 58, in recognition of the fact that these facilities are complex and application of NFPA 58 could be difficult. Safety is frequently achieved in these facilities by a combination of design and operating features that, by their nature, might be unnecessarily prohibited in NFPA 58. API 2510 provides valuable information on these facilities. The American Petroleum Institute also publishes API 2510A, *Fire Protection Considerations for the Design and Operation of Liquefied Petroleum Gas (LPG) Storage Facilities*, which supplements API 2510.

(c) LP-Gas (including refrigerated storage) at utility gas plants. NFPA 59, *Standard for the Storage and Handling of Liquefied Petroleum Gases at Utility Gas Plants*, shall apply.

The LP-Gas application covered by this exclusion is storage (which can be both refrigerated and nonrefrigerated) of LP-Gas, vaporization, and often its mixing with air to form a mixture having a heating value similar to that of natural gas in a facility operated by a gas utility. A utility is a private or governmental body specifically designated as a public utility by an action of a governmental entity, e.g., state or local government.

Such facilities have limited firesafety impact upon general consumers and industry. However, because they are covered by an NFPA standard, NFPA 59 is adopted by reference in NFPA 58.

In recent years, the size and nature of an increasing number of industrial nonrefrigerated installations have approached that of many utility gas plants. In 1966, a reorganization of the NFPA Technical Committee structure in the field of gases resulted in assignment of NFPA 59 to the same committee responsible for NFPA 58 (and the committee membership was augmented accordingly). This was changed in 1993 by the separation of NFPA 59 to its own technical committee, the Technical Committee on Liquefied Petroleum Gases at Utility Gas Plants. The two committees have overlapping membership. As a result, both standards are continuously monitored to maintain their provisions consistently, and this process will continue.

(d) Chemical plants where specific approval of construction and installation plans, based on substantially similar requirements, is obtained from the authority having jurisdiction.

This exclusion acknowledges the use of LP-Gas as a chemical reactant (feedstock), rather than as a fuel, in a chemical plant, and the unique and complex fire hazard problems that often exist. There are, however, many uses of LP-Gas as a fuel in such facilities, and the storage hazards are often not unique. Moreover, there is no standard definition of a "chemical plant"; facilities in which few or no chemical reactions are carried out are often called chemical plants—for example, aerosol product manufacturing facilities using LP-Gas only as a propellant. The term "chemical plant" includes all facilities owned by chemical companies where LP-Gas is used for other than a fuel, including pilot plants, which are facilities used for large scale testing and production of commercially small quantities of materials. In practice, the chemical industry uses NFPA 58 extensively.

(e) LP-Gas used with oxygen. NFPA 51, *Standard for the Design and Installation of Oxygen-Fuel Gas Systems for Welding, Cutting, and Allied Processes*, and ANSI Z49.1, *Safety in Welding and Cutting*, shall apply.

The burning of LP-Gas with oxygen rather than air is most commonly associated with the flame cutting of metal in industry and construction. The use of oxygen introduces another dimension to the fire hazards. These operations are widespread and the two standards referenced cover them. Because both NFPA 51 and ANSI Z49.1 are national consensus standards, their adoption in NFPA 58 is mandatory.

Unfortunately, the exclusion regarding use with oxygen, and the specific scopes of NFPA 51 and ANSI Z49.1, technically have left some areas of firesafety uncovered. These include small jewelry and glass-forming operations (often in mercantile occupancies having considerable public exposure) and the use of oxy-propane torches by plumbers and do-it-yourselfers. Because manufacturers of these torches customarily obtain product listings, this has not proven to be a significant problem; however, regulatory authorities have had difficulty with small mercantile operations. Used with judgment, many of the provisions in NFPA 51, NFPA 58, and ANSI Z49.1 are appropriate, especially those concerned with LP-Gas and oxygen storage.

(f) Those portions of LP-Gas systems covered by NFPA 54 (ANSI Z223.1), *National Fuel Gas Code,* where NFPA 54 (ANSI Z223.1), *National Fuel Gas Code,* is adopted, used, or enforced.

NOTE: Several types of LP-Gas systems are not covered by the *National Fuel Gas Code* as noted in 1.1.1(b) therein. These include, but are not restricted to, most portable applications; many farm installations; vaporization, mixing, and gas manufacturing; temporary systems, e.g., in construction; and systems on vehicles. For those systems within its scope, the *National Fuel Gas Code* is applicable to those portions of a system downstream of the outlet of the first stage of pressure regulation.

The *National Fuel Gas Code* (NFPA 54/ANSI Z223.1) and its predecessors (NFPA 54 and 54A and ANSI Z21.30 and Z83.1) were developed originally to cover the installation of fuel gas piping and appliances in buildings. In the early days, the fuel gas was manufactured gas and the buildings were residential and commercial in character. After World War II, natural gas replaced most manufactured gas, and by the early 1960s industrial systems were included.

In the 1950s, NFPA 52, *Liquefied Petroleum Gas Piping and Appliance Installations in Buildings,* was developed to provide similar coverage for undiluted LP-Gas piping and appliances in residential and commercial buildings. It was soon recognized, however, that with the exception of the piping materials and testing of piping for leaks, NFPA 52 and 54 were similar. Differences were resolved in the 1959 edition of NFPA 54, NFPA 52 was withdrawn, and the scope of NFPA 54 was expanded to include LP-Gas.

Still, with very few minor exceptions, NFPA 54 is restricted by its scope to the installation of fixed-in-place appliances and other gas-consuming equipment connected to a building's gas piping system. Furthermore, many specialized types of farm equipment are not covered, even though they may be fixed in place in a building and connected to a piping system in that building. These systems are customarily LP-Gas systems rather than natural gas systems.

While the great majority of LP-Gas piping and appliance installations in completed buildings require the application of the *National Fuel Gas Code* to part of the system, it is necessary to refer to 1.1.1, "Applicability," in that code to determine this. In this respect, the note to 1-1.3.1(f) should be considered only as a rough guide.

To all LP-Gas systems covered by the *National Fuel Gas Code* that are supplied from on-site storage, it is necessary also to apply NFPA 58. The point of demarcation between these two documents is the outlet of the first-stage pressure regulator. Many residential LP-Gas systems have only one regulator installed very close to the storage container(s) or a second regulator a few feet away. However, some installations involving multiple buildings, such as apartment complexes, may have second regulators downstream of the first that may be hundreds of feet away from the storage. The piping between them is within the scope of the *National Fuel Gas Code.*

This provision was modified in the 1995 edition to limit the exclusion to areas where the National Fuel Gas Code has been adopted. This was the result of a severe problem in the Sierra Mountains of Northern California during the winter of 1992–1993. Exceptionally heavy snowfall covered the area resulting in several accidents in which propane was released and ignited causing multiple deaths. The accidents, resulting from the weight of the snow—which was as high as 30 ft—broke piping, allowing propane gas to migrate under the snow and enter buildings.

The amended requirement recognizes that some areas of the United States do not adopt the NFPA 54, *National Fuel Gas Code*, for gas piping downstream of the first-stage regulator, but use a "model" mechanical code. The model codes begin at the building entrance, leaving a portion of the piping system not required to meet a code or standard. The piping failures occurred in this section between the first-stage pressure regulator and the building. The lack of standards coverage, and local practices, were considered factors in the failures. This new requirement extends coverage of NFPA 58 beyond its normal termination at the discharge of the first-stage regulator where other codes do not begin at that point, to where they begin. Local requirements have been addressed via a model ordinance recommended by the propane industry.

(g) Transportation by air (including use in hot air balloons), rail, or water under the jurisdiction of the U.S. Department of Transportation.

This exclusion acknowledges the primary jurisdiction in the United States of the U.S. Department of Transportation in transportation of LP-Gas where that jurisdiction is valid. This is discussed in more detail in Chapter 6.

Hot air ballooning is a popular sport in the United States that uses propane to heat the air in the balloon. The use of propane cylinders in a balloon in flight, or in flight preparation, is covered by the U.S. Department of Transportation. The storage of propane cylinders used in hot air balloons (when not in flight or preparation thereof) falls under NFPA 58. The same rules apply to the storage of these cylinders as any other propane cylinder.

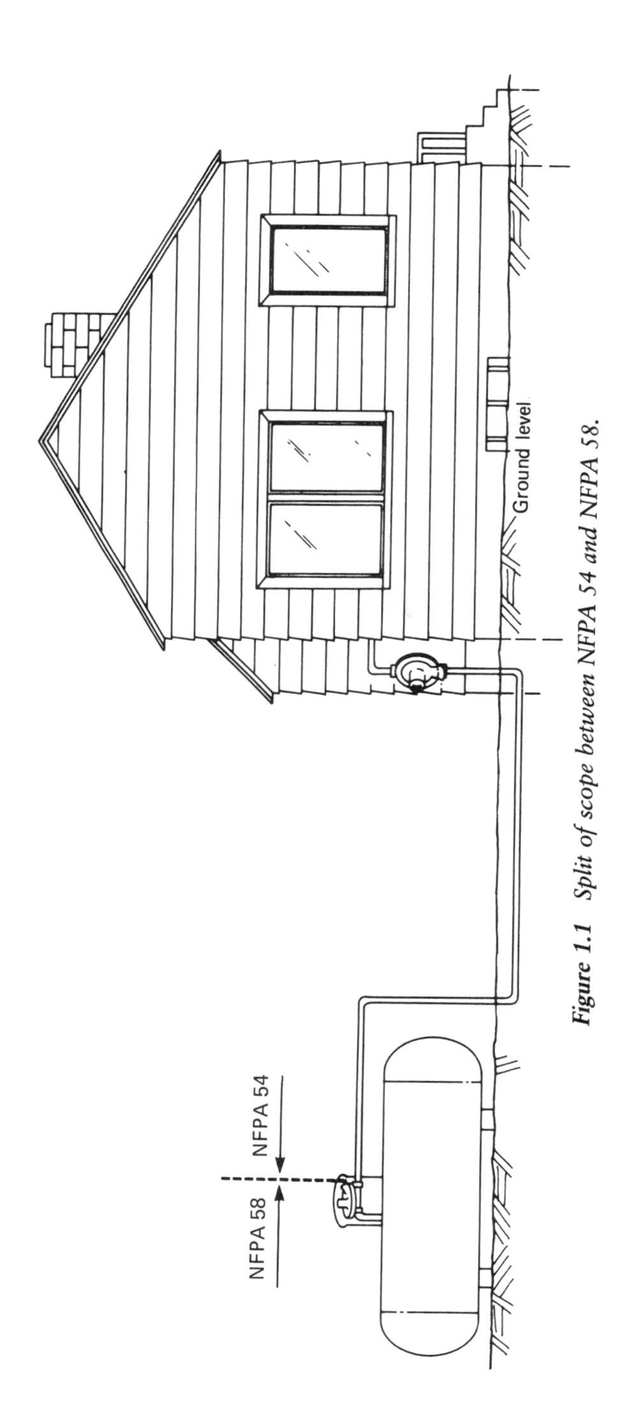

***Figure 1.1** Split of scope between NFPA 54 and NFPA 58.*

(h) Marine fire protection. NFPA 302, *Fire Protection Standard for Pleasure and Commercial Motor Craft*, shall apply.

LP-Gas has become a popular fuel for galley stoves, cabin heaters, and other appliances on motor vessels within the scope of NFPA 302 [less than 300 gross tons (849 m^3) and used for pleasure and commercial purposes]. This use is extensive enough to represent substantial public safety exposure and, therefore, its adoption in NFPA 58 is mandatory.

(i) Refrigeration cycle equipment and LP-Gas used as a refrigerant in a closed cycle.

This exception was added in the 1995 edition following a report by the LP-Gas Committee on the use of propane and other LP-Gases as a refrigerant to replace the nonflammable, ozone depleting R-12 (see commentary following the Foreword). The Committee concluded that LP-Gas in a closed cycle refrigeration system be excluded from the scope of NFPA 58. This was done because the Committee had no expertise on refrigeration, and recognized that it would be more appropriate for the product standards for these devices (such as refrigerators and air conditioners) to be modified to recognize the additional hazard of substituting a flammable refrigerant for a nonflammable one. At this writing, the standards for appliances that may use flammable refrigerants have not been revised to address this hazard.

1-1.4 Alternate Materials, Equipment, and Procedures. The provisions of this code are not intended to prevent the use of any material, method of construction, or installation procedure not specifically prescribed by this code, provided any such use is acceptable to the authority having jurisdiction (*see Section 1-6, Approved*). The authority having jurisdiction shall require that sufficient evidence be submitted to substantiate any claims made regarding the safety of such alternate use.

This provision was added in the 1992 edition by the Committee as part of the project to make the standard more usable, adoptable, and enforceable. It is the intent of the Committee to empower the authority having jurisdiction to be able to permit the use of items that have not been thought of or developed when the standard was written. The Committee recognizes that technology continues to evolve and that there are cases where new technology permits prohibited materials, techniques, and equipment to be installed safely. The requirement for demonstrating that equivalent safety is achieved is placed on the proponent, who must demonstrate safety to the satisfaction of the authority having jurisdiction. Of course, the Committee encourages all involved with such situations to advise the Committee by the use of a public proposal for the next edition of NFPA 58. Forms are included in the standard and their use is encouraged.

1-1.5 Retroactivity. The provisions of this standard are considered necessary to provide a reasonable level of protection from loss of life and property from fire and explosion. They reflect situations and the state of the art prevalent at the time the standard was issued.

Unless otherwise noted, it is not intended that the provisions of this document be applied to facilities, equipment, appliances, structures, or installations that were in existence or approved for construction or installation prior to the effective date of the document, except in those cases where it is determined by the authority having jurisdiction that the existing situation involves a distinct hazard to life or adjacent property. Equipment and appliances include stocks in manufacturers' storage, distribution warehouses, and dealers' storage and showrooms in compliance with the provisions of this standard in effect at the time of manufacture.

The matter of retroactivity in a standard that is as widely adopted by public safety regulatory agencies and used in litigation as is NFPA 58 is of considerable importance. Over the more than 60 years of its existence, the technical committees responsible for NFPA 58 have consistently taken the approach that when an amendment should be applied retroactively, the amendment will so state. Typically, this has taken the form of stipulating a certain date by which something must be accomplished (*see 3-2.8.9, for example*). The date is retained in subsequent editions of the standard for many years—even though it may predate the year of a specific edition—because of its potential importance in, for example, litigation.

Over the years, such retroactive provisions have been few. While many amendments do provide an additional degree of safety, the functional and economic burdens of complying with them are commensurate with a more gradual phasing-in, or no phasing-in, when compared with the additional degree of safety to be achieved.

The expression "distinct hazard to life or adjoining property" is rather obtuse and subject to a good deal of judgment. In practice, prior to an incident, what constitutes this hazard can be determined only by argument between parties or perhaps by regulatory directive. After an incident, especially a serious one, the opportunity for differing views is somewhat lessened. While many NFPA committees have tried to express this thought in more specific terms, the above phrase remains the best they can come up with.

The retroactivity statement was revised in the 1992 edition to more positively state its objective and limitation that a reasonable level of protection from loss of life and property from fire and explosion is provided in the standard, and that it represents the state of the art at the time of writing.

1-2 Acceptance of Equipment and Systems.

1-2.1 Systems, or components assembled to make up systems, shall be approved (*see Section 1-6, Approved*) as specified in Table 1-2.

1-2.2 Acceptance applies to the complete system, or to the individual components of which it is comprised, as specified in Table 1-2.

Table 1-2

Containers Used	Capacity in Water Gal (m^3)	Approval Applies to:
DOT Cylinders	Up to 120 (0.454) (1,000 lb, 454 kg)	1. Container Valves and Connectors 2. Manifold Valve Assemblies 3. Regulators and Pressure Relief Devices
ASME Tanks	2,000 (7.6 m^3) or less	1. Container System* including Regulator, or 2. Container Assembly* and Regulator separately
ASME Tanks	Over 2,000 (7.6 m^3)	1. Container Valves 2. Container Excess Flow Valves, Back Flow Check Valves, or alternate means of providing this protection such as remotely controlled Manual or Automatic Internal Valves 3. Container Gauging Devices 4. Regulators and Container Pressure Relief Devices

*Where necessary to alter or repair such systems or assemblies in the field in order to provide for different operating pressures, change from vapor to liquid withdrawal, or the like, such changes shall be permitted to be made by the use of approved components.

Approval of systems using DOT cylinders includes:

(a) The cylinder itself, which must meet DOT specs and be labeled with a DOT number that indicates which specification it meets and the date of manufacture or retest (*see Appendix C for information on requalification of DOT cylinders*).

(b) Cylinder valves, which must be in accordance with DOT regulations and American National, Canadian, and CGA Standard Compressed Gas Cylinder Valve outlet and inlet connections (ANSI B57.1/CSA B 96/CGA V-1).

(c) Manifold valve assemblies, regulators, and pressure relief devices, which must also be approved.

Approval of systems using ASME tanks (cylinders) of 2,000 gal (7.6 m^3) or less includes:

(a) The container (tank), which must be designed, fabricated, and stamped in accordance with the ASME Code, and its appurtenances, which may be approved as a unit, or
(b) The container, which may be approved separately, in which case the pressure regulator must also be approved separately.

Approval of systems using ASME tanks (containers) of more than 2,000 gal (7.6 m^3) includes:

(a) Container valves.
(b) All other container appurtenances as listed in the table above.

NOTE: On these larger systems utilizing containers of more than 2,000 gal (7.6 m^3) capacity, the entire system does not have to be approved as a system; instead, the individual parts must be approved. The reason for this is that there are many application variations, and approval of the entire unit as a system would be very difficult, possibly requiring many different approvals to cover all possible applications.

1-3 LP-Gas Odorization.

1-3.1* All LP-Gases shall be odorized prior to delivery to a bulk plant by the addition of a warning agent of such character that the gases are detectable, by a distinct odor, to a concentration in air of not over one-fifth the lower limit of flammability.

Exception: Odorization, however, shall not be required if harmful in the use of further processing of the LP-Gas or if such odorization will serve no useful purpose as a warning agent in such further use or processing.

This provision specifies where odorization must take place, that is, at some point in its distribution chain prior to a bulk plant (*see definition*). This is based on research that has identified that ethyl mercaptian, the most widely used LP-Gas odorant, can be oxidized by oxygen or oxides (including iron oxide, or rust) in LP-Gas containers, and a number of incidents where failure to smell leaking LP-Gas was a factor.

There has been extensive research on odorants and odorization of LP-Gas. Two symposiums have been held on the topic, in April 1989 and October 1990, where much information was disseminated. Copies of the proceedings are available from the Gas Processors Association in Tulsa, Oklahoma. Additional research has been sponsored by the U.S. Consumer Product Safety Commission. Those interested in odorization are encouraged to review the latest available work in the field.

A-1-3.1 It is recognized that no odorant will be completely effective as a warning agent in every circumstance.

It is recommended that odorants be qualified as to compliance with 1-3.1 by tests or experience. Where qualifying is by tests, such tests should be certified by an approved laboratory not associated with the odorant manufacturer. Experience has shown that ethyl mercaptan in the ratio of 1.0 lb (0.45 kg) per 10,000 gal (37.9 m^3) of liquid LP-Gas has been recognized as an effective odorant. Other odorants and quantities meeting the provisions of 1-3.1 may be used. Research on odorants has shown that thiophane (tetrahydrothiophene) in a ratio of at least 6.4 lb (2.9 kg) per 10,000 gal (37.9 m^3) of liquid LP-Gas may satisfy the requirements of 1-3.1.

NOTE: Odorant research includes *A New Look at Odorization Levels for Propane Gas*, BERC/RI-77/1, United States Energy Research & Development Administration, Technical Information Center, September, 1977.

This provides additional information to the user of the standard on the important subject of odorization. It must be remembered that despite the specific reference to odorization levels in this appendix, it is the responsibility of the deliverer of LP-Gas to ensure that the gas is odorized, as stated in 1-3.1.

1-3.2* If odorization is required, the presence of the odorant shall be determined by sniff-testing or other means and the results shall be documented:

A-1-3.2 Another method of determining the presence of odorant is the stain tube test. This method uses a small handheld pump to draw a sample across a filled glass tube and reading the length of color change. For additional information see GPA Standard 2188-89 and CAN/CGSB-3.0 No. 18.5-M89. At the time of the preparation of this standard additional analytical methods are under development.

(a) Whenever LP-Gas is delivered to a bulk plant; and
(b) When shipments of LP-Gas bypass the bulk plant.

This provision, added in the 1989 edition, resulted from a recommendation by the U.S. Consumer Product Safety Commission (CPC) in their Status Report on LP-Gas Residential Heating Equipment, July 24, 1986, which stated:

"NFPA Pamphlet 58 should be revised to assign responsibility for testing and certification of odorant testing at some point in the distribution chain." This was based on technical work done for the CPSC. Testing for the presence of odorant is required to verify that the odorization has taken place, and that the odorant has not "faded" or been otherwise rendered ineffective.

The LP-Gas Committee, in considering this recommendation, evaluated the possible locations at which this testing might be required, including arrival at the bulk plant, departure from the bulk plant, and upon delivery from a bobtail (delivery truck).

Additional information on odorization of LP-Gases may be found in the Fuel Gas Odorization supplement to the *National Fuel Gas Code Handbook*.

The odorization of gas, started as early as 1880 in Europe and shortly after World War I in the United States, was used to prevent injuries or fatalities resulting from escaping gas. The first reported odorant used was ethyl mercaptan, a predominant compound still used to odorize LP-Gas. This odorant has a characteristic smell often described as that of garlic or rotten eggs—essentially a sulfurous odor.[2]

Previous wording in NFPA 58 specified that odorization was to indicate positively, by a distinctive odor, the presence of gas. The implication was that all must be warned by the odorant. This was an impossible requirement, for it is known that some persons have suffered partial or complete loss of smell and were not warned by any level of any odorant system. Factors that can compromise the sense of smell in humans include:

- Certain diseases that cause permant loss of smell
- Certain diseases that cause temporary loss of smell, i.e., colds
- Allergies
- Intoxication
- Eating certain foods[2]

Thus, regardless of dosage levels, it is not possible to warn 100 percent of the population of gas leaks by use of odorant systems. This is pointed out in A-1-3.1. Despite being doomed to less than perfection, odorization is a valuable practice as it does serve as a useful warning in the vast majority of cases. In a study conducted by inserting a "scratch and sniff" card in its magazine, *National Geographic* Magazine reported that gas was one of the most recognized odors.[3]

A significant research study on odorization of LP-Gas was conducted at the Bartlesville Energy Research Center (BERC) in 1977, referenced in A-1-3.1. This research showed that defects of the nasal anatomy and psychological factors may affect olfactory responses. In addition, unfamiliarity with a given environment, as well as anxieties or mental distractions, can produce reduced awareness to odorants intended to warn individuals of the presence of LP-Gas.[2]

[2] U.S. Department of Energy. Technical Information Center. *A New Look at Odorization Levels for Propane Gas*, by Whisman ML, Goetziner JW, Cotton FO, Brinkman DW, and Thompson CJ. BERC/RI-77/1. Bartlesville, Okla. Sept 1977.

[3] Gilbert AN and Wysocki CJ. "The Smell Survey Results." *National Geographic* 1987;172(4): 514–526.

1-4 Notification of Installations.

1-4.1 Stationary Installations. Plans for stationary installations utilizing storage containers of over 2,000 gal (7.6 m^3) individual water capacity, or with aggregate water capacity exceeding 4,000 gal (15.1 m^3), shall be submitted to the authority having jurisdiction before the installation is started. [*See also 3-4.9.1(e).*]

In terms of the number of installations actually made, installations of the sizes cited are much less numerous, to the knowledge of the editor. Requiring plans for these installations to be submitted would place a burden on the authority having jurisdiction (usually a fire department) that could not be justified for these small, relatively simple and standardized installations. However, it is not uncommon for those authorities to require approval of small installations in heavily populated or congested areas. In such instances, they will either amend 1-5.1 or adopt specific wording in the enabling document.

1-4.2 Temporary Installations. The authority having jurisdiction shall be notified of temporary (not to exceed six months) installations of the sizes covered in 1-4.1 before the installation is started.

It is important that such installations be brought to the attention of the authority having jurisdiction so that the fire service can be made aware of them. Such notice is important so that plans can be made for emergency response, if required. Six months is the maximum installation period for temporary installations, which are found at construction sites, where temporary heating is needed, at fairs for cooking, and other locations. The intent of the Committee in using the term "temporary" is just that. Installations that are of a onetime nature, or if recurring, are such that permanent installations are not practical. Therefore, a 5-month installation at a cabin used only in the winter could not be considered temporary beyond the first year as it can be reasonably expected that an identical installation will be required in future years. Fairs require propane for cooking, and many are held at the same location for many years. The installations could be considered temporary if the specific locations vary within the fairgrounds each year. Judgment on the part of the authority having jurisdiction is required in such cases. Specific requirements for containers in temporary service are found in 2-2.5.4 and 3-2.4.2(a)(2)(b).

1-5 Qualification of Personnel.

In the interest of safety, all persons employed in handling LP-Gases shall be trained in proper handling and operating procedures, which the employer shall document.

This requirement for training in proper handling and operation procedures has been in NFPA 58 since the first edition in 1932. In the 1992 edition, a requirement for documentation of training and written certification to be carried by employees was added to permit verification that training had been accomplished. The requirement of carrying written certification was deleted in the 1995 edition by the Committee because it is important that the documentation be available, not that it be carried by the employee.

Considerable educational information is available for users handling LP-Gas. Some states provide training, and material is available from the National Propane Gas Association, including the Certified Employee Training Program, which is a curriculum for a training program given by LP-Gas companies, LP-Gas associations, and others such as community colleges.

It is the opinion of the editor that all new employees should be trained in the properties of LP-Gas, as those properties may differ from the properties of other flammable materials the employees have experienced. Many users and enforcers of the standard new to propane have had experience with gasoline, and assume that they should be treated similarly. This is not true as gasoline is stored in tanks open to the atmosphere with air above the liquid level, while propane and butane are stored in closed, purged containers with no air present. The avoidance of sparks in gasoline tanks is imperative for preventing fires although not needed in propane and butane tanks, as ignition of propane liquid and vapor is not possible without air or oxygen present.

The incidental use of LP-Gas in the performance of job requirements is beyond the scope of 1-5, and documentation of training is not needed. It does, however, apply to all employees engaged in liquid transfer and bulk storage of LP-Gas.

1-6 Definitions, Glossary of Terms, and Abbreviations.

Actuated Liquid Withdrawal Excess-Flow Valve. An excess-flow valve for liquid withdrawal applications where the valve is in a closed position until actuated by a pipe nipple or adapter, as recommended by the manufacturer, and is used with a shutoff valve attached to the actuator.

This definition was added in the 1995 edition to describe the special purpose valve required in 2-2.3.3, which has been required since the 1961 edition with no definition or name. The valve is specified by its function in 2-2.3.3. Since 1961 at least three manufacturers have made a special purpose valve to accomplish the requirements of 2-2.3.2, under the names of "Check-Lok" and "Check-Mate." In the past few years, questions have been raised over possible uses of the valve other than its mandated use. This has been addressed in replies to interpretation requests and was first covered in the previous edition of this handbook. The discussions on the uses of the valve led to the recognition that a definition was needed.

While some readers may come to the conclusion that this is just another example of the Committee adding to the size of the standard with no benefit to the user, please keep in mind that it permits reference to a valve with specific function without having to use brand names. While the Committee could have used "Check-Lok" or an equivalent, there are no specific means to determine what is equivalent, and it implies a preference. Further, to provide a list of the three major manufacturers would imply that others (whether available at any time or not) are not acceptable and would discriminate against or discourage manufacturers from producing a valve.

AGA. American Gas Association.

The American Gas Association is a trade association comprised largely of the gas utilities and gas transmission pipeline companies in the United States These utilities (the local "gas company") distribute natural gas primarily but often use LP-Gas in their supply and standby operations [*see commentary on 1-2.3.1(c)*]. The AGA is not specifically cited elsewhere in NFPA 58. The primary role of the AGA in the context of NFPA 58 is its sponsorship of the ANSI Z21 and Z83 series of standards covering residential, commercial, and industrial gas appliances and equipment. These standards cover both natural gas and LP-Gas appliances. (*See commentary on* 2-6.1.2.)

Anodeless Riser. A transition assembly where polyethylene pipe or tubing is permitted to be installed and terminated aboveground outside of a building. The polyethylene pipe or tubing is piped from at least 12 in. (300 mm) below grade to an aboveground location inside a protective steel casing and terminates in either a factory-assembled transition fitting or a field-assembled, service head, adapter-type transition fitting. The gas-carrying portion of the anodeless riser after the transition fitting shall be at least Schedule 40 steel pipe.

The definition was added in the 1995 edition, along with a corresponding definition of Service Head Adapter, to define components used with polyethylene pipe to provide a termination below grade. (Polyethylene pipe is permitted only outdoors, for underground service.) The riser is a purchased component (see 3-2.8.6) which connects to the polyethylene pipe underground and terminates in a metal thread connection aboveground.

ANSI. American National Standards Institute.

The American National Standards Institute is the overall focus of all private-sector standards-making in the United States The institute does not develop its own standards, but does adopt standards prepared by others provided they are developed by ANSI procedures. Because these procedures are compatible with NFPA procedures, the ANSI standards cited in Chapter 11 have been adopted as mandatory in NFPA 58. In fact, each edition of NFPA 58 is also submitted to ANSI for adoption both to support ANSI and for the additional status obtained from ANSI approval.

API. American Petroleum Institute.

The American Petroleum Institute is a trade association comprised largely of U.S. oil companies. Many oil companies also produce and distribute LP-Gas because it is a product of the refining of crude oil as well as the processing of natural gas. A number of oil companies have LP-Gas marketing subsidiaries that market LP-Gas and install and service LP-Gas installations.

The API produces many standards. While very useful, their adoption is not considered mandatory in NFPA 58 unless they are adopted by ANSI and written in a mandatory format that can be enforced.

API-ASME Container (or Tank). A container constructed in accordance with the pressure vessel code jointly developed by the American Petroleum Institute and the American Society of Mechanical Engineers (*see Appendix D*).

Approved. Acceptable to the authority having jurisdiction.

NOTE: The National Fire Protection Association does not approve, inspect, or certify any installations, procedures, equipment, or materials; nor does it approve or evaluate testing laboratories. In determining the acceptability of installations, procedures, equipment, or materials, the authority having jurisdiction may base acceptance on compliance with NFPA or other appropriate standards. In the absence of such standards, said authority may require evidence of proper installation, procedure, or use. The authority having jurisdiction may also refer to the listings or labeling practices of an organization concerned with product evaluations that is in a position to determine compliance with appropriate standards for the current production of listed items.

ASME. American Society of Mechanical Engineers.

ASME Code. *The Boiler and Pressure Vessel Code* (Section VIII, "Rules for the Construction of Unfired Pressure Vessels") of the American Society of Mechanical Engineers. Division I of Section VIII of the ASME Code is applicable in this standard except UG-125 through UG-136 shall not apply. Division 2 is also permitted.

The purpose of using the ASME Code was to adopt a nationally recognized code for LP-Gas containers other than those covered by DOT (ICC) regulations. As most states adopt the ASME Code, this results in national standardization, which has a safety value as well as being practical from an economic standpoint.

The definition was revised in the 1995 edition to drop the limitation to Division I and to identify that Division II was acceptable. Division II permits a smaller safety factor to be used provided an extensive and rigorous structural analysis is made. Prior to the last few years, there was no interest in the use of Division II for LP-Gas containers. This changed with several requests from designers and fabricators and resulted in the current text, as no reason could be found to continue to prohibit Division II. Those using Division II, must ensure that the inspections required as a part of design and fabrication to Division II be done over the life of the container.

A major exception to the ASME Code in the application of NFPA 58 is the exclusion of paragraphs UG-125 through UG-136. The ASME Code applies to pressure vessels in all kinds of service with, as the membership of the ASME Committee indicates, emphasis on vessels in air, steam, and other nonflammable products service. Paragraphs UG-125 through UG-136 in Division I of the ASME Code cover pressure relief devices. The possible consequences of operation of the pressure relief devices discharging these products are obviously different from those discharging LP-Gas. Therefore, the provisions of NFPA 58—e.g., start-to-discharge pressure settings—are different from those in the ASME Code. This does not mean that no standards apply to pressure relief valves. See 2-3.2.3, which adopts UL Standard 132 by reference.

Other differences between NFPA 58 and the ASME Code involve: the use of pilot-operated relief valves (UG-126) (*see commentary on 2-3.2.3*); the use of frangible disk devices (UG-127) (such devices discharge the entire contents of a container and reduce the pressure to atmospheric rather than control the pressure as a spring-loaded device does—a circumstance believed to be undesirable in installations covered by NFPA 58); marking (UG-129) (if functions are different, markings will be also); and certification (UG-131 and UG-132). NFPA 58 covers certification, in the contexts of "approved" and listing by Underwriters Laboratories Inc., Factory Mutual, or another agency acceptable to the authority having jurisdiction. These agencies, as noted in 2-3.2.3, have developed criteria especially for LP-Gas service that are believed to be more appropriate in NFPA 58 than are the ASME Code provisions.

These exceptions from the ASME Code are in the large view only small details and do not lessen the respect of the LP-Gas Committee for the ASME Code.

ASME Container (or Tank). A container constructed in accordance with the ASME Code. (*See Appendix D.*)

ASTM. American Society for Testing and Materials.

The American Society for Testing and Materials promulgates the largest body of private-sector standards in the United States. As noted in Chapter 11, many ASTM standards concerned with piping and castings are adopted by mandatory reference in NFPA 58.

Authority Having Jurisdiction. The organization, office, or individual responsible for approving equipment, an installation, or a procedure.

NOTE: The phrase "authority having jurisdiction" is used in NFPA documents in a broad manner, since jurisdictions and approval agencies vary, as do their responsibilities. Where public safety is primary, the authority having jurisdiction may be a federal, state, local, or other regional department or individual such as a fire chief; fire marshal; chief of a fire prevention bureau, labor department, or health department; building official; electrical inspector; or others having statutory authority. For insurance purposes, an insurance inspection department, rating bureau, or other insurance company representative may be the authority having jurisdiction. In many circumstances, the property owner or his or her designated agent assumes the role of the authority having jurisdiction; at government installations, the commanding officer or departmental official may be the authority having jurisdiction.

NFPA 58 has been adopted as the basis of public safety regulations by most states in the United States and by cities or counties in states where not adopted, by federal agencies, and in many other countries. In most states, the adopting and enforcing agency is the Office of the State Fire Marshal. The chief officer of each fire department in the state is considered an agent of the state fire marshal. In some states, an agency concerned only with LP-Gas has been established to adopt and enforce the regulations. In others, only the provisions concerned with certain applications have been adopted—for example, those concerned with vehicle applications or employee safety—and these involve other agencies, such as labor and industry departments and insurance commissions.

Since its formation, the U.S. Occupational Safety and Health Administration has adopted those provisions of the 1969 edition of NFPA 58 within its jurisdiction in its regulations [29 *Code of Federal Regulations* (CFR) 1910.110]. This has the impact of making OSHA the AHJ, but only when OSHA uses the authority of its enabling legislation via a citation. OSHA has a policy of issuing only a de minimus citation when an installation is found that does not comply with the 1969 edition but complies with the current edition.

These authorities vary in their ability to update their adoptive documents with new editions of NFPA 58. It is the exception where the edition of record is the current NFPA edition. This can present problems, although most authorities are amenable to use of the current edition, especially if the reason for an amendment is available to them.

As observed in the note to the definition, nongovernmental bodies can be the authority having jurisdiction by virtue of the conditions of a private contract, e.g., an insurance policy.

Bulk Plant. A facility, the primary purpose of which is the distribution of gas, that receives LP-Gas by tank car, tank truck, or piping, distributing this gas to the end user by portable container (package) delivery, by tank truck, or through gas piping. Such plants have bulk storage [2,000 gal (7.6 m^3) water capacity or more] and usually have container-filling and truck-loading facilities on the premises. Normally, no persons other than the plant management or plant employees have access to these facilities. A facility that transfers LP-Gas from tank cars on a private track directly into cargo tanks is also in this category.

This term was added to the standard in 1992 as part of the revisions updating dispensation of propane as an engine fuel, and was revised editorially in the 1995 edition. It replaced "distributing plant." At the same time, the definition of "distributing point" was also deleted and the terms "dispenser, vehicle fuel" and "dispensing station" were defined for the first time. Bulk plant is the term used in the propane distribution industry when referring to these facilities.

The terms "bulk plant" and "dispenser, vehicle fuel" and "dispensing station" indicate a basic distinction between the two types of facilities with respect to public access. This distinction enables the organization of provisions in the standard that differ depending on whether there is close involvement of the public with the facility. Particular attention is given to the protection of the public in such matters as ignition source control, protection against vehicular damage, container valve protection, separation distances, etc. This has become necessary because of a large increase in recent years of consumer-owned containers that need to be refilled, such as those used for gas-fired outdoor grills (*see commentary following 4-2.2.2*), recreational vehicles, and vehicular propulsion fuel. Generally, the public is expected to be in and around a dispensing station and vehicle fuel dispenser facility, but not in and around liquid transfer areas in a bulk plant. The public may also go to bulk plants to have their cylinders filled, but public movement is controlled so that only employees are permitted in and around the filling operations. A bulk plant may also have a dispensing station or vehicle fuel dispenser in connection with the plant facility.

Bulk plants have bulk storage facilities of 2,000 gal (7.6 m^3) water capacity or larger. A typical propane bulk plant distributes LP-Gas to its customers by delivering cylinders to them and by filling tank trucks, or bobtails, that take propane to tanks located on consumers' property. A bulk plant may also be a facility serving a number of customers, such as in a residential area, by piping to each user. However, if the latter is classified as a public utility it would be subject to NFPA 59, *Standard for the Storage and Handling of Liquefied Petroleum Gases at Utility Gas Plants.* A bulk plant serving customers via piping to the users would also be subject to the federal pipeline safety regulations, Part 192, CFR 49, if the system served ten or more customers. This may not be the case, however, if the storage facility serves this type of customer from a master meter. In this case only that part of the system downstream from the master meter would be subject to federal regulations.

Bureau of Explosives (B of E). An agency of the Association of American Railroads.

Cargo Tank. (Primarily a DOT designation.) A container used to transport LP-Gas over a highway as liquid cargo, either mounted on a conventional truck chassis or as an integral part of a transporting vehicle in which the container constitutes in whole, or in part, the stress member used as a frame. Essentially, it is a permanent part of the transporting vehicle.

CGA. Compressed Gas Association, Inc.

In 1913, a nonprofit service organization was incorporated in New York as the Compressed Gas Manufacturers' Association to promote, develop, represent, and coordinate technical and standardization activities in the compressed gas industries in the interest of safety and efficiency. Among its members are companies and individuals producing compressed, liquefied, and cryogenic gases and their containers (including cargo containers), container appurtenances, and other system components.

From its inception, a major activity of CGA has been the development of standards through the efforts of more than 40 technical committees. The 1932 edition of what is now NFPA 58 (*see History of NFPA 58*) was based upon a draft prepared by CGA's Test and Specification Committee in 1930 and 1931. Much of this work has now passed on to the National Propane Gas Association (NPGA); CGA continues to provide valuable technical assistance to the NFPA LP-Gas Committee.

Charging. See Fill, Filling.

Compressed Gas. Any material or mixture having, when in its container, an absolute pressure exceeding 40 psia (an absolute pressure of 276 kPa) at 70°F (21.1°C) or, regardless of the pressure at 70°F (21.1°C), having an absolute pressure exceeding 104 psia (an absolute pressure of 717 kPa) at 130°F (54.4°C).

This is a very specific definition developed primarily to meet the needs of the U.S. Department of Transportation in regulating the safety of hazardous materials in transportation. All of the LP-Gases covered by NFPA 58 are compressed gases in accordance with this definition. In addition, they are liquefied compressed gases. Many compressed gases, such as oxygen, argon, and nitrogen are stored and transported solely in the gaseous state.

Container. Any vessel, including cylinders, tanks, portable tanks, and cargo tanks, used for the transporting or storing of LP-Gases.

As noted in the definition, there are several kinds of pressure containers used to store and transport LP-Gas, and each must comply with specific fabrication criteria. The word "container" is a generic description and is used by itself in the standard whenever it is unnecessary to cite a specific fabrication type. Otherwise, it is qualified, e.g., DOT container, ASME container.

Container Appurtenances. Items connected to container openings needed to make a container a gastight entity. These include, but are not limited to, pressure relief devices; shutoff, backflow check, excess-flow check, and internal valves; liquid level gauges; pressure gauges; and plugs.

Container Assembly. An assembly consisting essentially of the container and fittings for all container openings. These include shutoff valves, excess-flow valves, liquid level gauging devices, pressure relief devices, and protective housings.

Cylinder. A portable container constructed to DOT (formerly ICC) cylinder specifications or, in some cases, constructed in accordance with the ASME Code of a similar size and for similar service. The maximum size permitted under DOT specifications is 1,000 lb (454 kg) water capacity.

Design Certification. The process by which a product is evaluated and tested by an independent laboratory to affirm that the product design complies with specific requirements.

Direct Gas-Fired Tank Heater. A gas-fired device that applies hot gas from the heater combustion chamber directly to a portion of the container surface in contact with LP-Gas liquid.

Dispenser, Vehicle Fuel. A device or system designed to measure and transfer volumes of LP-Gas into permanently mounted fuel containers on vehicles. (This serves the same purpose as the gasoline dispenser in a gasoline filling station.)

Dispensing Station. Fixed equipment where LP-Gas is stored and dispensed into portable containers. The public can be permitted access to the dispensing station area.

The terms "dispenser, vehicle fuel" and "dispensing station" were added in the 1992 edition as part of the revisions updating the dispensation of propane as an engine fuel. These definitions replace "distributing point" and "dispensing device." Prior to the 1992 edition, a dispensing station was called a distributing point. The definition of dispensing station was further revised in the 1995 edition by deleting the restriction of filling noncargo containers on vehicles so that all requirements of dispensers are the same.

This definition was added as part of a complete revision of the way the dispensing of LP-Gas was addressed in the 1992 edition of NFPA 58. Text was extracted from NFPA 30A, *Automotive and Marine Service Station Code*, for consistency with this widely used code.

The fundamental feature differentiating vehicle fuel dispensers and dispensing stations from bulk plants or industrial plants is that persons other than management or employees also have access to the facility (*see commentary on the definition of bulk plant*). In the 1969 edition of NFPA 58 there was a chapter devoted solely to service stations. The distinct provisions for dispensing fuel in this type of facility are now included in 3-9.

Dispensing stations include cylinder dealers whose distribution is on a smaller scale and open to the public (for a facility at a distributing plant), cylinder refueling facilities at recreational vehicle parks, facilities at hardware, equipment rental, and sporting goods stores for filling gas grill cylinders, and facilities at marinas. Refueling of industrial truck cylinders could be at a bulk plant where exchange cylinders are refilled or at an industrial plant where either exchange cylinders or containers mounted on the industrial truck are refilled. An installation where industrial trucks are refilled at an industrial plant [and the storage facilities are less than 2,000 gal (7.6 m^3) water capacity] is not considered to be a dispensing station since only employees have access to this operation.

DOT. U.S. Department of Transportation.

DOT Cylinder. See Cylinder.

Emergency Shutoff Valve. A shutoff valve incorporating thermal and manual means of closing that also provides for remote means of closing.

Excess-Flow Valve (also called Excess-Flow Check Valve). A device designed to close when the liquid or vapor passing through it exceeds a prescribed flow rate as determined by pressure drop.

Fill, Filling. Transferring liquid LP-Gas into a container.

Filling by Volume. See Volumetric Filling.

Filling by Weight. See Weight Filling.

Fixed Liquid Level Gauge. A type of liquid level gauge using a relatively small positive shutoff valve and designed to indicate when the liquid level in a container being filled reaches the point at which this gauge or its connecting tube communicates with the interior of the container.

Fixed Maximum Liquid Level Gauge. A fixed liquid level gauge that indicates the liquid level at which the container is filled to its maximum permitted filling limit.

Fixed Piping System. Piping, valves, and fittings permanently installed in a location to connect the source of the LP-Gas to the utilization equipment.

Flexible Connector. A short [not exceeding 36 in. (0.91 m) overall length] component of a piping system fabricated of flexible material (such as hose) and equipped with suitable connections on both ends. LP-Gas resistant rubber and fabric (or metal), or a combination of these, or metal only are used. Flexible connectors are used where there is the need for, or the possibility of, greater relative movement between the points connected than is acceptable for rigid pipe.

Float Gauge. A gauge constructed with a float inside the container resting on the liquid surface that transmits its position through suitable leverage to a pointer and dial outside the container, indicating the liquid level. Normally the motion is transmitted magnetically through a nonmagnetic plate so that no LP-Gas is released to the atmosphere.

Gallon. U.S. Standard. 1 U.S. gal = 0.833 Imperial gal = 231 cu in. = 3.785 liters.

Gas. Liquefied Petroleum Gas in either the liquid or vapor state. The more specific terms "liquid LP-Gas" or "vapor LP-Gas" are normally used for clarity.

Gas-Air Mixer. A device, or system of piping and controls, that mixes LP-Gas vapor with air to produce a mixed gas of a lower heating value than the LP-Gas. The mixture thus created normally is used in industrial or commercial facilities as a substitute for another fuel gas. The mixture can replace another fuel gas completely, or can be mixed to produce similar characteristics and then be mixed with the basic fuel gas. Any gas-air mixer that is designed to produce a mixture containing more than 85 percent air is not subject to the provisions of this standard.

GPA. Gas Processors Association.

ICC. U.S. Interstate Commerce Commission.

ICC Cylinder. See Cylinder.

Ignition Source. See Sources of Ignition.

Industrial Occupancy. Includes factories that manufacture products of all kinds and properties devoted to operations such as processing, assembling, mixing, packaging, finishing or decorating, and repairing.

Industrial Plant. An industrial facility that utilizes gas incident to plant operations, with LP-Gas storage of 2,000 gal (7.6 m^3) water capacity or more, and that receives gas by means of tank car, truck transport, or truck lots. Normally LP-Gas is used through piping systems in the plant but also can be used to fill small containers, such as for engine fuel on industrial (i.e., forklift) trucks. Since only plant employees have access to these filling facilities, they are not considered to be distributing points.

As with bulk plants, the public does not have access to liquid transfer operations in an industrial plant facility, as only employees are involved and the LP-Gas storage capacity is 2,000 gal (7.6 m^3) water capacity or more. Actually, storage capacities can be rather large (in some instances these may be industrial tank farms), but not those facilities covered by API Standard 2510. [*See commentary on 1-2.3.1(b).*]

Internal Valve. A primary shutoff valve for containers that has adequate means of actuation and that is constructed in such a manner that its seat is inside the container and that damage to parts exterior to the container or mating flange will not prevent effective seating of the valve.

Labeled. Equipment or materials to which has been attached a label, symbol, or other identifying mark of an organization that is acceptable to the authority having jurisdiction and concerned with product evaluation that maintains periodic inspection of production of labeled equipment or materials and by whose labeling the manufacturer indicates compliance with appropriate standards or performance in a specified manner.

See definition of Listed.

Liquefied Petroleum Gas (LP-Gas or LPG). Any material having a vapor pressure not exceeding that allowed for commercial propane composed predominantly of the following hydrocarbons, either by themselves or as mixtures: propane, propylene, butane (normal butane or isobutane), and butylenes.

Listed. Equipment or materials included in a list published by an organization acceptable to the authority having jurisdiction and concerned with product evaluation that maintains periodic inspection of production of listed equipment or materials and whose listing states either that the equipment or material meets appropriate standards or has been tested and found suitable for use in a specified manner.

NOTE: The means for identifying listed equipment may vary for each organization concerned with product evaluation, some of which do not recognize equipment as listed unless it is also labeled. The authority having jurisdiction should utilize the system employed by the listing organization to identify a listed product.

The authorities having jurisdiction will customarily accept or even require (whether or not NFPA 58 requires) the use of listed equipment whenever it is available. In the United States and Canada, the major LP-Gas equipment listing agencies customarily recognized as such by the authorities having jurisdiction are Underwriters Laboratories Inc., Underwriters' Laboratories of Canada, Inc., the Factory Mutual Engineering Corporation, the American Gas Association Laboratories, and the Canadian Gas Association Laboratories. Many other organizations perform such services, however, and equipment listed by them is acceptable provided the organization is acceptable to the authority having jurisdiction.

A point of some confusion is that not all the above organizations use the term "listed." For example, AGA and CGA Laboratories use the term "certified" and Factory Mutual uses the term "approved." Both are synonymous with "listed" in this definition.

Load, Loading. See Filling.

LPG. See Liquefied Petroleum Gas.

LP-Gas. See Liquefied Petroleum Gas.

LP-Gas System. An assembly consisting of one or more containers with a means for conveying LP-Gas from the container(s) to dispensing or consuming devices (either continuously or intermittently) and that incorporates components intended to achieve control of quantity, flow, pressure, or state (liquid or vapor).

Magnetic Gauge. See Float Gauge.

Mobile Containers. Containers that are permanently mounted on a vehicle and are connected for uses other than engine fuel.

This definition has been added to the 1992 edition in conjunction with the revisions to 2-3.3.2 to define permanently mounted non-engine fuel containers on vehicles, and is used in the new Table 2-3.3.2(a), Container Connection and Appurtenance Requirements.

Mounded Container. An ASME container designed for underground service installed above the minimum depth required for underground service and covered with earth, sand, or other material, or an ASME container designed for aboveground service installed above grade and covered with earth, sand, or other material.

Movable Fuel Storage Tenders, Including Farm Carts. Containers not in excess of 1,200 gal (4.5 m^3) water capacity, equipped with wheels to be towed from one location to another. These are basically non-highway vehicles but can occasionally be moved over public roads or high ways for short distances for use as fuel supplies for farm tractors, construction machinery, and similar equipment.

Multipurpose Passenger Vehicle. A motor vehicle with motive power, with the exception of a trailer, designed to carry 10 or fewer persons that is constructed on a truck chassis or with special features for occasional off-road operations.

This definition and an amendment to 8-2.2(a)(6) comprised TIA 58-83-1 (issued October 8, 1982) originally. Such vehicles are identified by a label affixed to the vehicle in accordance with federal regulations.

NFPA. National Fire Protection Association.

NPGA. National Propane Gas Association.

The National Propane Gas Association, originally the Bottled Gas Association of America and later the National LP-Gas Association, is the national association for the liquefied petroleum gas industry. Among its membership are producers of LP-Gas, wholesale and retail marketers, gas appliance and equipment manufacturers and distributors, tank and cylinder fabricators, transport firms, and others. NPGA has nearly 4,000 members throughout the United States and 33 other countries in 1995.

One of the primary reasons for the establishment of the Association in 1932 was to afford all interested divisions of the industry full opportunity to aid in making the rules that safeguard their industry. The association was first affiliated with the Compressed Gas Manufacturers' Association (now the Compressed Gas Association). This was a natural beginning, as the CGMA was the only technical association in existence that could cope with the problems involving gases in cylinders, the primary method of handling LP-Gases at that time.

Today the NPGA is a separate and distinct organization and maintains a liaison with CGA, but carries on its own technical activities through its technical committees. One of these, the NPGA Technology and Standards Committee, has engineering and technical representation from all phases of the industry. In addition, there are Advisory Members representing other related industry associations, testing laboratories, regulatory officials, insurance organizations, and others, who provide much assistance with the Technical Committee's standards development work. Many recommendations for new provisions and amendments of existing provisions in NFPA 58 are initiated by this technical committee as industry-sponsored recommendations. These, in turn, are considered by the NFPA Committee on LP-Gases, which reflects a broader interest. Because the Technology and Standards Committee has at its disposal a large amount of technical talent, the NFPA Committee on LP-Gases has at times referred matters to this group for initial study and recommendations.

Overpressure Shutoff Device. A device that shuts off the flow of LP-Gas vapor when the outlet pressure of the regulator reaches a predetermined maximum allowable pressure.

This definition was added in the 1995 edition as part of a change to require protection against excessive pressure in propane piping systems in buildings. The changes limit the outlet pressures of a first-stage pressure regulator to 10.0 psi and a second-stage pressure regulator to 14 in WC. They also require a feature (which can be an overpressure shutoff device) to prevent the pressure at the discharge of the second stage regulator from exceeding 2.0 psi.

Permanent Installation. See Stationary Installation.

Piping, Piping Systems. Pipe, tubing, hose, and flexible rubber or metallic hose connectors with valves and fittings made into complete systems for conveying LP-Gas in either the liquid or vapor state at various pressures from one point to another.

Point of Transfer. The location where connections and disconnections are made or where LP-Gas is vented to the atmosphere in the course of transfer operations.

The release of a small quantity of LP-Gas in either vapor or liquid form cannot be avoided during liquid transfer operations when filling a container. In all instances liquid is trapped between the valve on the end of the transfer hose and the filler valve on the container or piping. When the connection is disengaged, a small amount of liquid escapes to the atmosphere. If the filling level is being gauged with a fixed liquid level, fixed maximum liquid level, rotary, or slip tube gauge (*see definitions*), vapor and liquid are released during the operation. Therefore, the formation of a flammable mixture is unavoidable in the immediate vicinity of the points where the LP-Gas is released, and it is necessary that there be space for the gas to dissipate without ignition.

In most residential and commercial fixed installations, the filling connection is fastened to the container (it is, in fact, a "container appurtenance"), and the container siting distances are adequate to control the hazard of ignition. Where large containers are installed, it would be unsafe to require the delivery person to climb to the top to connect and disconnect the hose, so the connection is made to a filler pipe located near the ground some distance from the container. Even though container siting criteria would not necessarily address this hazard, established good practice in these large installations generally has provided an adequate safeguard.

With the growing general public ownership of portable containers associated with recreational applications, however, the location of the points where this release was occurring was no longer controlled by container siting criteria or by the expertise brought to bear in large installations. Moreover, the actual liquid transfer was being performed by individuals not employed by LP-Gas industry firms and to whom this operation was incidental to their principal occupations. Experience soon showed that the standard needed to incorporate a means by which the location of these release points could be better controlled.

The point-of-transfer concept was incorporated into NFPA 58 for the first time in 1972.

Portable Container. A container designed to be moved readily, as distinguished from containers designed for stationary installations. Portable containers, designed for transportation, filled to their maximum filling limit include "cylinders," "cargo tanks," and "portable tanks," all three of which are separately defined. Containers designed to be readily moved from one usage location to another, but substantially empty of product, are "portable storage containers" and are separately defined.

Portable Storage Container. A container similar to, but distinct from, those designed and constructed for stationary installation, designed so that it can be moved readily over the highways, substantially empty of liquid, from one usage location to another. Such containers either have legs or other supports attached, or are mounted on running gear (such as trailer or semitrailer chassis) with suitable supports that can be of the fold-down type, allowing them to be placed or parked in a stable position on a reasonably firm and level surface. For large-volume, limited-duration product usage (such as at construction sites and normally for 12 months or less), portable storage containers function in lieu of permanently installed stationary containers.

Portable Tank (also called Skid Tank). A container of more than 1,000 lb (454 kg) water capacity used to transport LP-Gas handled as a "package," that is, filled to its maximum permitted filling limit. Such containers are mounted on skids or runners and have all container appurtenances protected in such a manner that they can be safely handled as a "package."

Pressure Relief Device. A device designed to open to prevent a rise of internal fluid pressure in excess of a specified value due to emergency or abnormal conditions.

Pressure Relief Valve. A type of pressure relief device designed to both open and close to maintain internal fluid pressure. Pressure relief valves are further characterized as follows:

This definition identifies the different types of relief valves in use. Note that the standard contains only a list of the different pressure relief valves. The descriptions and drawings are included in Appendix A, which means that they are advisory material included for information purposes only.

*External Pressure Relief Valve.** A relief valve that is located entirely outside the container connection except the threaded portion, which is screwed into the container connection, and that has all of its parts exposed to the atmosphere.

*Flush-type Full Internal Pressure Relief Valve.** A full internal relief valve in which the wrenching section is also within the container connection, except for pipe thread tolerances on make up.

*Full Internal Pressure Relief Valve.** A relief valve in which all working parts are recessed within the container connections, and the spring and guiding mechanism are not exposed to the atmosphere.

*Internal Spring-type Pressure Relief Valve.** A relief valve in which only the spring and stem are within the container connection, and the spring and stem are not exposed to the atmosphere. The exposed parts of the relief valve have a low profile.

*Sump-type Full Internal Pressure Relief Valve.** A relief valve in which all working parts are recessed within the container connection, but the spring and guiding mechanism are exposed to the atmosphere.

A-1-6 **External Pressure Relief Valve.** Describes the type of relief valves used on older domestic tanks, relief valve manifolds, and piping protection.

Flush-type Full Internal Pressure Relief Valve. Describes the type of relief valve being required on cargo vehicles in most states.

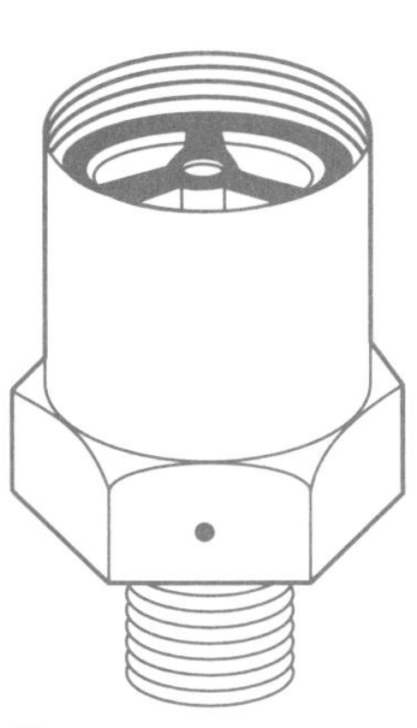

Figure A-1-6(a) *External relief valve.*

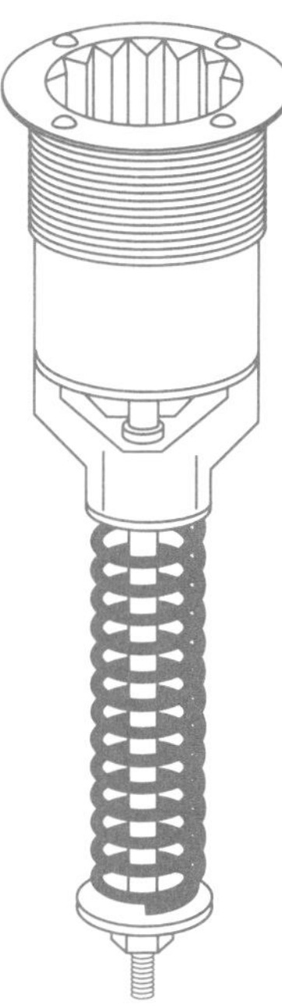

Figure A-1-6(b) *Flush-type full internal relief valve.*

Full Internal Pressure Relief Valve. Describes the type of relief valve being converted to for engine fuel use.

Internal Spring-type Pressure Relief Valve. Describes the type of relief valve used on modern domestic tanks; looks similar to full internal relief valve but has seat and poppet above the tank connection.

Sump-type Full Internal Pressure Relief Valve. Describes the type of relief valve used on older engine fuel tanks.

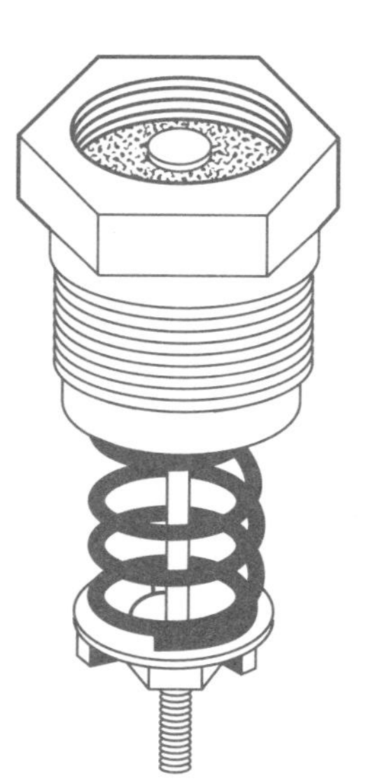

Figure A-1-6(c) *Full internal relief valve.*

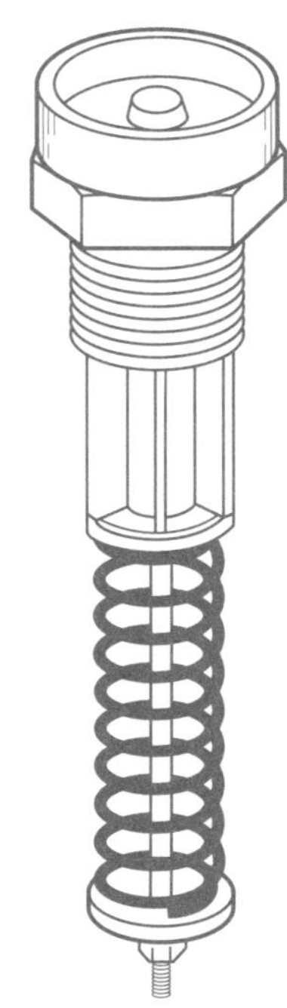

Figure A-1-6(d) *Internal spring-type relief valve.*

PSI, PSIG, and PSIA. Pounds per square inch, pounds per square inch gauge, and pounds per square inch absolute, respectively.

Quick Connectors. Devices used for quick connections of the acme thread or lever-cam types. This does not include devices used for cylinder-filling connections.

Quick connectors, in the context of this definition and as used in 2-4.6.2, are generally end fittings such as hose couplings on transfer hose used to make connections for loading and unloading cargo tank vehicles, railroad tank cars, etc. They may also be used as connectors on swivel-type transfer piping. There are other quick connectors used in LP-Gas systems, such as a quick-closing couplings for engine fuel refueling and cylinder filling, quick-disconnect devices

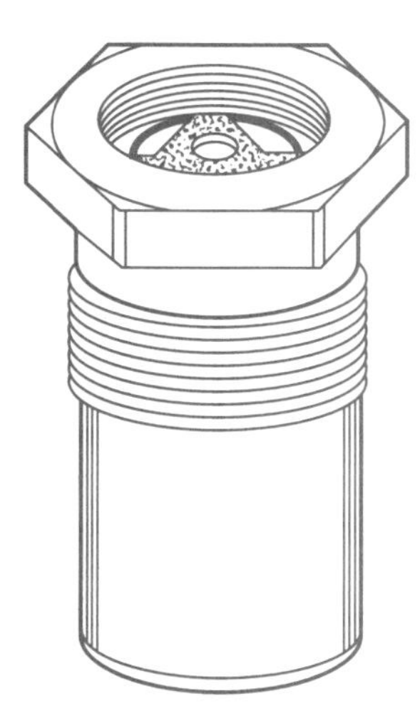

Figure A-1-6(e) *Sump-type full internal relief valve.*

for gas-fired barbecue grills, and connecting gas systems between two sections of mobile homes. This definition, however, is limited only to the two basic types used for transfer of LP-Gas as cargo.

Regulator, Automatic Changeover. An integral two-stage regulator that combines two high pressure regulators and a second-stage regulator into a single unit. It incorporates two inlet connections and a service-reserve indicator and is designed for use with dual or multiple cylinder installations. The system automatically changes the LP-Gas vapor withdrawal from the designated service cylinder(s) when depleted to the designated reserve cylinder(s) without interruption of service. The service reserve indicator gives a visual indication of the cylinder(s) that are supplying the system.

Regulator, First Stage. A pressure regulator for LP-Gas vapor service designed to reduce pressure from the container to 10.0 psi (69 kPa) or less.

Regulator, High Pressure. A pressure regulator for LP-Gas liquid or vapor service designed to reduce pressure from the container to a lower pressure in excess of 1.0 psi (6.9 kPa).

Regulator, Integral Two Stage. A pressure regulator that combines a high pressure regulator and a second-stage regulator into a single unit.

Regulator, Second Stage. A pressure regulator for LP-Gas vapor service designed to reduce first-stage regulator outlet pressure to 14 in. W.C. (4.0 kPa) or less.

Regulator, Single Stage. A pressure regulator for LP-Gas vapor service designed to reduce pressure from the container to 1.0 psi (6.9 kPa) or less.

These definitions were added in the 1995 edition as part of changes to provide a means to prevent excess pressure from reaching appliances. A maximum of 2.0 psi is permitted, and manufacturers of appliance valves and controls design their products not to fail at pressures up to 2.0 psi. Note that these changes are restrictive and do take some flexibility away from the LP-Gas installer. With the limit of first-stage outlet pressure to 10.0 psi, the installer can no longer increase the capacity of an existing interstage piping system by increasing the pressure to between 10 and 20 psi if appliances are added. The piping system now must be changed in that case. This restriction is done to improve safety, and some flexibility must be sacrificed for this worthwhile cause. See also the new definitions of Overpressure Shutoff Device, Two-Stage Regulator System, and Sections 2-5.8 and 3-2.6.

Rotary Gauge. A variable liquid level gauge consisting of a small positive shutoff valve located at the outer end of a tube, the bent inner end of which communicates with the container interior. The tube is installed in a fitting designed so that the tube can be rotated with a pointer on the outside to indicate the relative position of the bent inlet end. The length of the tube and the configuration to which it is bent are suitable for the range of liquid levels to be gauged. By means of a suitable outside scale, the level in the container at which the inner end begins to receive liquid can be determined by the pointer position on the scale at which a liquid-vapor mixture is observed to be discharged from the valve.

A rotary gauge is built with an open-ended bent tube within the container so that the open end of the tube can be rotated to determine the point of liquid level in the container. The gauge can be rotated 360° and is generally made for containers up to 130 in. (3.3 m) internal diameter of the container.

The gauge can be end- or side-mounted, and can be used either on stationary or mobile storage containers. On straight-bodied delivery trucks, it is generally mounted in the head of the cargo tank because the length of the average unit is such that being at the end does not throw off the gauging accuracy to a great extent when the truck is not on a level surface. On the other hand, on longer semitrailer transports, rotary gauges must be side-mounted so as to read accurately the actual liquid level in such containers. As a result, the proper side location for the rotary gauge is at the center longitudinal point; the liquid level can then be gauged at the average level when the transport unit is not on a level surface. The length of the tube within the container must be suitable for the particular container in which it is installed.

To determine the liquid level in the container, the rotary gauge should be set with the tube in the vapor space and then slowly rotated toward the liquid level with the vent valve open. When in the vapor space there will be a vapor discharge from the vent but, when the gauge reaches the liquid level, liquid (a white fog) will start spewing out of the vent opening. It is then good practice to move the rotary gauge back to the vapor space slowly and again to the liquid state so as not to get an erroneous reading. The percentage full point is indicated on the dial on the outside of the gauge to determine the content level.

On large transports, a heavy metal hanger should be welded to the inside top of the container and attached to the rotary gauge stem to give it support so that in the movement over the road the bent tube will not be bent out of proper position. This also prevents or lessens vibration, which could eventually crack or break the tube.

The rotary gauge is not a gauge to be used for accurate determination of the contents of a container, but only for general use in determining the approximate quantity in the container. The rotary gauge should never be used to determine the maximum filling level of a container; this should be done by a fixed liquid level gauge set in accordance with 2-3.4.2.

Service Head Adapter. A transition fitting for use with polyethylene pipe or tubing that is recommended by the manufacturer for field assembly and installation at the aboveground termination end of an anodeless riser (*see definition*). This fitting makes the transition from polyethylene pipe or tubing to a gas-carrying Schedule 40 steel fitting.

Skid Tank. See Portable Tank.

Slip Tube Gauge. A variable liquid level gauge in which a relatively small positive shutoff valve is located at the outside end of a straight tube, normally installed vertically, that communicates with the container interior. The installation fitting for the tube is designed so that the tube can be slipped in and out of the container and so that the liquid level at the inner end can be determined by observing when the shutoff valve vents a liquid-vapor mixture.

Sources of Ignition. Devices or equipment that, because of their modes of use or operation, are capable of providing sufficient thermal energy to ignite flammable LP-Gas vapor-air mixtures when introduced into such a mixture or when such a mixture comes into contact with them, and which will permit propagation of flame away from them.

Although the standard has contained provisions to control ignition sources since its inception, this definition was adopted first in the 1979 edition. It became necessary because of the existence of devices or equipment that could ignite a flammable LP-Gas–air mixture under certain conditions although not in the manner in which they were normally used. In addition, the definition clarifies the fact that other devices might produce ignition but would not permit a flame to propagate away from them, in which case they are not a hazard. Certain Class I, Division 1, electrical devices using gap flame quenching principles and devices equipped with flame arrestors are examples of such devices.

Some sources of ignition are:

1. Devices or equipment—such as gas- or oil-burning appliances, torches, matches, and lighters—producing, or capable of producing, flame
2. Devices or equipment—such as motors, generators, switches, lights, and wiring and welding machines—producing, or capable of producing, electrical arcs or sparks of sufficient energy. Such devices or equipment listed for

Class I, Group D, locations are not sources of ignition when properly used and maintained in the classified areas prescribed in this standard.

3. Devices or equipment having, or capable of having, surfaces at temperatures exceeding 700°F (371°C)

4. Devices or equipment—such as grinders, metal saws, chipping hammers, and flint lighters—producing, or capable of producing, mechanical (struck) sparks of sufficient energy. Unpowered hand tools used only by one individual at a time and for their intended purposes are not considered sources of ignition.

5. Cigarettes and other smoking materials

6. Because they constitute a readily communicating pathway, gravity or mechanical air intakes and exhaust terminals of ventilation systems for structures housing the ignition sources described. These include air intakes and exhaust terminals for gas- or oil-burning appliances, including those for direct-vent appliances, and air intakes and exhaust terminals for air conditioners.

Operating spark-ignition internal combustion engines do not appear, based on experience, to be an ignition source; however, they have been an ignition source during the act of being started. Operating diesel engines have been identified as the ignition source in fire reports, usually as the result of burning carbon being blown out of their exhausts. Such engines tend to overspeed when LP-Gas is drawn into their intake manifolds.

Special Protection. A means of limiting the temperature of an LP-Gas container for purposes of minimizing the possibility of failure of the container as the result of fire exposure.

Where required in this standard, special protection consists of any of the following: applied insulating coatings, mounding, burial, water spray fixed systems, or fixed monitor nozzles, meeting the criteria specified in this standard (*see 3-10.3*), or by any means listed (*see definition of Listed*) for this purpose.

In the early 1970s, the United States was subjected to an outbreak of Boiling Liquid-Expanding Vapor-Explosions (BLEVEs) of LP-Gas railroad tank cars as a result of derailments. Most of these were caused by container failure resulting from overheating of the container metal (steel) through contact with flames from flammable or combustible materials released from other cars.

While this hazard was beyond the scope of NFPA 58 and, in fact, the circumstances involved control problems not found in containers covered by the standard, these accidents resulted in widespread fear of all large LP-Gas containers. Notable in this respect was an increasing reluctance by firefighters to approach an LP-Gas container exposed to fire in order to limit the container temperature by the application of water from hose streams. As this procedure was, and is, a fundamental and universal BLEVE prevention safeguard, the NFPA Technical Committee on LP-Gases and the industry were concerned that a reduction in its use would increase the frequency and severity of BLEVEs.

A joint task force of the NFPA Technical Committee and the NPGA's Technology and Standards Committee (augmented by representatives of railroad and U.S. DOT groups studying the tank car problem) concluded that additional provisions were needed to reduce the chances for BLEVEs caused by fire exposure. Two new provisions were first included in the 1976 edition of NFPA 58.

The first provision was for the emergency shutoff valves described in 2-4.5.4 and 3-2.8.10. These valves were considered important enough to require retroactive installation in all of the larger facilities as prescribed in 3-2.8.10.

The second provision was for special protection as stipulated in 3-10.2.3, Exception No. 2. It is noted, however, that the need for this protection is predicated upon the results of a required fire safety analysis.

The five modes of special protection cited in the above definition represent those currently recognized and for which standards exist. However, the definition is broad enough to recognize other means, provided they are listed for the purpose.

Stationary Installation ("Permanent" Installation). An installation of LP-Gas containers, piping, and equipment for indefinite use at a particular location; an installation not normally expected to change in status, condition, or place.

Two-Stage Regulator System. An LP-Gas vapor delivery system that combines a first-stage regulator and a second-stage regulator(s), or an integral two-stage regulator.

UL. Underwriters Laboratories Inc.

Founded in 1894, Underwriters Laboratories Inc. is chartered as a not-for-profit, independent organization performing testing for public safety. It maintains laboratories for the examination and testing of devices, systems, and materials to determine their relation to life, fire and casualty hazards, and crime prevention.

UL-listed materials within the scope of NFPA 58 are included in the following UL directories: *Hazardous Location Equipment; Marine Products; Automotive, Burglary Protection, and Mechanical Equipment; and Gas and Oil Equipment.*

Universal Cylinder. A DOT cylinder specification container, constructed and fitted with appurtenances in such a manner that it can be connected for service with its longitudinal axis in either the vertical or the horizontal position, and so that its fixed maximum liquid level gauge, pressure relief device(s), and withdrawal appurtenance will function properly in either position.

Vaporizer. A device other than a container that receives LP-Gas in liquid form and adds sufficient heat to convert the liquid to a gaseous state.

This was editorially revised in the 1986 edition to facilitate the use of standardized classifications for the many types of liquefied gas vaporizers in several NFPA standards.

Vaporizer, Direct-Fired. A vaporizer in which heat furnished by a flame is directly applied to some form of heat exchange surface in contact with the liquid LP-Gas to be vaporized. This classification includes submerged-combustion vaporizers.

Some have considered submerged-combustion vaporizers to be indirect-fired because they contain a noncombustible heat transfer medium. For purposes of this standard, however, they are treated as direct-fired. The second sentence was added in the 1986 edition to make this clear.

Vaporizer, Electric. A unit using electricity as a source of heat.

Because electric vaporizers can be designed so that they are not a source of ignition (unlike a vaporizer using fuel-fired burners), they can be treated differently and thus need to be defined separately. Electric vaporizers are further separated into two types, each of which may be designed not to be a source of ignition.

Direct Immersion Electric Vaporizer. A vaporizer wherein an electric element is immersed directly in the LP-Gas liquid and vapor.

Indirect Electric Vaporizer. An immersion type wherein the electric element heats an interface solution in which the LP-Gas heat exchanger is immersed or heats an intermediate heat sink.

Vaporizer, Indirect (also called Indirect-Fired). A vaporizer in which heat furnished by steam, hot water, the ground, surrounding air, or other heating medium is applied to a vaporizing chamber or to tubing, pipe coils, or other heat exchange surface containing the liquid LP-Gas to be vaporized; the heating of the medium used being at a point remote from the vaporizer.

Vaporizer, Waterbath (also called Immersion Type). A vaporizer in which a vaporizing chamber, tubing, pipe coils, or other heat exchange surface containing liquid LP-Gas to be vaporized is immersed in a temperature-controlled bath of water, water-glycol combination, or other noncombustible heat transfer medium that is heated by an immersion heater not in contact with the LP-Gas heat exchange surface.

Although the inference was that the heat transfer medium was not capable of burning, the word "noncombustible" was added in the 1986 edition to make this clear.

Vaporizing-Burner (also called Vaporizer-Burner and Self-Vaporizing Liquid Burner). A burner containing an integral vaporizer that receives LP-Gas in liquid form and that uses part of the heat generated by the burner to vaporize the liquid in the burner so that it is burned as a vapor.

Variable Liquid Level Gauge. A device to indicate the liquid level in a container throughout a range of levels. See Float, Rotary, and Slip Tube Gauge.

Volumetric Filling. Filling a container by determination of the volume of LP-Gas in the container. Unless a container is filled by a fixed maximum liquid level gauge, correction of the volume for liquid temperature is necessary.

Gauging the quantity of liquid in a container by means of its volume rather than its weight is a practical necessity on ASME containers, cargo vehicles, tank cars, and larger DOT containers, and is a convenience on smaller DOT containers.

Slip tube, rotary, and fixed tube gauges are used. Because these gauges must be manipulated during gauging and, with the exception of the fixed maximum liquid level gauge, the temperature of the liquid during filling must be taken into consideration, the volumetric method may be somewhat less precise than the weight method in practice. The degree of accuracy has seldom been a problem. However, the smaller the container, the greater the opportunity for overfilling, whether filling by volume or weight.

Appendix F discusses at considerable length the factors involved in volumetric filling.

Volumetric Loading. See Volumetric Filling.

Many years ago the U.S. Department of Transportation (DOT) drew up regulations that limit the amount of gases that may be charged into containers. NFPA 58 provisions are based on these regulations. With respect to LP-Gas, which is a liquefied compressed gas, the filling limitation is primarily for the purpose of preventing excess hydrostatic pressures if liquid expansion results in a liquid-full container. A certain outage is left in the container to account for this expansion. DOT regulations use a filling-density concept to designate the maximum amount of liquid LP-Gas that may be placed into a container and, thus, what the outage should be. Filling density is a ratio of the quantity of LP-Gas in a container to the water capacity of the container. Water capacity of a container is a measure of its total volume and is the only way to express its gross capacity.

Water capacity may be expressed in gallons or pounds of water. Where expressed in pounds, it reflects both the feasibility and traditional practice of filling by weight using a scale. This is obviously limited to containers small enough to be moved by a person—usually up to about 240 lb (109 kg) water capacity and generally constructed in accordance with DOT requirements. Where capacity is expressed in gallons, it usually applies to containers larger than 240 lb (109 kg) water capacity. In recent years, however, it has become more common to find containers as small as 50 lb (23 kg) water capacity filled by volume.

Water Capacity. The amount of water, in either lb or gal, at 60°F (15.6°C) required to fill a container liquid full of water.

Weight Filling. Filling containers by weighing the LP-Gas in the container. No temperature determination or correction is required, as a unit of weight is a constant quantity regardless of temperature.

References Cited in Commentary

The following publications are available from the National Fire Protection Association, 1 Batterymarch Park, P.O. Box 9101, Quincy, MA 02269-9101.

NFPA 30A, *Automotive and Marine Service Station Code*, 1990 edition.

NFPA 51, *Standard for the Design and Installation of Oxygen-Fuel Gas Systems for Welding, Cutting, and Allied Processes,* 1992 edition.

NFPA 54, *National Fuel Gas Code,* 1988 edition.

NFPA 302, *Fire Protection Standard for Pleasure and Commercial Motor Craft,* 1989 edition.

The following publication is available from the American National Standards Institute, 11 West 42nd Street, New York, NY 10036.

ANSI Z49.1-88, *Safety in Welding and Cutting.*

For information on the ANSI Z21 and Z83 Series, see the reference list following Chapter 2.

The following publication is available from the American Society of Mechanical Engineers, 345 East 47th Street, New York, NY 10017.

ANSI/ASME B31.8-89, *Gas Transmission and Distribution Piping Systems.*

The following publication is available from the American Society for Testing and Materials, 1916 Race Street, Philadelphia, PA 19103.

ASTM D 1835, *Standard Specification for Liquefied Petroleum (LP) Gases.*

The following publication is available from the Compressed Gas Association, Inc., 1235 Jefferson Davis Highway, Arlington, VA 22202.

ANSI B57.1/CGA V-1/CSA B 96, *Standard Compressed Gas Cylinder Valve Outlet and Inlet Connections.*

The following publication is available from the Gas Processors Association, 1812 First National Building, Tulsa, OK 74103.

GPA Standard 2410-88, *Liquefied Petroleum Gas Specifications for Test Methods.*

The following publications are available from the National LP-Gas Association, 1600 Eisenhower Lane, Lisle, IL 60521.

A Study of Cleaning Methods for LP-Gas Transport.

Certified Employee Training Program.

The following publications are available from Underwriters Laboratories Inc., 333 Pfingsten Road, Northbrook, IL 60062.

Automotive, Burglary Protection, and Mechanical Equipment, Directory, 1991.

Gas and Oil Equipment, Directory, 1991.
Hazardous Location Equipment, Directory, 1991.
Marine Products, Directory, 1991.

The following publication is available from the U.S. Government Printing Office, Washington, DC. Parts 171–190 are also available from the Association of American Railroads, American Railroads Building, 1920 L Street, NW, Washington, DC 20036 and the American Trucking Associations, Inc., 2201 Mill Road, Alexandria, VA 22314.

Code of Federal Regulations, Title 49, Parts 100–199.

2

LP-Gas Equipment and Appliances

2-1 Scope.

2-1.1 This chapter includes the basic provisions for individual components, or for such components shop-fabricated into subassemblies, container assemblies, or complete container systems.

The intent of Chapter 2 is to include all information needed by the designer and fabricator of LP-Gas system components, with the exception of cargo tank vehicles, which are covered by Chapter 6.

2-1.2 The field assembly of components, subassemblies, container assemblies, or complete container systems into complete LP-Gas systems is covered by Chapter 3. (*See definition of LP-Gas System.*)

2-2 Containers.

2-2.1 General.

2-2.1.1 This section includes design, fabrication, and marking provisions for containers, and features normally associated with container fabrication, such as container openings, appurtenances required for these openings to make the containers gastight entities, physical damage protecting devices, and container supports attached to, or furnished with, the container by the manufacturer.

Other sections in Chapter 2 cover items which are needed to make the container a complete system to safely contain LP-Gases:

- 2-3 Container Appurtenances
- 2-4 Piping, (including Hoses, Fittings and Valves)
- 2-5 Equipment

The final section, 2-6, covers appliances.

2-2.1.2 Refrigerated containers shall comply with Chapter 9.

Because of the inherent difference between refrigerated and pressurized storage containers, requirements for the less common refrigerated containers are placed separately in NFPA 58, in Chapter 9. Refrigerated storage is normally used for storing a large quantity of LP-Gas (1,000,000 gal or more) as a liquid at its normal atmospheric pressure boiling point (about –40°). At that temperature, the liquid can be stored at a very low pressure, usually less than 5 psi.

2-2.1.3* Containers shall be designed, fabricated, tested, and marked (or stamped) in accordance with the Regulations of the U.S. Department of Transportation (DOT), the ASME *Boiler and Pressure Vessel Code*, "Rules for the Construction of Unfired Pressure Vessels," Section VIII, or the API-ASME *Code for Unfired Pressure Vessels for Petroleum Liquids and Gases* applicable at the date of manufacture; and as follows: (*See Appendices C and D.*)

In the 1995 edition, this paragraph was revised (with a corresponding change to the definition of ASME Code) by removal of the restriction that pressure vessels be constructed according to ASME Code, Section VIII, Division 1, and now permits construction to Section VIII, Division 1 or 2. Refer to the commentary following the definition of ASME Code for more information on Division 2.

Containers that have passed their useful service life due to corrosion, mechanical damage, or lack of a nameplate must be removed from service. As most ASME containers are the property of a propane marketer, the marketer will usually accumulate a number of containers destined for scrap and ship them to a scrap metal dealer. The marketer is advised to take steps to insure that these containers are in fact destroyed. The editor recently received a phone call from, and later met with, a state's fire marshal and chief pressure vessel inspector who discovered a number of propane tanks that had recently been sold for, and were being used in, propane service that did not meet the ASME Code requirements. Investigation determined that the tanks had been sold to a scrap dealer and had subsequently been "repaired" by the use of fiberglass patches, by sanding corroded areas and repainting. The tanks were sold to propane consumers who were persuaded that they would be able to purchase propane at a lower price if they owned their own container. To the editor's knowledge, this case is being treated as a criminal matter.

There is nothing in NFPA 58 to prevent a consumer from purchasing his or her own LP-Gas tank. When this is done, the owner is not relieved of the requirements of the standard that the container be maintained. It is the responsibility of the propane marketer to take reasonable steps to verify that a container is fit for service before filling the container. This applies equally to a 20-lb gas grill cylinder, a 500-gal tank at a residence, and a 30,000-gal container at an industrial facility.

(a) Adherence to applicable ASME Code Case Interpretations and Addenda shall be considered as compliance with the ASME Code.

There has been some confusion as to the intent and application of this paragraph, particularly in regard to containers built to ASME-API specifications U-200 and U-201, and ASME specifications U-68 and U-69 constructed to codes in existence prior to 1949. As these containers have a very long service life when properly maintained, many remain in use and are sometimes relocated and reinstalled. It is the Committee's intent that use, relocation, and reinstallation of such containers be permitted to continue provided the requirements of 1-2.5, Retroactivity, are met; 1-2.5 states that the provisions of the current edition of NFPA 58 not be applied to "equipment . . . in compliance with the provisions of this standard in effect at the time of manufacture." The reason for allowing these old containers to continue to be used is simply that there has been no evidence that there are any flaws in the design and construction of these containers which become evident after over 40 years of use. At the time these containers were built they incorporated the then state of the art in design, metallurgy and fabrication techniques. While it is true that these have improved significantly over the years, there is no evidence that these containers are less safe that when they were built, where they are properly maintained.

There are a number of nondestructive tests that can be used to determine whether the container is in compliance with the standard in effect at the time of manufacture. These include hydrostatic testing, ultrasonic metal thickness testing and others. The person performing the evaluation must be familiar with the code to which the container was constructed and the test method to be used. Each test method has advantages and disadvantages. Hydrostatic testing, for example, will verify that a container is suitable for continued service; however, because water is heavier than propane it must first be determined that both the container and its foundations can support the additional weight without causing damage. This is a real concern in larger containers.

(b) Containers fabricated to earlier editions of regulations, rules, or codes listed in 2-2.1.3 and the ICC *Rules for Construction of Unfired Pressure Vessels*, prior to April 1, 1967, shall be permitted to be continued in use in accordance with 1-1.4.

A-2-2.1.3 Prior to April 1, 1967, regulations of the U.S. Department of Transportation were promulgated by the Interstate Commerce Commission. In Canada, the regulations of the Canadian Transport Commission apply. Available from the Canadian Transport Commission, Union Station, Ottawa, Canada.

Construction of containers to the API-ASME Code has not been authorized after July 1, 1961.

To the editor's knowledge, all pressurized LP-Gas containers in the U.S., Canada, and other countries served by marketers based in the U.S. and Canada comply with 2-2.1.3. Many other countries that use NFPA 58 have their own container requirements. Furthermore, a safe container can be built using other criteria, and this flexibility is desirable in the standard.

Containers originally used for LP-Gas were Interstate Commerce Commission (ICC) cylinders. DOT/ICC restricts cylinders to 1,000 lb (454 kg) of water capacity and less. In the early days, there were some ICC specifications, such as ICC-26 and others, but these have not been used for many years. The current basic container for reusable service is a DOT 4BA-240 cylinder (steel construction) and a DOT 4E specification (aluminum construction). There are also "throw-away" cylinders, such as the hand torch cylinders, which have different DOT specifications (or old ICC specifications)—currently DOT 39.

Even though the responsibilities of the U.S. Department of Transportation and the Canadian Transport Commission are restricted to the transportation environment while much of the application of NFPA 58 is concerned with container usage and storage, there are compelling reasons why the standard must recognize these containers. From a safety point of view, it would be extremely hazardous to require that LP-Gas be transferred from a DOT shipping container to another use or storage container on the premises of the user or storer. Such a procedure would also be adverse from an economic standpoint. However, the NFPA Technical Committee continuously monitors DOT and CTC standards to ensure that they are adequate for the applications covered by NFPA 58. For example, the Committee did not permit the 4E (aluminum) cylinder for several years until its use and storage safety had been demonstrated by tests and experience.

While all DOT/ICC/CTC containers (cylinders) are portable in the sense that they are designed to be transportable full of product, when it is considered that such containers commonly hold from 1 lb (0.5 kg) to 420 lb (190 kg) of propane, the degree of portability varies widely. In practice, cylinders in portable service are those that can be moved with reasonable ease by a strong individual, or those that contain up to about 100 lb (45 kg) of LP-Gas [about 200 lb (90 kg) total weight of cylinder and product]. These containers are usually filled at a location other than where they are installed or used, although it is common to see 100-lb cylinders "permanently" installed at residential and commercial locations being filled on site. When this is done the standard cylinder valve

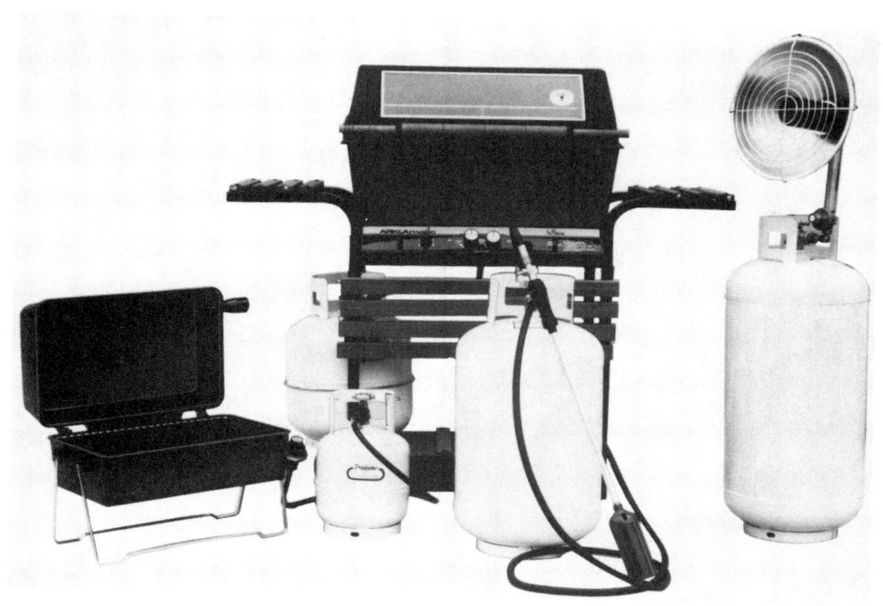

Figure 2.1 DOT/ICC/CTC portable cylinders. (Photo courtesy of Manchester Tank Company.)

is replaced with a multiport valve so that the cylinder can be refilled without dismantling the piping. Containers larger than these usually are filled at the installed location and are considered to be in stationary service.

A DOT/ICC cylinder is required to be marked with certain information—usually on the cylinder itself or on its neck ring or collar. This information is especially vital in determining the suitability of the cylinder for continued service indicated by the retest date.

The second type of container is the "bulk container," usually an unfired pressure vessel built to the ASME Code. The original LP-Gas bulk containers were riveted, but there was great difficulty in keeping them leak-free around the rivets and at the seams. The industry then began to forge-weld containers, and in the 1930s began to use the fusion-welded pressure vessels. At that time the ASME Code specifications were U-68 and U-69, which had a safety factor of 5 to 1 (design operating pressure to theoretical burst pressure). The normal working pressure of these containers was 200 psi (1.4 MPa). NFPA 58 allowed the pressure relief valve setting to be 125 percent of the working pressure, or 250 psi (1.7 MPa). This was done to ensure that there would be no pressure relief of the flammable product into the atmosphere as a result of ambient air temperatures or solar radiation, which would create a hazard. Prior to 1949, there were also ASME specifications U-200 and U-201, which were developed in conjunction with the American Petroleum Institute. These were known as the API-ASME Code specifications. Such containers,

Figure 2.2 *DOT/ICC/CTC industrial truck motor (engine) fuel cylinders. (Photo courtesy of Manchester Tank Company.)*

however, were restricted in use to installations such as refineries, gas processing plants, and tank farms.

In 1949, ASME adopted the current ASME Code precept with a safety factor of 4 to 1. At that time the requirement for the pressure relief valve setting was changed to the working pressure of a container, which generally was 250 psi (1.7 MPa). In effect, the previous 4 to 1 safety factor and the 5 to 1 safety factor were comparable from an ultimate burst standpoint because 200 × 5 and 250 × 4 both equal 1,000. Many pre-1949, 200-psi (1.4 MPa) ASME tanks are still in service.

With the exception of those used in vehicular systems, ASME Code containers are usually found in stationary installations.

At times there are ASME Code Case Interpretations and Addenda to the base code; they are also applicable to ASME containers. See Appendices C and D for further description of DOT/CTC and ASME Code containers.

Figure 2.3 DOT/ICC/CTC stationary and portable cylinders. (Photo courtesy of Manchester Tank Company.)

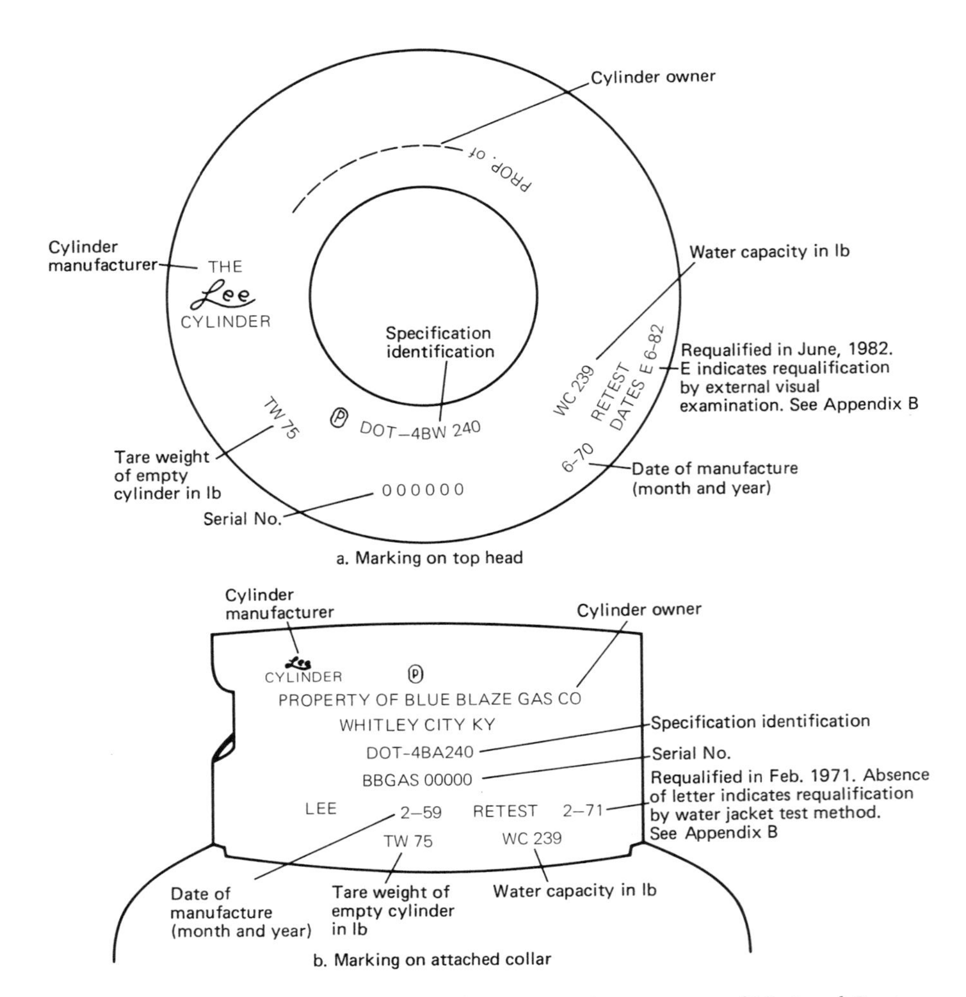

Figure 2.4 Typical DOT cylinder marking. (Drawing courtesy of National Propane Gas Association.)

Table 2.1 DOT/ICC/CTC Cylinder Applications.

Type of Service	Typical Use	Propane Capacity		Water Capacity		Common DOT Mfg. Code
		(lb) (kg)	(gal)	(lb) (kg)	(gal)	
Stationary	Homes, Business	420 (191)	99	1,000 (454)	119	4B, 4BA, 4BW
Stationary	Homes, Business	300 (136)	71	715 (324)	86	4B, 4BA, 4BW
Stationary	Homes, Business	200 (91)	47	477 (216)	57	4B, 4BA, 4BW
Stationary	Homes, Business	150 (68)	35	357 (162)	43	4B
Exchange	Homes, Business	100 (45)	24	239 (108)	29	4B, 4BA, 4BW
Exchange	Home, Business	60 (27)	14	144 (65)	17	4B, 4BA, 4BW
Motor Fuel	Tractor	100 (45)	24	239 (108)	29	4B, 4BA, 4BW
Motor Fuel	Tractor	60 (27)	14	144 (65)	17	4B, 4BA, 4BW
Motor Fuel	Forklift	43.5 (19.7)	10	104 (47)	12	4B, 4BA, 4BW, 4E
Motor Fuel	Forklift	33.5 (15.2)	8	80 (36)	9.6	4B, 4BA, 4BW, 4E
Motor Fuel	Forklift	20 (9)	4.7	48 (22)	5.7	4B, 4BA, 4BW, 4E
Motor Fuel	Forklift	14 (6.4)	3.3	34 (15.4)	4.1	4B, 4BA, 4BW, 4E
Portable	Rec. Vehicles	40 (18)	9.5	95 (43)	11	4B, 4BA, 4BW, 4E
Portable	Rec. Vehicles	30 (13.6)	7.1	72 (32.7)	8.6	4B, 4BA, 4BW, 4E
Portable	Rec. Vehicles	25 (11.3)	5.9	59.5 (27)	7.1	4B, 4BA, 4BW
Portable	Rec. Vehicles, Grills	20 (9)	4.7	48 (22)	5.7	4B, 4BA, 4BW, 4E
Portable	Rec. Vehicles & Sm. Ind.	10 (4.5)	2.4	23.8 (10.8)	2.8	4B, 4BA, 4BW, 4E
Portable	Indoors, Trailers	5 (2.3)	1.2	12 (5.4)	1.4	4B, 4BA, 4BW, 4E
Portable	Torches, RV	.93 lb (1 lb) (.42)	0.2	2.2 (1)	0.3	39 (disposable) 4B240 (refillable)

Table 2.2 Typical Stationary ASME Container Applications.

Type of Service	Water Capacity (gal)	LP-Gas Capacity (gal*)	LP-Gas Capacity (lb)
Domestic	100 (379 L)	80 (301 L)	338
Domestic	125 (473 L)	100 (379 L)	423
Domestic	150 (568 L)	120 (454 L)	508
Domestic	250 (946 L)	200 (757 L)	848
Domestic	325 (1 230 L)	260 (984 L)	
Domestic	500 (1 893 L)	400 (1 514 L)	
Domestic	1,000 (3.8 m^3)	800 (3 m^3)	
Industrial/Agricultural/Commercial	1,000–5,000 (3.8–19 m^3)	800–4,500 (3–17 m^3)	
Service Stations	1,000–6,500 (3.8–24.6 m^3)	800–5,850 (3–22 m^3)	
Bulk Plant or Standby Storage	12,000–18,000 (45.4–68 m^3)	10,800–16,200 (41–61 m^3)	
Bulk Plant or Standby Storage	20,000–30,000 (76–114 m^3)	18,000–27,000 (45.4–102 m^3)	
Bulk Plant or Standby Storage	30,000–60,000 (114–227 m^3)	27,000–54,000 (102–204 m^3)	
Bulk Plant or Standby Storage	60,000–120,000 (227–454 m^3)	48,000–96,000 (182–364 m^3)	

*Based on propane specific gravity of 0.508 at 60°F (15.6°C). Actual quantity depends upon actual specific gravity.

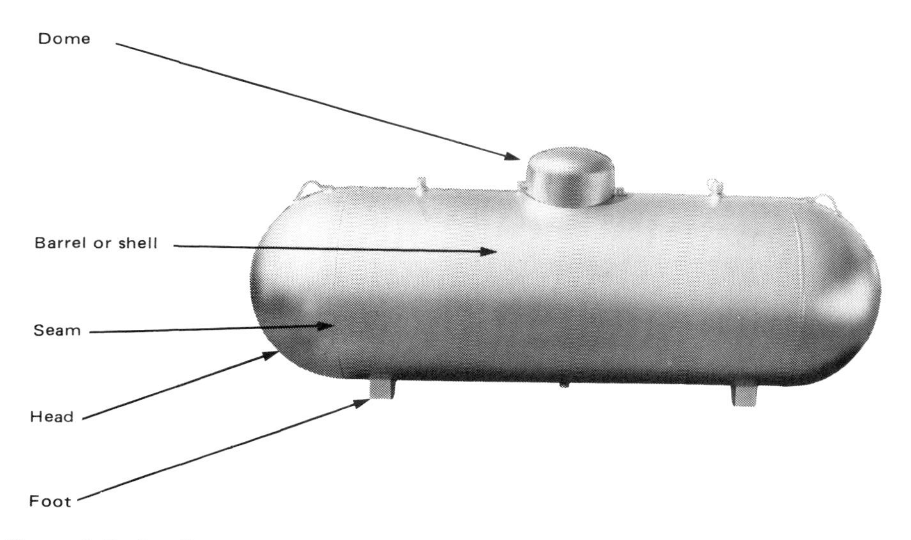

Figure 2.5 *Small ASME container. Used commonly in residential and commercial applications. (Photo courtesy of American Welding and Tank Company.)*

2-2.1.4 Containers complying with 2-2.1.3 shall be permitted to be reused, reinstalled, or continued in use as follows:

(a) A container shall not be filled if it is not suitable for continued service.

(b) DOT cylinders shall not be refilled, continued in service, or transported unless they are properly qualified or requalified for LP-Gas service in accordance with DOT regulations.

(c) Containers that have been involved in a fire and show no distortion shall be requalified for continued service before being used or reinstalled as follows:

1. DOT containers shall be requalified by a manufacturer of the type of cylinder to be requalified or by a repair facility approved by DOT.

Exception: DOT 4E specification (aluminum) cylinders shall be permanently removed from service.

2. ASME or API-ASME containers shall be retested, using the hydrostatic test procedure applicable at the time of original fabrication.
3. All appurtenances shall be replaced.

Maintenance of the container is extremely important. Containers may be subjected to conditions at the point of installation that can create corrosion. In addition, the handling of cylinders can result in cuts, gouges, and dents, which can weaken them and defeat protective coatings (paint). It is the responsibility of the container's owner to see that it is properly maintained and requalified, that the container is repaired if such can be done in a safe manner and, if not, that the container is withdrawn from service and scrapped. The U.S. Department of Transportation and Canadian Transport Commission have regulations for retest or requalification of DOT and CTC cylinders. Appendix C provides an overview of the DOT rules for the requalification of cylinders.

Corrosion is a major factor in the failure of DOT cylinders, especially where the wearing ring is attached to the bottom of the cylinder. This area is especially vulnerable to the elements, and there will be cycles of drying and wetting that accelerate corrosion. The cylinder has to be upended in order for this area to be checked, and it has to be done very thoroughly or small areas of heavy corrosion can be missed. Many times there will be deep corrosion under scale on the surface and the deep corrosive spot will not be observed. Then when the cylinder is put back into use, it can fail at such a point. All scale, rust, and corrosion must be thoroughly removed at the time of retest or requalification; this is especially the case with the area where the wearing ring is attached to the bottom of the container. After the rusted or corroded area is properly cleaned, the cylinder surface should be well coated with a protective coating to prevent future rust and corrosion.

When a cylinder is removed from service and discarded, it should be cut up so that it cannot be repaired by a junkyard or others and put back into pressure service. Care should be used when it is cut up to ensure that it has been properly purged of all product, including vapors. Compressed Gas Association Pamphlet C-2, *Recommendations for the Disposition of Unserviceable Compressed Gas Cylinders,* is useful in these circumstances.

The retesting and requalification of DOT and CTC cylinders are best done by a qualified test organization or, if not, then by a thoroughly qualified and trained individual.

If a cylinder has been subjected to fire, the metal shell (usually steel or aluminum) can be weakened and can be requalified only by a properly qualified repair facility authorized by DOT or CTC. When an ASME Code container has been subjected to fire, it must be retested using the hydrostatic test procedure applicable at the time it was originally fabricated, because the fire heat can alter the properties of the steel and result in reduction of its ability to contain pressure.

In the 1995 edition, (c) was added to require replacement of container appurtenances where a container was exposed to fire. The Committee was advised that there were no procedures for requalification of container appurtenances where the container was exposed to fire, and agreed that replacement was needed.

2-2.1.5 ASME paragraph U-68 or U-69 containers shall be permitted to be continued in use, installed, reinstalled, or placed back into service. Installation of containers shall be in accordance with all provisions listed in this standard. (*See Section 2-2, Table 2-2.2.2, Table 2-3.2.3, and Appendix D.*)

This requirement was added in the 1995 edition to clearly and specifically state that these containers can be continued in service and reinstalled. See commentary following 2-2.1.3(a).

2-2.1.6 Containers showing serious denting, bulging, gouging, or excessive corrosion shall be removed from service.

This includes ASME and API/ASME containers as subject to inspection. Prior to the 1986 edition, only DOT/CTC containers were addressed in this context. ASME containers are being used more widely as bulk distribution has replaced cylinder exchange as the preferred method of distributing propane. AMSE containers are also used in vehicular applications, where they are more subject to physical damage and corrosion than stationary containers.

2-2.1.7 Repair or alteration of containers shall comply with the regulations, rules, or code under which the container was fabricated. Other welding is permitted only on saddle plates, lugs, or brackets attached to the container by the container manufacturer.

The heat of welding directly to the container can affect the strength of the metal and change its characteristics so that it does not meet the code. Such heating can set up localized stressing, which can affect materially the strength of the metal. Welders holding certification for repair of pressure vessels can perform welding operations in a manner that will not weaken the container. For repairs to pressure vessels, an ASME "R" stamp is required.

2-2.1.8 Containers for general use shall not have individual water capacities greater than 120,000 gal (454 m^3). Containers in dispensing stations shall have an aggregate water capacity not greater than 30,000 gal (114 m^3). This capacity restriction shall not apply to LP-Gas bulk plants, industrial plants, or industrial applications.

This limits the size of containers that may be installed for general use and in service stations. The 30,000-gal (114-m^3) limit on total storage in dispensing stations intended to provide a limit on storage at locations where only portable cylinders are filled. The term "general use," which is not defined in the standard, is intended to mean installations where LP-Gas is not the primary or sole product or raw material used. Thus an LP-Gas bulk plant where most employees can be expected to be aware of the proper procedures for handling LP-Gas is not general use, while a container used as an alternate fuel supply at a manufacturing plant whose product is not related to LP-Gas is general use.

2-2.1.9 Heating or cooling coils shall not be installed inside storage containers.

Internal components of a container can only be inspected when the container is emptied, purged, and declared safe for entry. This may not be done over the normal lifetime of a container, and corrosion or other failure will not be observed. If there is leakage from heating or cooling coils installed inside the storage container, the gases can get out into the piping of the heating or cooling system because the LP-Gas pressure is normally higher. This may result in a pressure that can rupture or create leaks in the heating or cooling system. Even if this doesn't occur, a flammable product will be present in the heating or cooling system and may escape near an ignition source.

The user of an LP-Gas vapor withdrawal system who finds that the system will not deliver sufficient pressure during cold weather might elect to heat a container with internal coils to maintain continued operation. This is prohibited, however, and alternate methods must be used, such as vaporizers and external heating, including electric or circulating fluid coils mounted external to the container. At the other extreme, in very hot areas, overheating of a container may result in relief valve opening. Again, with internal coils for cooling being prohibited, options exist such as shading or insulating the container, and external coils for cooling fluids. Reducing the filling levels may also be effective.

2-2.2 Container Design or Service Pressure.

2-2.2.1 The minimum design or service pressure of DOT specification containers shall be in accordance with the appropriate regulations published under Title 49, *Code of Federal Regulations.*

See Appendix C.

2-2.2.2 The minimum design pressure for ASME containers shall be in accordance with Table 2-2.2.2.

Table 2-2.2.2 was revised by adding more columns for API-ASME and older ASME Codes. The current table was formerly located in Appendix D. The term "Container Type" is not used in NFPA 58, but is needed when using the older ASME and API-ASME Codes. Note that the last two rows in Table 2-2.2.2 cover gases with a vapor pressure not exceeding 215 psi at 100°F, which is the vapor pressure of commercial propane. Most containers require a design pressure of 250 psi. The higher pressure of 312.5 psi (125 percent of 250) is required for motor fuel containers installed in industrial trucks, buses (including school buses), and multipurpose passenger vehicles. Refer to commentary following 8-2.2(a)(6) for additional information on containers installed in interior spaces of vehicle.

See Appendix D for additional information on container design.

2-2.2.3 In addition to the applicable provisions for horizontal ASME storage containers, vertical ASME storage containers over 125 gal (0.5 m^3) water capacity shall comply with the following:

Table 2-2.2.2 Vapor Pressures and Design Pressures

Maximum Vapor Press. in psi (MPa) at 100°F (37.8°C)	Design Press. in psi (MPa) Present ASME Code[1]	Design Press. in psi (MPa) Earlier Codes	
		API-ASME	ASME[2]
80 (.6)	100 (.7)	100 (.7)	80 (.6)
100 (.7)	125 (.9)	125 (.9)	100 (.7)
125 (.9)	156 (1.1)	156 (1.1)	125 (.9)
150 (1.0)	187 (1.3)	187 (1.3)	150 (1.0)
175 (1.2)	219 (1.5)	219 (1.5)	175 (1.2)
215 (1.5)	250 (1.7)	250 (1.7)	200 (1.4)
215 (1.5)	312.5 (2.0)	312.5 (2.2)	—

[1]ASME Code edition for 1949, Par. U-200 and U-201 and all later editions (*See D-2.1.5*).

[2]All ASME Codes up to the 1946 edition and paragraphs U-68 and U-69 of the 1949 edition (*See D-2.1.5*).

[3]See 8-2.2(a)(6) for certain service conditions that require a higher pressure relief valve start-to-leak setting.

[4]See Appendix D for information on earlier ASME or API-ASME Code.

(a) Containers shall be designed to be self-supporting without the use of guy wires and shall satisfy proper design criteria taking into account wind, seismic (earthquake) forces, and hydrostatic test loads.

(b) Design pressure (*see Table 2-2.2.2*) shall be interpreted as the pressure at the top head with allowance made for increased pressure on lower shell sections and bottom head due to the static pressure of the product.

(c) Wind loading on containers shall be based on wind pressures on the projected area at various height zones above ground in accordance with ASCE 7, *Design Loads for Buildings and Other Structures*. Wind speeds shall be based on a Mean Occurrence Interval of 100 years.

(d) Seismic loading on containers shall be based on forces recommended in the ICBO *Uniform Building Code*. In those areas identified as zones 3 and 4 on the Seismic Risk Map of the United States, Figures 1, 2, and 3 of Chapter 23 of the UBC, a seismic analysis of the proposed installation shall be made that meets the approval of the authority having jurisdiction.

(e) Shop-fabricated containers shall be fabricated with lifting lugs or some other suitable means to facilitate erection in the field.

When a container is installed in a vertical rather than a horizontal position, in addition to the pressure factors involved there is also an increased weight load in the vertical container. As a result, some changes may have to be made in the design specifications for the vertical container as opposed to those of the horizontal container. Also, the methods of support of the vertical container must be designed to take into account the greater concentration of stresses. These methods are addressed further in 2-2.5.3.

Lifting lugs are not designed to carry the load of lifting the container when it is full of LP-Gas or water (the latter for testing or purging). Note that only shop-fabricated containers require lifting lugs.

2-2.3 Container Openings.

2-2.3.1 Containers shall be equipped with openings suitable for the service for which the container is to be used. Such openings shall be permitted to be either in the container proper or in the manhole cover, or partially in each location.

The number of openings in the container is not limited by the standard. Prior to the 1989 edition of the standard no more than two plugged openings were permitted in a 30- to 1,999-gal (0.1- to 7.6-m^3) container. The provision (2-2.3.3 in the 1986 edition) was deleted because the Technical Committee could see no reason why a properly installed plug, cap, or blind flange would not provide an effective seal and could find no other reason to continue the restriction.

2-2.3.2 Containers of more than 30 gal (0.1 m^3) and less than 2,000 gal (7.6 m^3) water capacity, designed to be filled volumetrically, and manufactured after December 1, 1963, shall be equipped for filling into the vapor space.

The reason for filling into the vapor space only and not through the liquid is to achieve equilibrium more rapidly between the vapor and the liquid in the container. It has been demonstrated repeatedly that lower container pressure results when filling into the vapor space, compared to filling into the liquid space. This is a matter of real, practical interest because the time it takes to fill cylinders, especially in the summer, can be lengthy when the vapor and liquid differ significantly in temperature. A valve manufacturer has reported a 27 percent increase in the filling rate of a 100-lb (45-kg) cylinder when the cylinder valve was equipped with a spray filler that sprayed the liquid into the cylinder rather than admitting a solid stream of liquid.

2-2.3.3 Containers of 125 gal (0.5 m^3) through 2,000 gal (7.6 m^3) water capacity manufactured after July 1, 1961, shall be provided with an opening for an actuated liquid withdrawal excess-flow valve with a connection not smaller than 3/4-in. National Pipe Thread.

This requirement was revised in the 1995 edition by substituting the term "actuated liquid withdrawal excess-flow valve" for text that described the function of the valve. A new definition of the term, an upper limit of 2,000 gal, revisions to Table 2-3.3.2(a), and a new 2-3.3.2(a)(2) were also added. See commentary following the definition of Actuated Liquid Withdrawal Valve. The new upper limit recognizes that the actuated liquid withdrawal valve is intended to eliminate the need to "roll" containers to empty them. This does not apply to containers over 2,000 gal, which are usually equipped with a bottom valve for liquid withdrawal or are in a location where they can be emptied, if needed, without presenting a potential hazard to the general public.

The provision for a liquid evacuation connection is both safety and utility oriented. From the safety standpoint, an LP-Gas container that has been damaged can become a severe hazard if there is no method of removing the liquid contents. This is important because a stationary container must contain less than 5 percent of its maximum contents prior to being transported. Liquid can, of course, be removed from a vapor connection by rolling or tipping the tank, but this can be a hazardous procedure.

Although the provision does not specifically state so, the intent is that the connection cited allows evacuation of the tank without moving the tank and with a device such as is shown in Figure 2.6 (a liquid withdrawal valve), specifically designed for this purpose and installed in the connection. This is the reason for stating that a plugged opening will not comply with the provision.

There are other reasons for evacuating the tank other than those caused by an accident. Note that 6-5.2.1 requires that containers of 125 gal (0.5 m^3) or more water capacity, which are designed for stationary service, must contain no more than 5 percent of their capacity in liquid during transportation. Five percent was selected for a number of reasons, not the least of which was that most gauges on domestic-size tanks, which are often transported with some liquid in them, would not read below 5 percent. However, it was believed that 5 percent left in the tank during transportation did not constitute a hazard. This enabled safe evacuation of the tank when the tank was to be moved from one location to another. Providing a method of removing most of the liquid in the tank prior to moving it reduces the danger of accident during such operation.

Liquid withdrawal valves are combination check and excess-flow valves that are actuated by threading an adapter or pipe nipple into the exposed threaded portion of the valve. This opens a No. 80 drill size hole and allows liquid propane to escape. When a valve is threaded into the adapter or nipple (and the valve is closed), pressure will build up and the excess-flow valve will open. The valve will remain open until the flow through it exceeds the valve design flow rate, or the threaded valve or adapter is removed.

A frequently asked question regards whether liquid withdrawal valves may be used for withdrawal from containers in liquid withdrawal service. This is a valid question and, unfortunately, the text of the standard provides no guidance in answering it. When the provision was added to the standard in the 1961 edition, the special fitting now in use had just been developed and its intended (and only foreseen) use was for evacuation. Since the evacuation fitting became a standard feature on containers larger than 125-gal (and especially on 500- and 1,000-gal containers used for many applications), propane marketers have used it to convert containers designed for vapor withdrawal to liquid withdrawal service. In doing so, it was believed that the safety requirements for liquid withdrawal connections were incorporated into the evacuation fitting. The matter has been complicated by confusion among the manufacturers of evacuation fittings who, at the time of publication, recommend that the fittings not be used for permanent liquid withdrawal.

The liquid withdrawal valve may be used for normal service if it meets all the requirements of the standard for container appurtenances (*see 2-3 and 3-2.5*). If there is doubt, the manufacturer of the liquid withdrawal valve should be contacted to determine if the valve is designed for continuous service or only for occasional use.

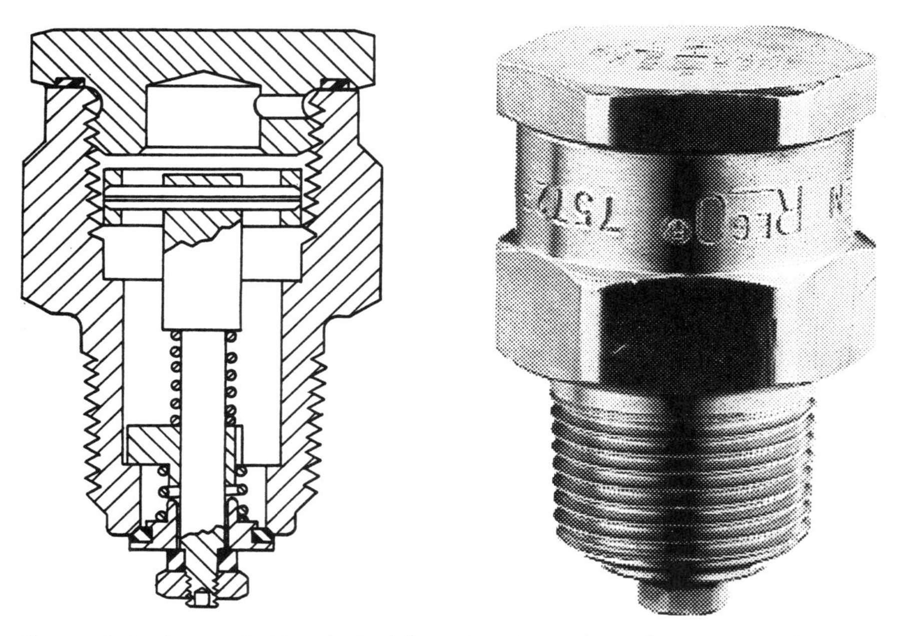

Figure 2.6 *Actuated Liquid Withdrawal Valve. The valve is a combination check valve and excess-flow check valve. The check valve is opened by insertion of an adapter (in place of caps) that opens the check valve against spring tension. The same functions can be achieved by using separate fittings. (Drawing and photo courtesy of Engineered Controls, Inc.)*

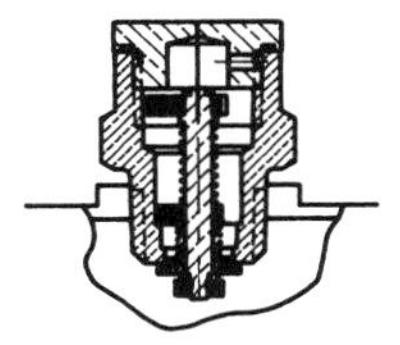

The special fitting in the capped position as the tank fabricator ships the tank.

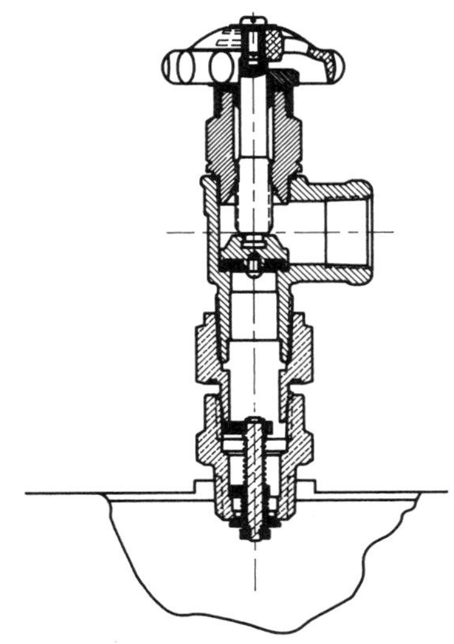

The special fitting with the cap removed and an adapter with transfer valve beginning to open the special fitting.

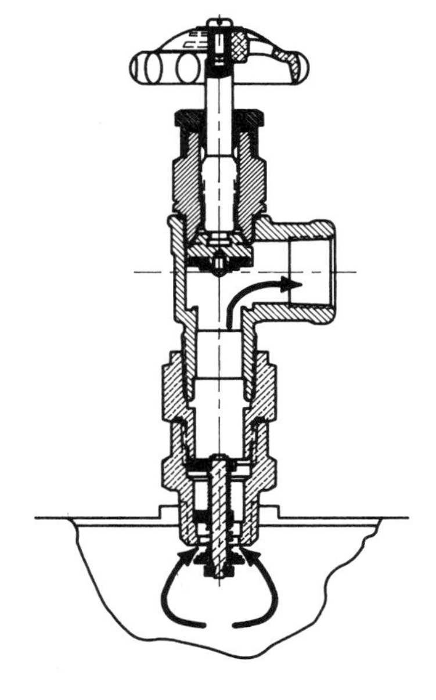

The special fitting/adapter/transfer valve assembly in the normal position. The excess-flow valve of the special fitting is in the open position, allowing flow.

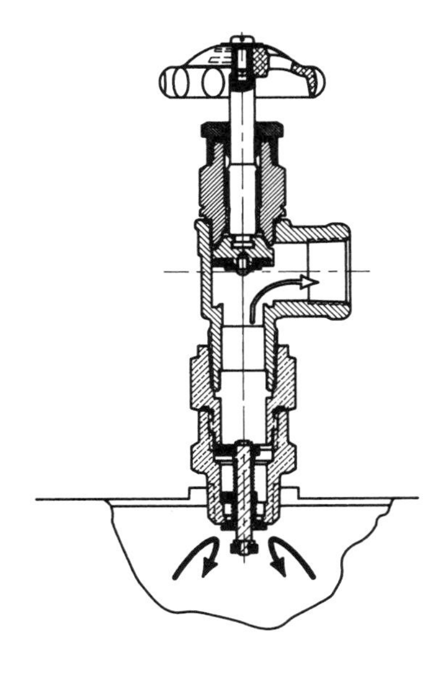

The excess-flow valve in the closed position, shutting off the flow of propane.

Figure 2.7 *Operation of an actuated liquid withdrawal valve. (Drawings courtesy of Engineered Controls, Inc.)*

2-2.3.4 Containers of more than 2,000 gal (7.6 m^3) water capacity shall be provided with an opening for a pressure gauge.

Knowledge of container pressure is useful on larger containers, especially during liquid transfer operations. When installing a pressure gauge, reference should be made to 2-3.5.2, which reminds the user of the standard of an important safety feature requiring either a maximum opening of a No. 54 drill size between the container and the pressure gauge or installation of an excess-flow check valve. This prevents discharge of significant amounts of LP-Gas in case the pressure gauge is broken off.

2-2.3.5 Connections for pressure relief valves shall be located and installed in such a way as to have direct communication with the vapor space, whether the container is in storage or in use.

(a) If located in a well inside the container with piping to the vapor space, the design of the well and piping shall permit sufficient pressure relief valve relieving capacity.

(b) If located in a protecting enclosure, design shall be such as to permit this enclosure to be protected against corrosion and to permit inspection.

(c) If located in any position other than the uppermost point of the container, the connection shall be internally piped to the uppermost point practical in the vapor space of the container.

This provision has two purposes. Because of the high liquid-to-vapor expansion ratio of LP-Gas (about 1:260 for propane), a liquid discharge represents a much greater quantity of gas that can flow through an opening, such as a relief valve into the atmosphere. If vapor is discharged, the liquid is vaporizing inside the container and drawing its heat of vaporization from the liquid in the container to do so. This limits the pressure inside the container. If liquid is discharged, this effect takes place outside the container, and container pressure is higher. This is significant if the valve is discharging as a result of fire exposure.

Note that the relief valve must either be located at the uppermost point of the container or be internally piped to the uppermost point practical in the vapor space of the container. This was added as a Tentative Interim Amendment to the 1986 edition of the standard and added in the 1989 edition when the Committee became aware that some containers (in motor fuel and recreational vehicle applications) were being manufactured with the relief valve located just above the 80 percent liquid level. The editor is aware of incidents that have occurred where these containers have released liquid LP-Gas upon relief valve operation. The provision that the relief valve be piped to the uppermost point practical in the container recognizes that in some containers, such as portable lift truck cylinders which are stored vertically and used horizontally, the relief valve cannot be piped to the uppermost point for both use positions.

2-2.3.6 Containers to be filled on a volumetric basis manufactured after December 31, 1965, shall be fabricated so that they can be equipped with a fixed liquid level gauge(s) capable of indicating the maximum permitted filling level(s) in accordance with 4-4.2.2.

Variable liquid level gauges (which indicate liquid level over most of the container's capacity) are not 100 percent reliable. The float may develop a leak, the float arm may sag, the mechanism may bind, or the gauge may otherwise indicate an inaccurate reading. Therefore, a fixed liquid level gauge is required to be used in filling the container. [Note that 4-4.3.3(c) requires that the variable gauge be checked for accuracy by comparison to the fixed maximum liquid level gauge.] In this way, overfilling can be prevented. If containers are filled only by weight, a fixed liquid level gauge is not required. Refer to Figures 2.26, 2.29, and 2.30 for illustrations of these gauges.

2-2.4 Portable Container Appurtenance Physical Damage Protection.

2-2.4.1 Portable containers of 1,000 lb (454 kg) [nominal 120 gal (0.5 m^3)] water capacity or less shall incorporate protection against physical damage to container appurtenances and immediate connections to these while in transit, storage, while being moved into position for use, and when in use except in residential and commercial installations, by:

(a) Recessing connections into the container so that valves will not be struck if the container is dropped on a flat surface, or,

(b) A ventilated cap or collar designed to permit adequate pressure relief valve discharge and capable of withstanding a blow from any direction equivalent to that of a 30-lb (14-kg) weight dropped 4 ft (1.2 m). Construction shall be such that the force of the blow will not be transmitted to the valve. Collars shall be designed so that they do not interfere with the free operation of the cylinder valve.

This cap or collar is required to provide physical protection for the valves and any other appurtenances connected to the cylinder. This recognizes that cylinders are portable and are normally transported from their normal use point to a filling point, and that a number of incidents where the cylinder is involved in a transportation incident, falls over, or is dropped is inevitable. Where a removable cap is used, it must be in place prior to moving the cylinder. The cap also provides a lifting point for the cylinder. The valve must never be used for lifting a cylinder. (*See Figures 2.1, 2.2, and 2.3, which show cylinders with collars.*) Ventilation is required to permit discharge from the pressure relief valve to dissipate.

2-2.4.2 Portable containers of more than 1,000 lb (454 kg) [nominal 120 gal (0.5 m^3)] water capacity, including skid tanks or for use as cargo containers, shall incorporate protection against physical damage to container appurtenances by recessing, protective housings, or by location on the vehicle. Such protection shall comply with the provisions under which the tanks are fabricated, and shall be designed to withstand static loadings in any direction equal to twice the weight of the container and attachments when filled with LP-Gas, using a safety factor of not less than four, based on the ultimate strength of the material to be used. (*See Chapters 3 and 6 for additional provisions applying to the LP-Gas system used.*)

2-2.5 Containers with Attached Supports.

2-2.5.1 Horizontal containers of more than 2,000 gal (7.6 m^3) water capacity designed for permanent installation in stationary service shall be permitted to be provided with nonfireproofed structural steel saddles designed to allow mounting the containers on flat-topped concrete foundations. The total height of the outside bottom of the container shell above the top of the concrete foundation shall not exceed 6 in. (152 mm).

2-2.5.2 Horizontal containers of 2,000 gal (7.6 m^3) water capacity or less, designed for permanent installation in stationary service, shall be permitted to be equipped with nonfireproofed structural steel supports and designed to allow mounting on firm foundations in accordance with the following:

(a) For installation on concrete foundations raised above the ground level by more than 12 in. (300 mm), the structural steel supports shall be designed so that the bottoms of the horizontal members are not less than 2 in. (51 mm), nor more than 12 in. (300 mm), below the outside bottom of the container shell.

(b) For installation on paved surfaces or concrete pads within 4 in. (102 mm) of ground level, the structural steel supports shall be permitted to be designed so that the bottoms of the structural members are not more than 24 in. (610 mm) below the outside bottom of the container shell. [*See 3-2.4.2(a)(3) for installation provisions for such containers, which are customarily used as components of prefabricated container-pump assemblies.*]

2-2.5.3 Vertical ASME containers over 125 gal (0.5 m^3) water capacity designed for permanent installation in stationary service shall be designed with steel supports designed to allow the container to be mounted on, and fastened to, concrete foundations or supports. Such steel supports shall be designed to make the container self-supporting without guy wires and shall satisfy proper design criteria, taking into account wind, seismic (earthquake) forces, and hydrostatic test load criteria established in 2-2.2.3.

The steel supports shall be protected against fire exposure with a material having a fire resistance rating of at least 2 hr.

Exception: Continuous steel skirts having only one opening of 18 in. (457 m) or less in diameter shall have such fire protection applied to the outside of the skirt.

2-2.5.4 Containers to be used as portable storage containers (*see definition*) for temporary stationary service (normally less than 6 months at any given location) and to be moved only when substantially empty of liquid shall comply with the following (this shall apply to movable fuel storage tenders including farm carts):

The time period constituting temporary service was changed in the 1992 edition from 12 months to 6 months as part of the Committee's efforts for consistency within NFPA 58. Notification of temporary installations not exceeding 6 months has been required in 1-4.2, and the Committee saw no reason to differ here. See commentary following 1-4.2 for additional information.

(a) If mounted on legs or supports, such supports shall be of steel and shall either be welded to the container by the manufacturer at the time of fabrication or shall be attached to lugs that have been so welded to the container. The legs or supports or the lugs for the attachment of these legs or supports shall be secured to the container in accordance with the code or rule under which the container is designed and built, with a minimum safety factor of four, to withstand loading in any direction equal to twice the weight of the empty container and attachments.

(b) If the container is mounted on a trailer or semitrailer running gear so that the unit can be moved by a conventional over-the-road tractor, attachment to the vehicle, or attachments to the container to make it a vehicle, shall comply with the appropriate DOT requirements for cargo tank service; except that stress calculations shall be based on twice the weight of the empty container. The unit shall also comply with applicable State and DOT motor carrier regulations and shall be approved by the authority having jurisdiction.

2-2.5.5 Portable tanks (*see definition*) shall comply with DOT portable tank container specifications as to container design and construction, securing of skids or lugs for the attachment of skids, and protection of fittings. In addition, the bottom of the skids shall be not less than 2 in. (51 mm) or more than 12 in. (300 mm) below the outside bottom of the container shell.

2-2.5.6 Movable fuel storage tenders, including farm carts, shall be secured to the trailer support structure for the service involved.

2-2.6 Container Marking and Recordkeeping.

2-2.6.1 Containers shall be marked as provided in the regulations, rules, or code under which they are fabricated and in accordance with the following:

(a) Where LP-Gas and one or more other compressed gases are to be stored or used in the same area, the containers shall be marked "Flammable" and either "LP-Gas," "LPG," "Propane," or "Butane." Compliance with marking requirements of Title 49 of the *Code of Federal Regulations* shall meet this provision.

(b) When being transported, portable DOT containers shall be marked and labeled in accordance with Title 49 of the *Code of Federal Regulations*.

2-2.6.2 Portable DOT containers designed to be filled by weight, including those optionally filled volumetrically but which may require check weighing, shall be marked with:

Figure 2.8 Label marking for a DOT container. (Drawing courtesy of the National Propane Gas Association.)

(a) The water capacity of the container in lb.

(b) The tare weight of the container in lb, fitted for service. The tare weight is the container weight plus the weight of all permanently attached valves and other fittings but does not include the weight of protecting devices removed in order to load the container.

See Figure 2.4.

2-2.6.3 ASME containers shall be marked in accordance with the following:

(a) The marking specified shall be on a stainless steel metal nameplate attached to the container, located to remain visible after the container is installed. The nameplate shall be attached in such a way to minimize corrosion of the nameplate or its fastening means and not contribute to corrosion of the container.

Exception: Where the container is buried, mounded, insulated, or otherwise covered so the nameplate is obscured, the information contained on the nameplate shall be duplicated and installed on adjacent piping or on a structure in a clearly visible location.

Note that the container nameplate must be stainless steel and must be attached to the container in such a way as to eliminate the possibility of corrosion to the nameplate, fasteners, or container. Since the ASME Code requires the nameplate to be attached to a container (or the container does not meet the ASME Code), the need for this provision is consistent with the long service life of propane containers. The exception which was editorially revised in the 1995 edition requires duplication of the information on the container nameplate for containers installed in such a way that the nameplate is not visible, and installation of the information nearby. This permits verification of the container nameplate information without digging up an underground container, which could damage the container, the protective coating, or a cathodic protection system (if used).

(b) Service for which the container is designed; i.e., underground, aboveground, or both.

(c) Name and address of container supplier or trade name of container.

(d) Water capacity of container in lb or U.S. gal.

(e) Design pressure in psi.

(f) The wording "This container shall not contain a product having a vapor pressure in excess of ________psi at 100°F." (*See Table 2-2.2.2.*)

(g) Tare weight of container fitted for service for containers to be filled by weight.

(h) Outside surface area in sq ft.

(i) Year of manufacture.

(j) Shell thickness ________head thickness.

(k) OL ________OD ________HD ________.

(l) Manufacturer's serial number.

(m) ASME Code symbol.

2-2.6.4* Effective January 1, 1993, a warning label shall be applied to all portable refillable LP-Gas cylinders of 100 lb (45.4 kg) LP-Gas capacity or less not filled on site. The label shall include information on the potential hazards of LP-Gas.

A-2-2.6.4 A sample warning label, which is intended for use with gas grill cylinders, is shown in Figure A-2-2.6.4.

This requirement was added in the 1992 edition of the standard and enforcement was delayed by one year to provide time to develop appropriate labels. The label shown in Figure A-2-2.6.4 was taken from ANSI Z21.58, *Outdoor Cooking Gas Appliances*, which requires it for cylinders sold for use with gas grills, and it was placed in Appendix A to provide an example to users of NFPA 58. Note that this specific labeling is not mandated by NFPA 58, and that an alternate label containing information on the potential hazards of LP-Gas would meet the requirement of 2-2.6.4. Cylinders used in service other than gas grills may need other information to convey the hazards of LP-Gas in the manner they are used.

The form of the label is not specified, only that it be "applied" to the cylinder. When the Committee added this requirement, the discussions covered

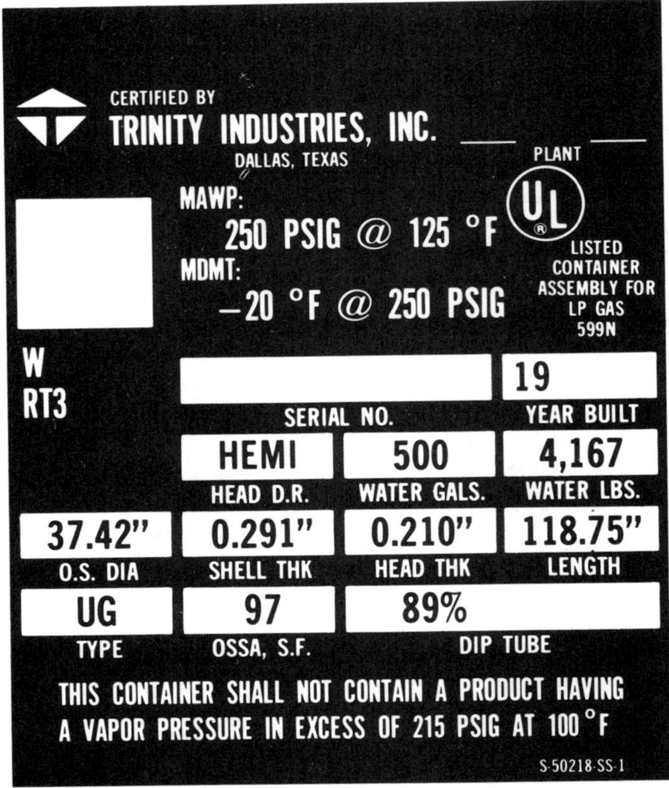

Figure 2.9 Marking required by 2-2.6.5. Container listing (in this example, by Underwriters Laboratories Inc. [UL]) is optional. (Nameplate courtesy of Trinity Industries.)

only labels applied with adhesive, with the gas grill cylinder label used as the mode. This has led to questions from users asking if the label must be glued to the cylinder, or if attachment by a mechanical means (such as a plastic strap) is acceptable. As the provision says "applied" and applied means "to put on," the label can therefore be glued or attached by a mechanical means. The Committee will certainly review the type and use of labels required, and changes may occur in future editions, if safety is being compromised.

2-2.6.5 Effective January 1, 1996, all ASME containers containing unodorized LP-Gas products shall be marked "NOT ODORIZED" in letters 4 in. (10 cm) in height with a contrasting background surrounded by a ½-in. (1.3-cm) rectangular border. The markings shall be located on either both sides or both ends of the container.

DANGER–**FLAMMABLE GAS UNDER PRESSURE**
LEAKING LP-GAS MAY CAUSE A FIRE OR EXPLOSION IF IGNITED
CONTACT LP-GAS SUPPLIER FOR REPAIRS, OR DISPOSAL OF THIS CYLINDER OR UNUSED LP-GAS
FOR OUTDOOR USE ONLY*
DO NOT USE OR STORE CYLINDER IN A BUILDING, GARAGE, OR ENCLOSED AREA

CUSTOMER WARNING:
- Know the odor of LP-Gas. If you hear, see, or smell leaking LP-Gas, immediately get everyone away from the cylinder and call the Fire Department. Do not attempt repairs.
- Caution your LP-Gas supplier to:
  Be certain cylinder is purged of trapped air prior to first filling.
  Be certain not to overfill the cylinder.
  Be certain cylinder requalification date is checked.
- LP-Gas is heavier than air and may settle in low places while dissipating.
- Contact with the liquid contents of cylinder will cause freeze burns to the skin.
- Do not allow children to tamper or play with cylinder.
- When not connected for use, keep cylinder valve turned off and, for 45 lb propane capacity or less, plug or cap valve outlet. Self-contained outdoor cooking appliances shall be limited to a cylinder of 20 lb capacity or less.
- Do not use, store, or transport cylinder where it would be exposed to high temperatures. Relief valve may open, allowing a large amount of flammable gas to escape.
- When transporting, keep cylinder secured in an upright position with cylinder valve turned off and tightly plugged or capped.

*EXCEPT AS AUTHORIZED BY ANSI/NFPA 58

WHEN CONNECTING FOR USE:
- Use only in compliance with applicable codes.
- Read and follow manufactures' instructions.
- If regulator has a coupling nut, the nut has left-handed **threads** and must be **wrench** tightened. If the regulator has a plug-in connection, pull the pull ring back, insert fully, and release ring. Be sure regulator **vent** is pointing down.
- Turn off all valves on the appliance.
- **Do not check for gas leaks with a match or open flame. Apply soapy water at areas marked "X." Open cylinder valve. If bubble appears, close valve and have LP-Gas service person make needed repairs. Also, check appliance valves and connections to make sure they do not leak before lighting appliance**.
- Light appliance(s) following manufacturers' instructions
- When appliance is not in use, keep the cylinder valve closed.

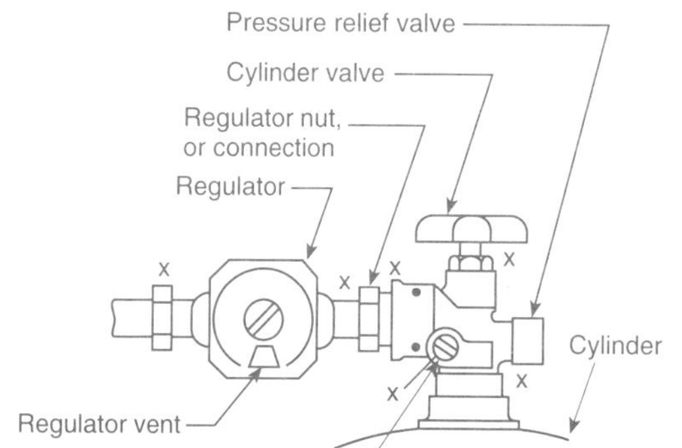

Figure A-2-2.6.4 *A sample cylinder warning label.*

This new marking requirement was added in the 1995 edition, providing important information to emergency responders. For example, odorized LP-Gas cannot be used in a few applications—where propane is used as a chemical feedstock and the odorant poisons a catalyst, or where butane is used as a foaming gas in the manufacture of styrene trays used to package meats for retail display or for filling aerosol cans as a propellant. Note that two labels are required and that the provision becomes effective on January 1, 1996. This does not mean that the new warning is not needed immediately, but recognizes the need to allow time for the standard to be adopted, for users to become aware of the new requirement, and for those affected to comply without being cited.

The lack of odorant in a storage container adds an additionally potential hazard that both users and emergency responders must face. By providing a visible marking, everyone working around stationary containers holding unodorized LP-Gas is made aware of the absence of odorant. Alternate sensing devices must be used, if leakage is suspected, to locate the leak.

2-2.6.6 The name and emergency service telephone number of the LP-Gas fuel suppliers shall be affixed and legibly maintained in a visible location on or near all stationary LP-Gas containers with an aggregate capacity of less than 4,000 gal (15.1 m^3) water capacity installed at consumer premises.

This additional labeling requirement was added in the 1995 edition by the Committee, and places requirements on propane suppliers who provide, and retain ownership of, containers to their customers. This is a very important matter for emergency responders who operate 24-hours a day, every day of the year. In an incident involving an LP-Gas container, emergency responders may either benefit from the assistance from the propane supplier, or need to notify the propane supplier that their container has been involved in an incident.

The new requirement provides flexibility in the means of compliance. The supplier can be available 24 hours a day for emergency calls; use a service that specializes in emergency response advice; install an answering machine that provides specific advice, such as to call the local fire department (with prior agreement of the local fire department); or provide another means so that a phone is answered at all times, and results in notifying the supplier of an incident involving their container.

(a) Effective January 1, 1996, the firm name and emergency telephone number on ASME containers shall be in text at least 1/2 in. (1.3 cm) in height.

(b) Effective January 1, 1996, the emergency telephone numbers shall be manned 24 hr a day to ensure that the LP-Gas supplier is available in the event of leakage or fire at the consumer premise. The use of an answering service or other means acceptable to the authority having jurisdiction during nonbusiness hours shall be permitted.

2-3 Container Appurtenances.

2-3.1 General.

2-3.1.1 This section includes fabrication and performance provisions for container appurtenances, such as pressure relief devices, container shutoff valves, backflow check valves, internal valves, excess-flow check valves, plugs, liquid level gauges, and pressure gauges connected directly into the container openings described in 2-2.3. Shop installation of such appurtenances in containers listed as container assemblies or container systems in accordance with 1-3.1 is a responsibility of the fabricator under the listing. Field installation of such appurtenances is covered in Chapters 3 and 8.

In the early days of distribution of LP-Gas in bulk quantities to an ASME container located on the consumer's premises, each container assembly of 1,200 gal (4.5 m^3) [and subsequently 2,000 gal (7.6 m^3)] water capacity or less had to be tested and listed or inspected and approved by the authority having jurisdiction. On larger containers each appurtenance and regulator had to be so tested and listed. Many LP-Gas distributing companies and tank fabricators had their own systems listed by Underwriters Laboratories Inc. Today, the individual appurtenances are tested and listed and either shop installed—e.g., by container fabricators—or field installed. This section sets out general, fundamental criteria for these appurtenances, which are used as a basis for testing laboratories, manufacturers, and installers.

2-3.1.2 Container appurtenances shall be fabricated of materials suitable for LP-Gas service and shall be resistant to the action of LP-Gas under service conditions. The following shall also apply:

(a) Pressure-containing metal parts of appurtenances, such as those listed in 2-3.1.1, except fusible elements, shall have a minimum melting point of 1,500°F (816°C) such as steel, ductile (nodular) iron, malleable iron, or brass. Ductile iron shall meet the requirements of ASTM A395, *Standard Specification for Ferritic Ductile Iron Pressure-Retaining Castings for Use at Elevated Temperatures*, or equivalent and malleable iron shall meet the requirements of ASTM A47, *Standard Specification for Ferritic Malleable Iron Castings*, or equivalent.

These criteria are intended to provide a degree of structural integrity in the event of fire exposure and application of water for fire control. It is recognized, however, that appurtenance configuration, mass, and location on the container affect the degree of hazard. The exception for liquid level gauges on smaller containers reflects this consideration.

Note that only metals are recognized for pressure-containing parts, and that the 1,500°F (816°C) requirement prohibits the use of all metals and alloys that have melting points below this point. As a general rule, metals and alloys become structurally unusable at temperatures approaching one-third to one-half their Fahrenheit melting points, making materials with melting points below 1,500°F unacceptable in fire exposure. The exception exempts listed or approved liquid level gauges on containers over 3,500 gal from the restrictions on use of metals melting below 1,500°F (816°C).

Exception: Approved or listed liquid level gauges used in containers of 3,500 gal (13.2 m^3) water capacity or less.

(b) Cast iron shall not be used.

Cast iron is subject to cracking and failure from mechanical shock and from severe thermal shock under fire conditions when used in container appurtenances and pipe fittings.

(c) Nonmetallic materials shall not be used for bonnets or bodies.

Although there have been considerable improvements in the temperature and pressure capabilities of plastic materials, these materials still do not approach the melting (or softening) point of metals. Because of the concern that valves maintain their integrity in fire as long as practical, nonmetallic materials are not permitted to be used for bonnets and bodies of appurtenances in LP-Gas service.

2-3.1.3 Container appurtenances shall have a rated working pressure of at least 250 psi (1.7 MPa).

2-3.1.4 Gaskets used to retain LP-Gas in containers shall be resistant to the action of LP-Gas. They shall be made of metal or other suitable material confined in metal having a melting point over 1,500°F (816°C) or shall be protected against fire exposure. When a flange is opened, the gasket shall be replaced.

Exception: Aluminum O-rings and spiral wound metal gaskets shall be acceptable. Gaskets for use with approved or listed liquid level gauges for installation on a container of 3,500 gal (13.2 m^3) water capacity or less shall be exempted from this provision.

This provision again demonstrates the Committee's concern that LP-Gas systems be able to maintain their integrity when subjected to fire. The requirement that gaskets maintain their integrity to a temperature of 1,500°F limits the use of gaskets to those that have the best chance of maintaining their integrity in a fire. Other gasket materials are permitted only when the flange containing the gasket is protected against fire exposure. This protection, while not specified, must safeguard the gasket from temperatures that would degrade it in a reasonably expected fire condition. An insulating material that would not be degraded in a fire situation could be used.

When a flange is opened, the gasket can be damaged and the damage may not be visible to the person opening the flange. Therefore, the replacement of gaskets, which are relatively inexpensive, is required.

The exemption of aluminum O-rings and spiral wound metal gaskets for use with approved or listed liquid level gauges on containers of 3,500 gal (13.2 m^3) water capacity or less is attributed to the long and successful experience with this equipment, which has been listed for many years. See also commentary following 2-3.1.2(a).

2-3.2 Pressure Relief Devices. *(See 2-4.7 for hydrostatic relief valves.)*

Editions of the standard prior to 1983 referred to "safety relief devices" to distinguish between those pressure relief devices normally in service with nonhazardous materials. The 1983 edition was revised to refer to "pressure relief devices" to bring the standard into agreement with ANSI B95.1, *Standard Terminology for Pressure Relief Devices.*

2-3.2.1 Containers shall be equipped with one or more pressure relief devices that, except as otherwise provided for in 2-3.2.2, shall be designed to relieve vapor.

Vapor discharge permits smaller valves to be used to obtain the needed relieving capacity under fire exposure conditions because some of the heat is used to vaporize the liquid inside the container. Also, the discharged vapor represents a lesser hazard than does discharged liquid.

2-3.2.2 The following shall apply to DOT containers:

(a) DOT cylinders complying with *Code of Federal Regulations,* Title 49, Part 178, Subpart C, shall be equipped with pressure relief valves or fusible plug devices as required by DOT regulations. *(See Appendix E for additional information.)*

DOT regulations [CFR 49, 173.34(d)] currently require that DOT containers be equipped with one or more pressure relief devices selected and tested in accordance with CGA Pamphlet S-1-1, *Pressure Relief Device Standards.* The pressure relief device system must be capable of preventing rupture of the normally charged cylinder when subjected to a fire test conducted in accordance with CGA Pamphlet C-14, *Procedures for Fire Testing of DOT Cylinder Safety Relief Device Systems.* For many years this testing was done by the Bureau of Explosives of the Association of American Railroads, but manufacturers may now conduct their own testing.

(b) DOT 2P and 2Q inside nonrefillable metal containers complying with *Code of Federal Regulations,* Title 49, Part 178, Subpart B, shall be equipped with a pressure relief device(s) or system(s) that will prevent rupture or propulsion of the container when the container is exposed to the action of fire.

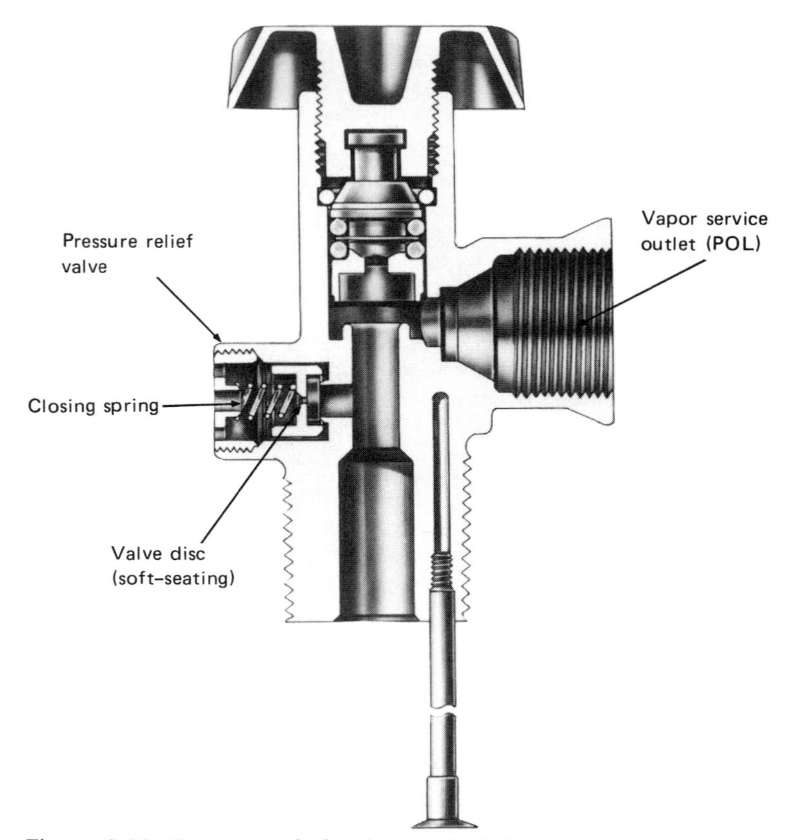

Figure 2.10 *Pressure relief valve on DOT/ICC/CTC cylinder. On these containers the relief valve is part of a fitting that includes the manual shutoff valve (also known as the "service valve"). Note that the relief valve is unaffected by the position of the manual valve and is, therefore, able to function whether the manual valve is open or closed. (Drawing courtesy of Engineered Controls, Inc.)*

DOT-specification 2P and 2Q inside nonrefillable metal containers are used primarily for aerosol products [*see Figure 3-31*]. They have also been used for LP-Gas, specifically butane, for portable butane-fired appliances and refills for butane cigarette lighters. With the proposals to permit portable butane cooking appliances fueled by these containers made to the 1992 edition of the standard, these requirements were added at that time based on test work done by Underwriters Laboratories Inc.

In 2-3.2.2 (a) all DOT cylinders are required to have relief devices. In accordance with the *Code of Federal Regulations,* Title 49, paragraph 173.304(d)(3)(ii), DOT 2P and 2Q inside nonrefillable metal containers are authorized for LP-Gas; however, depending upon pressure and quantity of gas, cylinders may not be required to contain a relief device. The LP-Gas Committee believes, though, that any container for LP-Gas must contain a relief device or system to prevent rupture or propulsion of the container. Small LP-Gas containers can be very dangerous in warehouse fires, as they have been known to "rocket" and spread the fire laterally. There have been warehouse fires involving aerosol containers that sprinkler systems have been unable to contain, resulting in loss of the structure. This requirement prevents such incidents involving butane containers.

The butane 2P/2Q containers being sold for use with portable cooking stoves contain a weakness at one seam which provides pressure relief without "rocketing." See Figure 3.31, where small dents at the top seam can be seen. These are the weakness points that accomplish pressure relief.

2-3.2.3 ASME containers for LP-Gas shall be equipped with direct spring-loaded pressure relief valves conforming with applicable requirements of the *Standard on Safety Relief Valves for Anhydrous Ammonia and LP-Gas*, UL 132; or other equivalent pressure relief valve standards. The start-to-leak setting of such pressure relief valves, with relation to the design pressure of the container, shall be in accordance with Table 2-3.2.3.

Exception: On containers of 40,000 gal (151 m^3) water capacity or more, a pilot-operated pressure relief valve in which the relief device is combined with and is controlled by a self-actuated, direct, spring-loaded pilot valve shall be permitted to be used provided it complies with Table 2-3.2.3, is approved (see definition), is inspected and maintained by persons with appropriate training and experience, and is tested for proper operation at intervals not exceeding 5 years.

Table 2-3.2.3

Containers	Minimum	Maximum
All ASME Codes prior to the 1949 Edition, and the 1949 Edition, paragraphs U-68 and U-69	110%	125%*
ASME Code, 1949 Edition, paragraphs U-200 and U-201, and all ASME Codes later than 1949	100%	100%*

*Manufacturers of pressure relief valves are allowed a plus tolerance not exceeding 10 percent of the set pressure marked on the valve.

This provision specifically states that these valves must comply with a UL standard, or its equivalent. It relates to the definition of ASME Code in Section 1-6 stating that Section VIII of that code is applicable to NFPA 58 except for the relief valve provisions in Sections UG-125 through UG-136 of the code. The exception was made in light of differences in this area between

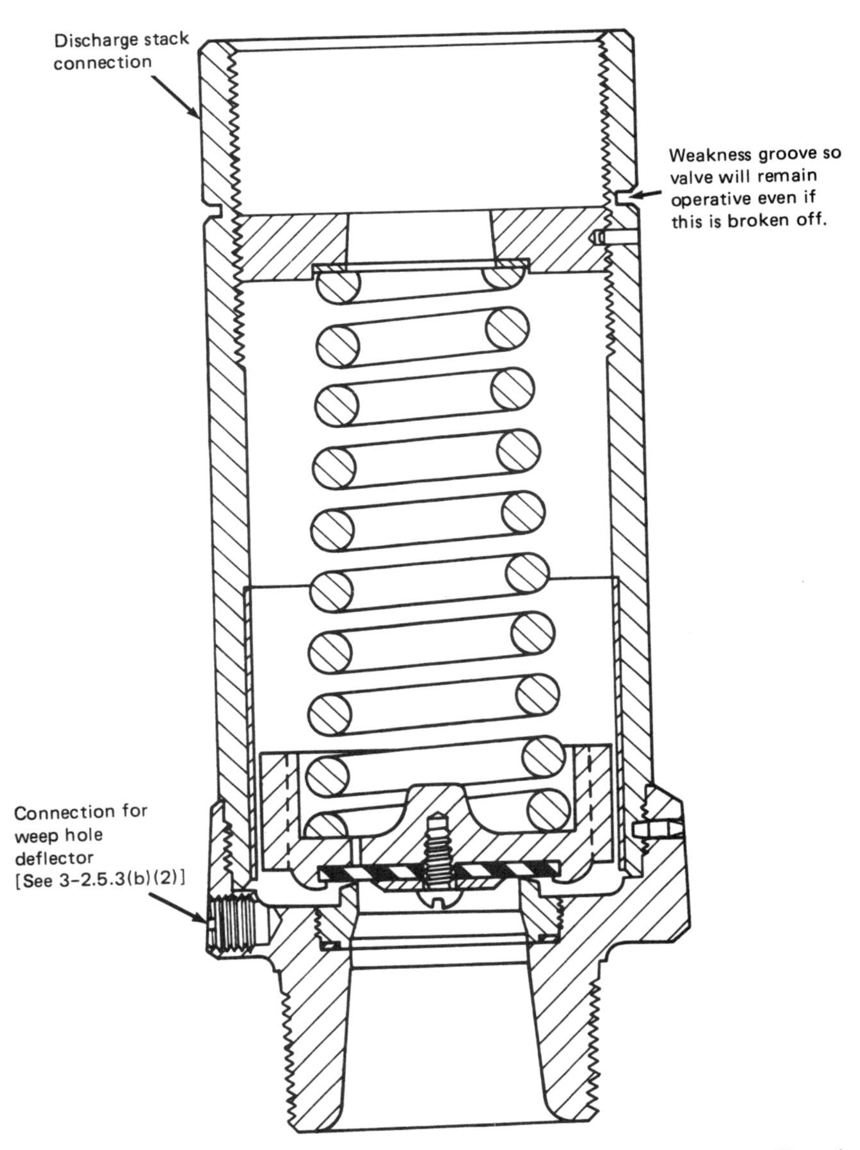

Figure 2.11 *Pressure relief valve for large stationary ASME container. (Drawing courtesy of Engineered Controls, Inc.)*

the ASME Code and longstanding practice within the LP-Gas and petroleum industries. Recognition is given to UL Standard 132 as the basic requirements for pressure relief valves installed on ASME containers.

One of the basic differences between the ASME Code and other standards was the tolerances for the start-to-leak setting of relief valves. The ASME Code

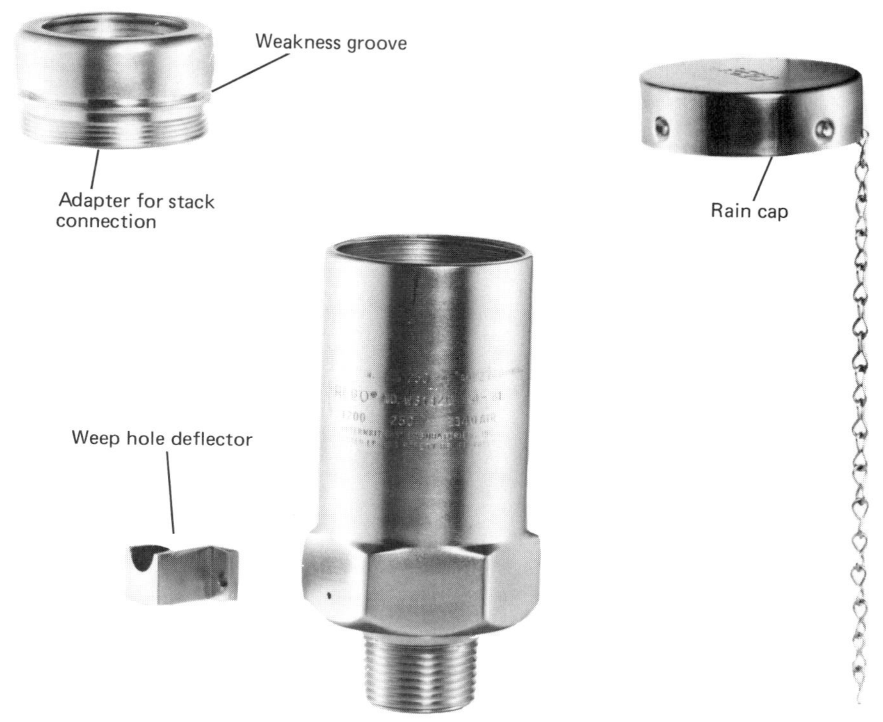

Figure 2.12 Pressure relief valve for small ASME container. (Photos courtesy of Engineered Controls, Inc.)

permitted a maximum of + 3 percent tolerance, whereas NFPA 58 set out a plus tolerance only, and one not exceeding 10 percent. The plus tolerance is based on the opinion that there should be no premature discharge of the valve and that the pressure relief valve should function only as a last resort. The ASME Code has subsequently been changed to agree with this concept.

There is apparently one remaining difference, and that is the official capacity of the valves. The ASME Code requires 10 percent derating of the average test capacities of three samples as the official capacity of the valve. NFPA 58 and UL do not require such derating.

Reference is made to "direct spring-loaded pressure relief valves" to describe accurately the type of relief valve desired on ASME containers used in installations covered by NFPA 58. Other types exist. More complex types are not functionally needed, are unlikely to receive adequate maintenance, and could result in undesirable frequent premature operation in facilities covered by NFPA 58. Simpler devices exist, such as rupture discs which fail catastrophically at the relief pressure and cannot reclose. They are not permitted, as the entire contents of the container would be released.

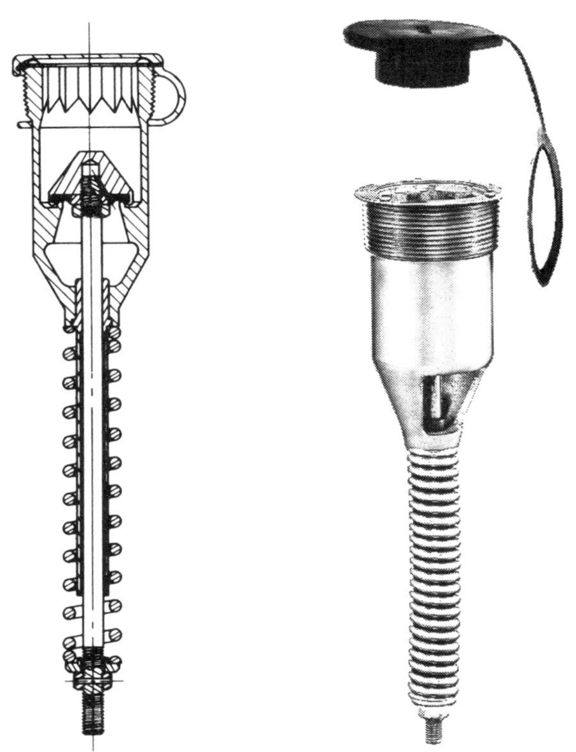

Figure 2.13 Internal-type pressure relief valves used on cargo containers. All working parts are within the containers or below the container shell. Stacks are not used. (Drawing and photo courtesy of the National Propane Gas Association.)

One exception to this provision is given for pilot-operated valves on 40,000 gal (151 m^3) water capacity or larger containers under certain conditions. This recognizes that direct spring-loaded pressure relief valves may not be readily available in the larger sizes required by larger containers. Also, large installations utilizing this size storage container are more likely to be able to comply with the testing and training conditions specified.

2-3.2.4 Pressure relief valves for ASME containers shall also comply with the following:

(a) Pressure relief valves shall be of sufficient individual or aggregate capacity to provide the relieving capacity in accordance with Appendix E for the container on which they are installed, and to relieve at not less than the rate indicated before the pressure is in excess of 120 percent of the maximum (not including the 10 percent referred to in the footnote of Table 2-3.2.3) permitted start-to-leak pressure

Figure 2.14 Internal pressure relief valve for motor fuel containers. See 8-2.3(a)(4). (Photo courtesy of Sherwood Selpac.)

setting of the device. This provision is applicable to all containers (including containers installed partially aboveground) except containers installed wholly underground in accordance with E-2.3.1.

While a pressure relief valve can function for other reasons, such as overfilling, the maximum relieving capacity needed is the result of fire exposure. In the fire exposure case, the calculation of relief valve capacity assumes that the container is exposed to fire on the entire surface area, and the vaporization rate is calculated from the heat transfer expected into the container. The Compressed Gas Association publications cited in E-2.1.1 describe the factors involved.

Because underground containers cannot be exposed to fire, their pressure relief valve capacities may be as small as 30 percent of those specified in Table E-2.2.2, if the container is not filled with LP-Gas prior to being installed and not completely emptied of LP-Gas prior to being uncovered for removal. If an underground container is installed aboveground, it is imperative that the relief valve be properly sized.

(b) Each pressure relief valve shall be plainly and permanently marked with:

It is standard practice to rate relief valves in flow capacity of air. This not only eases the testing requirements for the manufacturer, it provides a standard means to compare products from competing manufacturers. Table E-2.2.2 provides sizing information for LP-Gas container relief valves, giving flow rates of air from the surface area of the container, and is sufficient for most applications.

1. the pressure in psi at which the valve is set to start-to-leak;
2. rated relieving capacity in cu ft per min of air at 60°F (16°C) and 14.7 psia (at an absolute pressure of 0.1 MPa); and
3. the manufacturer's name and catalog number.

Example: A pressure relief valve is marked 250-4050 AIR. This indicates that the valve is set to start-to-leak at 250 psi (1.7 MPa) and that its rated relieving capacity is 4050 cfm (1.9 m^3/s) of air.

(c) Shutoff valves shall not be located between a pressure relief device and the container.

Exception: Where the arrangement is such that the relief device relieving capacity flow specified in 2-3.2.4(a) will be achieved through additional pressure relief devices that remain operative.

(d) Pressure relief valves shall be designed to minimize the possibility of tampering. Externally set or adjusted valves shall be provided with an approved means of sealing the adjustment.

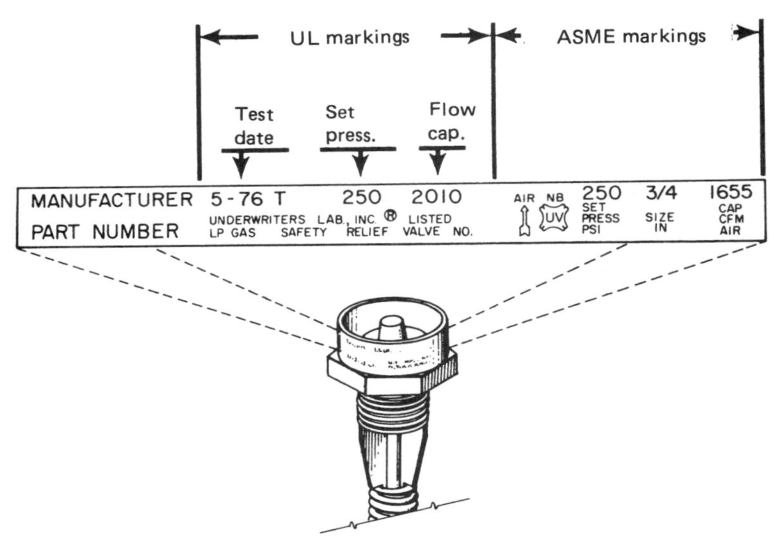

Figure 2.15 Markings on relief valves for ASME containers. The flow capacity marked on the relief valve is the volume of air that the valve can discharge when it is fully open. Again, this marking is based on a factory test when the relief valve is built. This capacity is usually measured in cubic feet of air per minute (cfm-air). Two different flow capacities are marked on the relief valve in Figure 2.15. On the left is the UL flow capacity rating (2,010). On the right is the ASME flow capacity rating (1,655). The UL rating is higher than the ASME rating because the ASME rating is at 90 percent of actual flow. (Drawing courtesy of the National Propane Gas Association.)

(e) Fusible plug devices, with a yield point of 208°F (98°C) minimum and 220°F (104°C) maximum, with a total discharge area not exceeding 0.25 sq in. (1.6 cm2), that communicate directly with the vapor space of the container shall be permitted to be used in addition to the spring-loaded pressure relief valves (as specified in Table 2-3.2.3) for aboveground containers of 1,200 gal (4.5 m^3) water capacity or less.

2-3.2.5 All containers used in industrial truck service (including forklift truck cylinders) shall have the container pressure relief valve replaced by a new or unused valve within 12 years of the date of manufacture of the container and every 10 years thereafter.

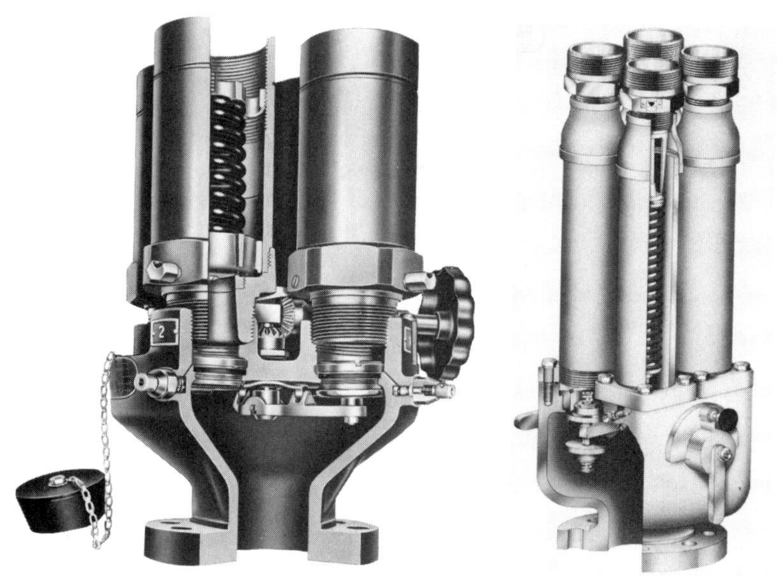

Figure 2.16 ASME container relief valve manifolds. The container requires three relief valves. The manifold contains four. By manipulating the handwheel or lever, an internal clapper-type valve can be rotated to isolate any one of the four relief valves for testing, maintenance, or replacement. (Photo courtesy of Engineered Controls, Inc.)

Pressure relief valves on DOT OR ASME containers used in industrial truck service are exposed to the environment that the truck operates in. If the truck operates in a dusty or corrosive environment as can be found in some industrial facilities, these conditions could prevent normal operation of a pressure relief valve if the valve becomes filled with dust or corrodes. The relief valve must initially be replaced 12 years after the date of manufacture of the container. This corresponds to the date of the first recertification of DOT cylinders, and it is appropriate to mark the cylinder to demonstrate that the valve has been replaced. Thereafter, the containers are normally recertified by the visual method every 5 years, and the relief valve is replaced on each second requalification; this should also be marked on the cylinder. When a permanently mounted ASME container is used on the truck, the valve replacement schedule is the same, and a mark should be made on the truck to indicate that the valve has been replaced.

2-3.3 Container Connections and Appurtenances.

2-3.3.1 Pressure relief devices, container shutoff valves, backflow check valves, internal valves, excess-flow check valves, plugs, liquid level gauges, and overfilling prevention devices that are used individually or in suitable combinations shall comply with 2-3.1.2 and 2-3.1.3 and with the following:

2-3.3.2 Container appurtenances shall be required as follows:

This section was completely revised in the 1992 edition. The revisions were proposed by the Technology and Standards Committee of the National Propane Gas Association, following several years of work, and the editor commends the Committee for their good work, which has enhanced use and understanding of the standard. The requirements are separated for containers of 2,000 gal and less and containers over 2,000 gal. This recognizes that the requirements for container appurtenances change at this size. The requirements for container appurtenances for containers of 2,000 gal and less used for domestic, commercial, industrial, engine fuel, and over-the-road uses are shown in Table 2-3.3.2 (a). This table, and the accompanying notes, have also been revised for the 1992 edition and the table is expanded to include all required valves and appurtenances for each category of container. This corrects several shortcomings of the table in previous editions. The new table includes 4 columns for DOT cylinders based on size, service (liquid or vapor withdrawal), and mode of use (portable vs. stationary); a separate column for DOT cylinders used for engine fuel and mobile applications; and 2 columns for ASME containers, both the stationary type (less than 2,000 gal only) and the engine fuel or mobile type. The concept of mobile containers is also added, and a new definition has been introduced to Section 1-7 defining these containers as those permanently mounted on a vehicle and connected for uses other than engine fuel.

The table does not include requirements for appurtenances for DOT cylinders of less than 2-lb LP-Gas. This is due to the fact that most cylinders of this small size are built to DOT exemptions and accomplish the necessary safety requirements by means other than those specifically required in NFPA 58. These containers include those used for butane lighter refills, portable butane and propane stoves, and similar applications.

Note that the requirements for shutoff valves on DOT cylinders differ for vapor service where a CGA 510 outlet is specified, and liquid service where a CGA 555 outlet is specified. These connections are defined in CGA

Standard V-1/ANSI B57.1, *Compressed Valve Cylinder Valve Outlet and Inlet Connections*. They are different so that accidental connection of liquid to vapor service applications are prevented. (There is also a CGA 600 connection for small torch-type cylinders.)

A new paragraph 2 was added in the 1995 edition covering actuated liquid withdrawal excess-flow valves. This is consistent with the new definition and the changes to 2-2.3.3, where these devices have been required for some time. The requirement is also placed here so that it will not be overlooked by users of the standard who use the comprehensive table for all container appurtenance requirements.

(a) For containers 2,000 gal (7.6 m^3) water capacity or less, see Table 2-3.3.2(a).

1. The requirement for internal spring-type pressure relief valves that are shown in Table 2-3.3.2(a) for stationary ASME containers up to and including 2,000 gal (7.6 m^3) water capacity shall not apply to underground containers where external pressure relief valves are permitted, or to containers that were originally equipped with external pressure relief valves.
2. Containers of 125 gal (0.5 m^3) or more water capacity shall be provided with an actuated liquid withdrawal excess-flow valve with a connection not smaller than ¾-in. National Pipe Thread. This valve shall not be connected for continuous service unless the valve is recommended by the manufacturer for such service.

(b) For containers of over 2,000 gal (7.6 m^3) water capacity.

This section specifies all appurtenances for containers over 2,000 gal water capacity. Containers of this size are used primarily in the distribution of LP-Gas and for industrial uses. Prior to the 1992 edition these requirements were included in Table 2-3.3.2(a). Note that a temperature gauge (thermometer) is required on all containers over 2,000 gal. This is needed so that these larger containers will not be overfilled. In order to determine proper fill level, the liquid temperature must be known (*see 4-4.2.2*).

1. For vapor and liquid withdrawal openings:

   a. A positive shutoff valve that is located as close to the tank as practical in combination with an excess-flow valve installed in the tank.
   b. An internal valve with an integral excess-flow valve or excess-flow protection.

2. For vapor and liquid inlet openings:

   a. A positive shutoff valve that is located as close to the tank as practical, in combination with either a backflow check valve or excess-flow valve installed in the tank.
   b. An internal valve with an integral excess-flow check valve or excess-flow protection.

Table 2-3.3.2(a) Container Connection and Appurtenance Requirements for Containers Used on Domestic, Commercial, Industrial, Engine Fuel, and Over-the-Road Mobile Applications (see Note 9)

Appurtenances	1. Portable DOT cylinders, 2-lb thru 100-lb (0.9- kg thru 45.4-kg) propane capacity for vapor service	2. Portable DOT cylinders, 2-lb thru 100-lb (0.9- kg thru 45.4-kg) propane capacity for liquid service	3. Portable DOT cylinders, 2-lb thru 100-lb (0.9-kg thru 45.4-kg) propane capacity for liquid and vapor service	4. DOT cylinders, 100-lb thru 420-lb (45.4-kg thru 190-kg) propane capacity filled on site	5. Stationary ASME containers thru 2,000-gal (7.6-m^3) water capacity	6. DOT Engine Fuel or Mobile containers (see Note 2)	7. ASME Engine Fuel or Mobile containers (see Note 2)
A. Manual Shutoff Valve (CGA 510 outlet) with Integral External Pressure Relief Valve	R√ (See Notes 4, 8, 10, 12)						
B. Manual Shutoff Valve (CGA 555 outlet) with Integral External Pressure Relief Valve and Internal Excess-Flow Valve		R√ (See Notes 4, 8, 10)					
C. Manual Shutoff Valve (CGA 555 outlet) with Internal Excess-Flow Valve for Liquid Service; Manual Shutoff Valve (CGA 510 outlet) with Integral External Pressure Relief Valve for Vapor Service			R√ (See Notes 4, 8, 10)				
D. Double Backflow Check Filler Valve		O√		R√	R√	O	R
E. Manual Shutoff Valve for Vapor Service	(see Notes 11, 12)			R√	R√		
F. Fixed Liquid Level Gauge (See Note 3)	O√	O√	O√	R√	R√	R	R
G. External Pressure Relief Valve (See Note 4)				R√			
H. Internal Spring-type Pressure Relief Valve (See Note 4)					R (See Note 1)		
I. Float Gauge	O√	O√	O√	O	R√	O	O
J. Backflow Check and Excess-Flow Vapor Return Valve					O√		

Table 2-3.3.2(a) Container Connection and Appurtenance Requirements for Containers Used on Domestic, Commercial, Industrial, Engine Fuel, and Over-the-Road Mobile Applications (see Note 9)

Appurtenances	1. Portable DOT cylinders, 2-lb thru 100-lb (0.9- kg thru 45.4-kg) propane capacity for vapor service	2. Portable DOT cylinders, 2-lb thru 100-lb (0.9- kg thru 45.4-kg) propane capacity for liquid service	3. Portable DOT cylinders, 2-lb thru 100-lb (0.9-kg thru 45.4-kg) propane capacity for liquid and vapor service	4. DOT cylinders, 100-lb thru 420-lb (45.4-kg thru 190-kg) propane capacity filled on site	5. Stationary ASME containers thru 2,000-gal (7.6-m^3) water capacity	6. DOT Engine Fuel or Mobile containers (see Note 2)	7. ASME Engine Fuel or Mobile containers (see Note 2)
K. Liquid Withdrawal Excess-Flow Actuated Valve (See Note 5)					R		
L. Manual Shutoff Liquid or Vapor Valve with Internal Excess-Flow Check Valve						R	R
M. Full Internal or Flush-type Full Internal Pressure Relief Valve						R	R (See Note 6)
N. Overfilling Prevention Device						R (See Note 7)	R (See Note 7)

O. R = Required. O = Optional. R√ = Required, but may be installed as part of a multipurpose valve. O√ = Optional, but may be installed as part of a multipurpose valve.

Notes to Table 2-3.3.2(a):

1. See 2-3.3.2(a)(1).
2. Mobile containers are containers that are permanently mounted on a vehicle and are connected for uses other than engine fuel (see Section 1-6, Definitions).
3. See 2-3.4.2.
4. See Appendix E for references on DOT and ASME pressure relief valve standards. See 1-6 for definitions of pressure relief valves.
5. A liquid withdrawal excess-flow check valve in combination with a manual shutoff valve, plugged, or internal valve with excess-flow protection, plugged, will also meet these requirements. Liquid withdrawal connections are not required on containers less than 125 gal (0.5 m^3) water capacity. (See 2-2.3.3.) Actuated liquid withdrawal excess-flow valves shall not be used for continuous service unless recommended for continuous service by the valve manufacturer.
6. Reference: 8-2.2(a)(6).
7. Reference: 8-2.3(a)(8).
   NOTE: An overfilling prevention device is not required for portable engine fuel containers used on industrial (and forklift) trucks powered by LP-Gas or for portable engine fuel containers used on vehicles having LP-Gas powered engines mounted on them (including floor maintenance machines, etc.).
8. Individual container valves meeting these requirements may be used instead of multipurpose valves.
9. Table 2-3.3.2(a) and paragraph 2-3.3.2(b) are not intended to restrict the utilization of additional container openings with appropriate appurtenances for use in other applications.
10. CGA outlets 510 and 555 are described in Compressed Gas Association Standards.
11. Excess-flow protection shall not be required when an approved regulator is directly attached or attached with a flexible connector to the outlet of the manual shutoff valve for vapor service and the controlling orifice between the container contents and the shutoff valve outlet does not exceed 5/16 in. (8 mm) in diameter.
12. Separate external excess-flow valve protection shall be provided for vapor service shutoff valves in DOT cylinders or ASME containers when used inside buildings, or on building roofs or exterior balconies.

3. Other required appurtenances:

a. Internal spring-type, flush-type full internal, or external pressure relief valve (*see Appendix E*).

b. Fixed liquid level gauge.

c. Float gauge, rotary gauge, or slip tube gauge, or a combination of these gauges.

d. Pressure gauge.

e. Temperature gauge.

2-3.3.3 The appurtenances specified in Table 2-3.3.2(a) and paragraph 2-3.2.2(b) shall comply with the following:

(a) Manual shutoff valves shall be designed to provide positive closure under service conditions.

(b) Excess-flow check valves shall be designed to close automatically at the rated flows of vapor or liquid specified by the manufacturer. Excess-flow valves shall be designed with a bypass that shall not exceed a No. 60 drill size opening to allow equalization of pressure.

This exception was added in the 1995 edition to require a smaller bypass opening in smaller pipes. A No. 60 drill size opening is 0.40 in. (1.02 mm) in diameter, which will permit a flow of about 300,000 Btu/hr (120 ft 3/hr) at 100 psi. The reduction to 10 ft^3/hr in piping smaller than ½ in. recognizes that these smaller piping systems should have a smaller bypass flow.

Exception: Excess-flow valves of less than ½ in. (1.3 cm) N.P.T. shall have a bypass that limits propane vapor flow to 10 scf/hr at 100 psi (690 kPa).

(c) Backflow check valves, which shall be permitted to be of the spring-loaded or weight-loaded type with in-line or swing operation, shall close when the flow is either stopped or reversed. Both valves of double backflow check valves shall comply with this provision.

(d) Internal valves (*see definition*), either manually or automatically operated and designed to remain closed except during operating periods, shall be considered positive shutoff valves. [*See 6-3.2.1 for special requirements for such valves used on cargo units.*]

An internal valve can be considered a substitute for an excess-flow check valve with the added feature that it remains in the closed position except when liquid is being transferred. An excess-flow check valve is in the open position except when a flow large enough to close the valve occurs.

Internal valves are categorized according to the means by which they are opened. Some are opened manually by means of a lever. Others are opened by liquid or gas pressure downstream of the valve. In the event of leakage, they close automatically, either by operation of an incorporated excess-flow

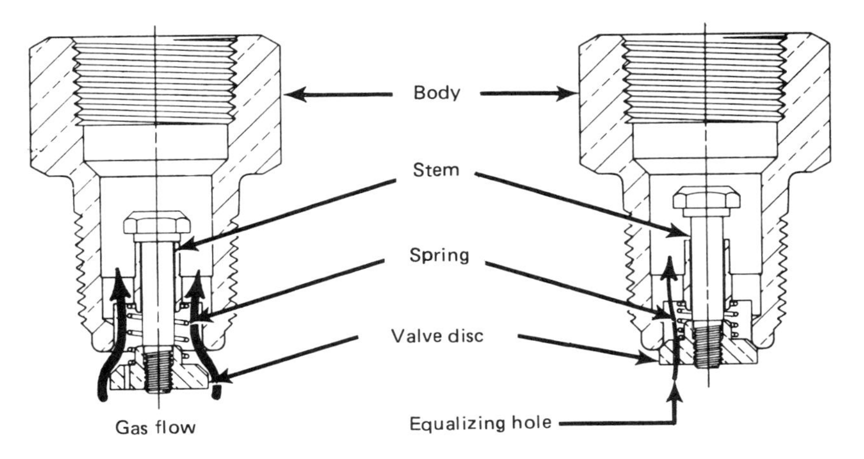

Figure 2.17 *Operation of excess-flow check valve. After flow has been stopped by closing the downstream valve (or other means), pressure on both sides will equalize through the equalizing hole and the spring will cause the valve to reopen. (Drawing courtesy of National Propane Gas Association.)*

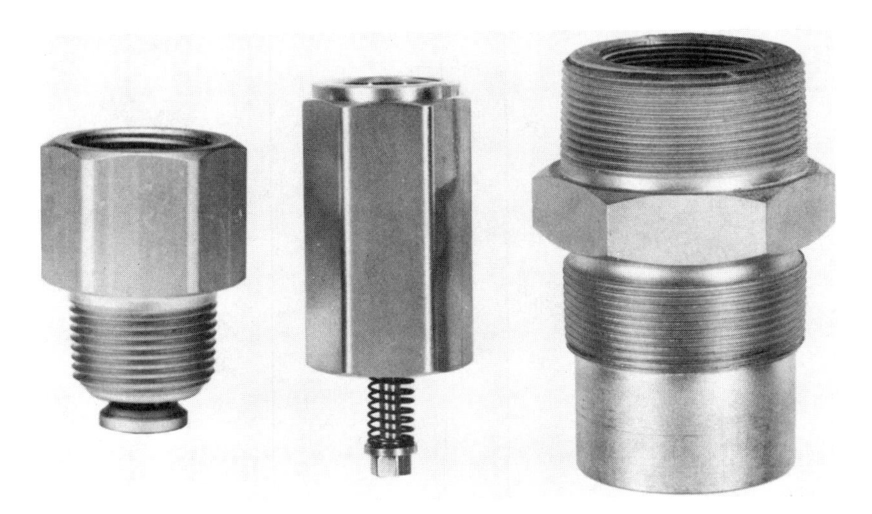

Figure 2.18 *Various excess-flow check valves. (Photo courtesy of Fisher Controls.)*

check valve or by lowering of downstream pressure. They also can be arranged for automatic closing through operation of fusible elements (fire exposure) and for remote manual closing.

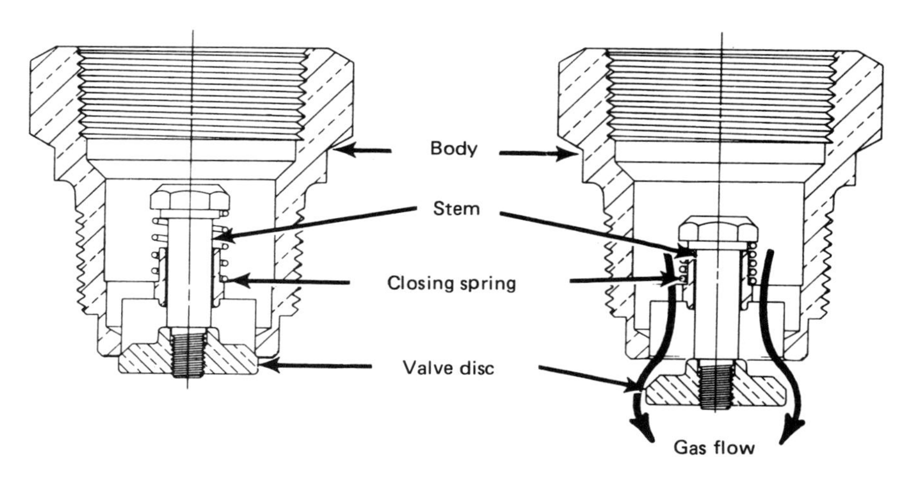

Figure 2.19 *Operation of backflow check valve. (Drawing courtesy of National Propane Gas Association.)*

Figure 2.20 *Lever-operated internal valve for threaded installation. (Photo courtesy of Engineered Controls, Inc.)*

Figure 2.21 *Lever-operated internal valve for flanged installation. (Photo courtesy of Engineered Controls, Inc.)*

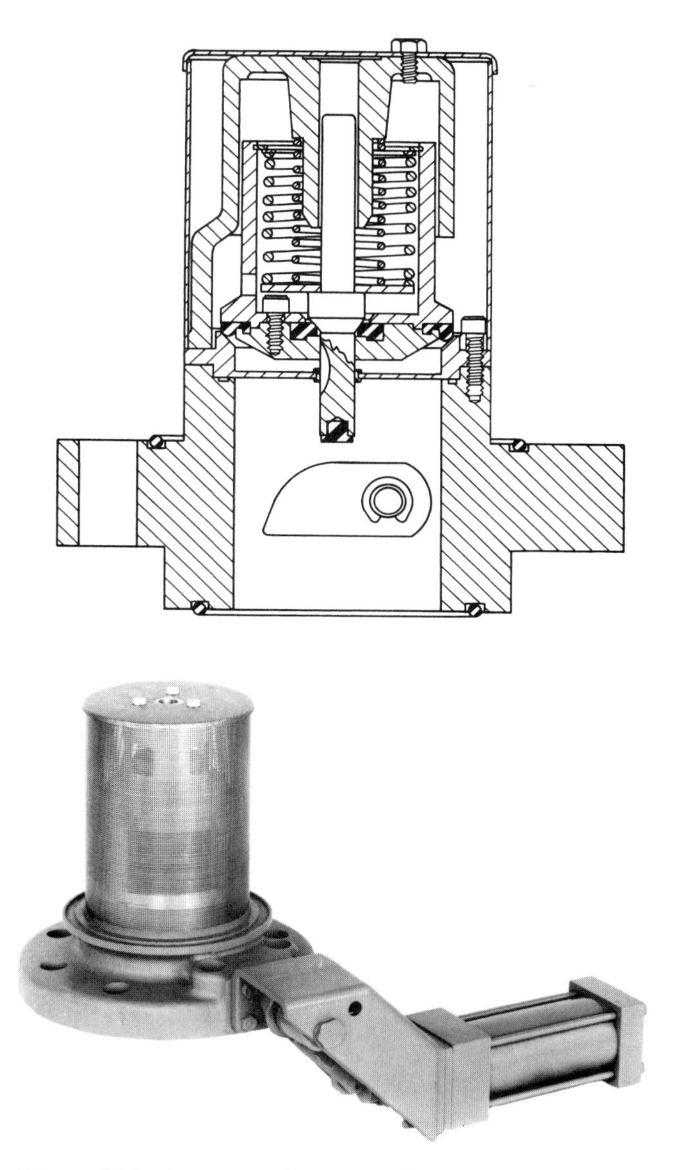

Figure 2.22 Pneumatically operated internal valve. (Photo and drawing courtesy of Engineered Controls, Inc.)

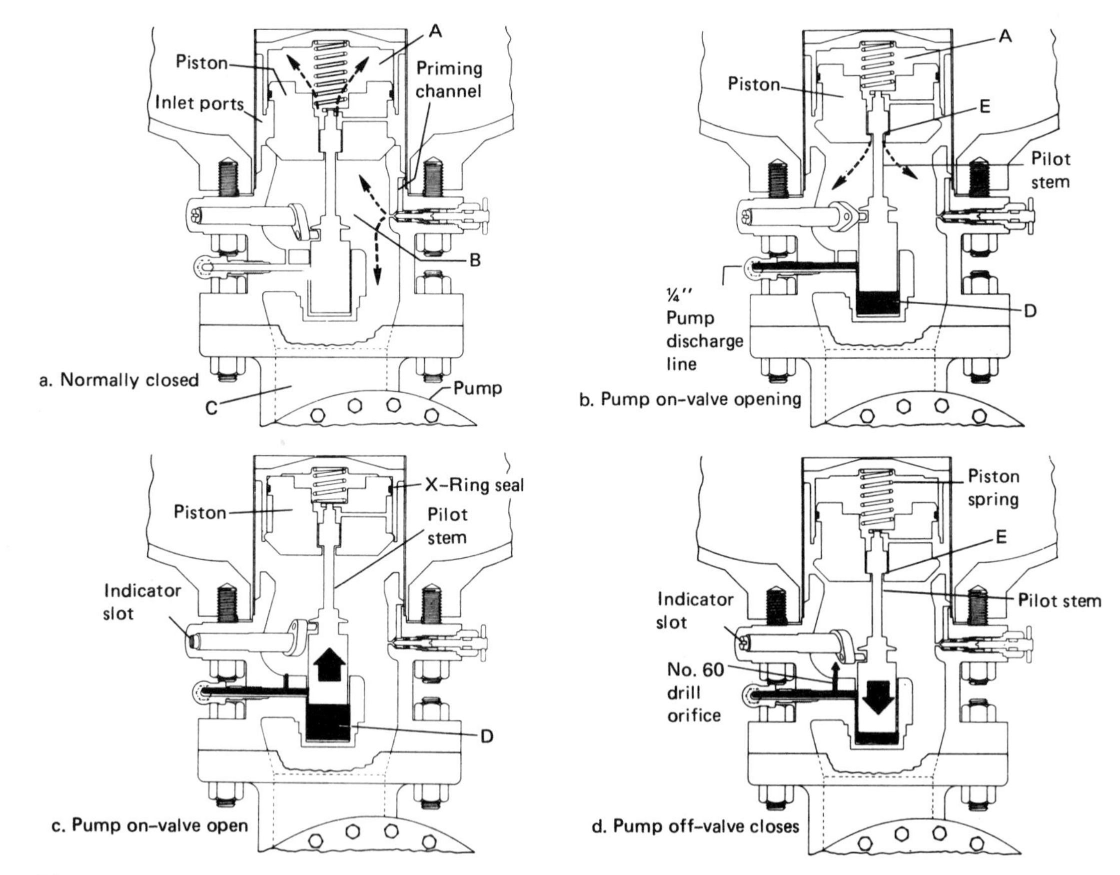

Figure 2.23 *Operation of pump pressure-operated internal valve. (Drawing courtesy of Engineered Controls, Inc.)*

2-3.4 Liquid Level Gauging Devices.

2-3.4.1 Liquid level gauging devices shall be provided on all containers filled by volume. Fixed level gauges or variable gauges of the slip tube, rotary tube, or float types (or combinations of such gauges) shall be permitted to be used to comply with this provision.

2-3.4.2 Every container constructed after December 31, 1965, designed to be filled on a volumetric basis shall be equipped with a fixed liquid level gauge(s) to indicate the maximum filling level(s) for the service(s) in which the container is to be used (*see 4-4.3.3*). This shall be permitted to be accomplished either by using a dip tube of appropriate length or by the position of the gauging device in the container. The following shall apply:

Because rotary or magnetic gauges can be inaccurate, it is necessary to have a fixed liquid level gauge available on all tanks to determine maximum filling levels in emergencies and to check the accuracy of these variable gauges to prevent inadvertent overfilling. This provision was incorporated in the 1965 edition of the standard. Fixed liquid level gauges are dip tubes of certain lengths installed in the container. The gauge discharge is invisible if vapor is emitted, but a fog (of condensed water vapor in the air) is created by refrigeration when vaporizing liquid is discharged. The criteria for their length is based on the specific gravity of the liquid at 40°F (4°C). Subparagraphs (a) through (d) provide marking requirements for containers equipped with fixed liquid level gauges to readily indicate the length of the dip tube or percentage fill indicated by the fixed level gauge.

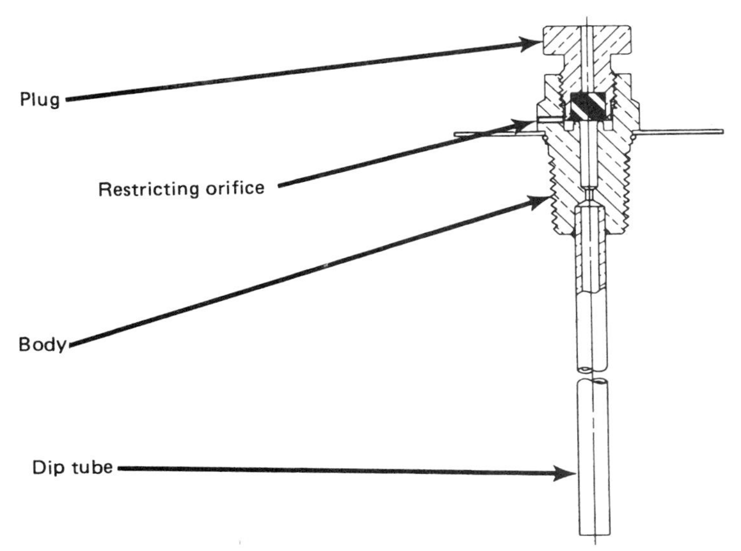

Figure 2.24 *Fixed liquid level gauge. (Photo courtesy of the National Propane Gas Association.)*

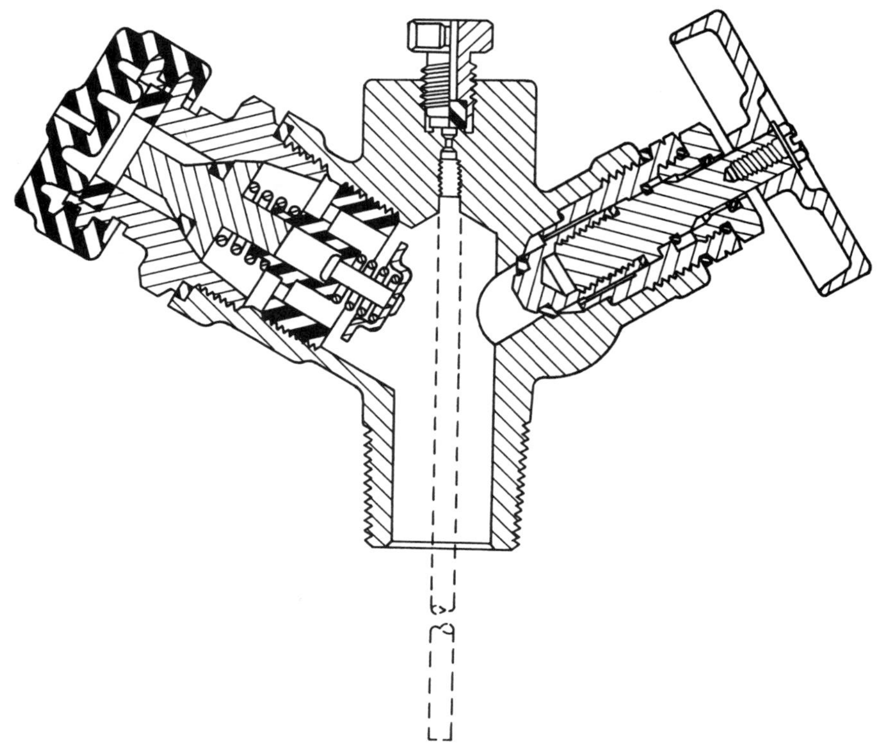

Figure 2.25 Fixed liquid level gauge incorporated into multiple function container valve assembly. (Drawing courtesy of Engineered Controls, Inc.)

(a) ASME containers manufactured after December 31, 1969, shall have permanently attached to the container adjacent to the fixed liquid level gauge, or on the container nameplate, markings showing the percentage full that is indicated by that gauge.

(b) Containers constructed to DOT cylinder specifications shall have stamped on the container the letters "DT" followed by the vertical distance (to the nearest tenth inch) from the top of the boss or coupling into which the gauge, or the container valve of which it is a part, is installed, to the end of the dip tube. [*See 2-3.4.2(c)(2) for DOT containers designed for loading in either the vertical or horizontal position.*]

(c) Each container manufactured after December 31, 1972, that is equipped with a fixed liquid level gauge for which the tube is not welded in place shall be permanently marked adjacent to such gauge or on a container nameplate as follows:

1. Containers designed to be filled in one position shall be marked with the letters "DT" followed by the vertical distance (to the nearest tenth inch) measured from the top center of the container boss or coupling into which the gauge is installed to the maximum permitted filling level.

2. Universal-type containers (*see definition*) shall be marked as follows:

a. *Vertical Filling*: With the letters "VDT" followed by the vertical distance (to the nearest tenth inch), measured from the top center of the container boss or coupling into which the gauge is installed to the maximum permitted filling level.

b. *Horizontal Filling*: With the letters "HDT" followed by the vertical distance (to the nearest tenth inch), measured from the top centerline of the container boss or coupling opening into which the gauge is installed to the inside top of the container when the container is in the horizontal position.

(d) Cargo tanks having several fixed level gauges positioned at different levels shall have stamped adjacent to each gauge the loading percentage (to the nearest 2/10 percent) of the container content indicated by that particular gauge.

Figure 2.26 *Fixed liquid level gauge. Marking on disk reads, "STOP FILLING WHEN LIQUID APPEARS." (Photo courtesy of Engineered Controls, Inc.)*

Figure 2.27 *Cylinder valve for 200-lb DOT cylinder. Note the fixed liquid level gauge (dip Tube) and top threaded connection for filling on site from a bobtail without disconnecting piping. (Photo courtesy of Sherwood Selpac.)*

2-3.4.3 Variable liquid level gauges shall comply with the following:

(a) Variable liquid level gauges shall be so marked that the maximum liquid level, in inches or percent of capacity of the container in which they are to be installed, is readily determinable. These markings shall indicate the maximum liquid level for propane, for 50/50 butane-propane mixtures, and for butane at liquid temperatures from 20°F (–6.7°C) to 130°F (54.4°C) and in increments not greater than 20°F (–6.7°C).

(b) The markings indicating the various liquid levels from empty to full shall be either directly on the system nameplate or on the gauging device or on both.

(c) Dials of magnetic float or rotary gauges shall indicate whether they are for cylindrical or spherical containers, and whether for aboveground or underground service.

(d) The dials of gauges for use only on aboveground containers of over 1,200 gal (4.5 m^3) water capacity shall be so marked.

2-3.4.4 Variable liquid level gauges shall comply with the provisions of 4-4.3.3(b) if they are used for filling containers.

2-3.4.5 Gauging devices requiring bleeding of product to the atmosphere, such as fixed liquid level, rotary tube, and slip tube gauges, shall be designed so that the bleed valve maximum opening to the atmosphere is not larger than a No. 54 drill size, unless they are equipped with excess-flow check valves.

2-3.5 Pressure Gauges.

2-3.5.1 Pressure gauges shall comply with 2-3.1.2 and 2-3.1.3.

2-3.5.2 Pressure gauges shall be attached directly to the container opening or to a valve or fitting that is directly attached to the container opening. If the effective opening into the container allows a flow greater than that of a No. 54 drill size, an excess-flow check valve shall be provided.

2-3.6 Other Container Connections. Container openings shall be equipped with one of the following:

(a) A positive shutoff valve in combination with either an excess-flow check valve or a backflow check valve, plugged.

(b) An internal valve, plugged.

(c) A backflow check valve, plugged.

(d) An actuated liquid withdrawal excess-flow valve, normally closed and plugged, with provision to allow for external actuation.

(e) A plug, blind flange, or plugged companion flange.

Exception No. 1: Pressure relief valves in accordance with 2-3.2.

Exception No. 2: Connections for flow controls in accordance with 2-3.3.

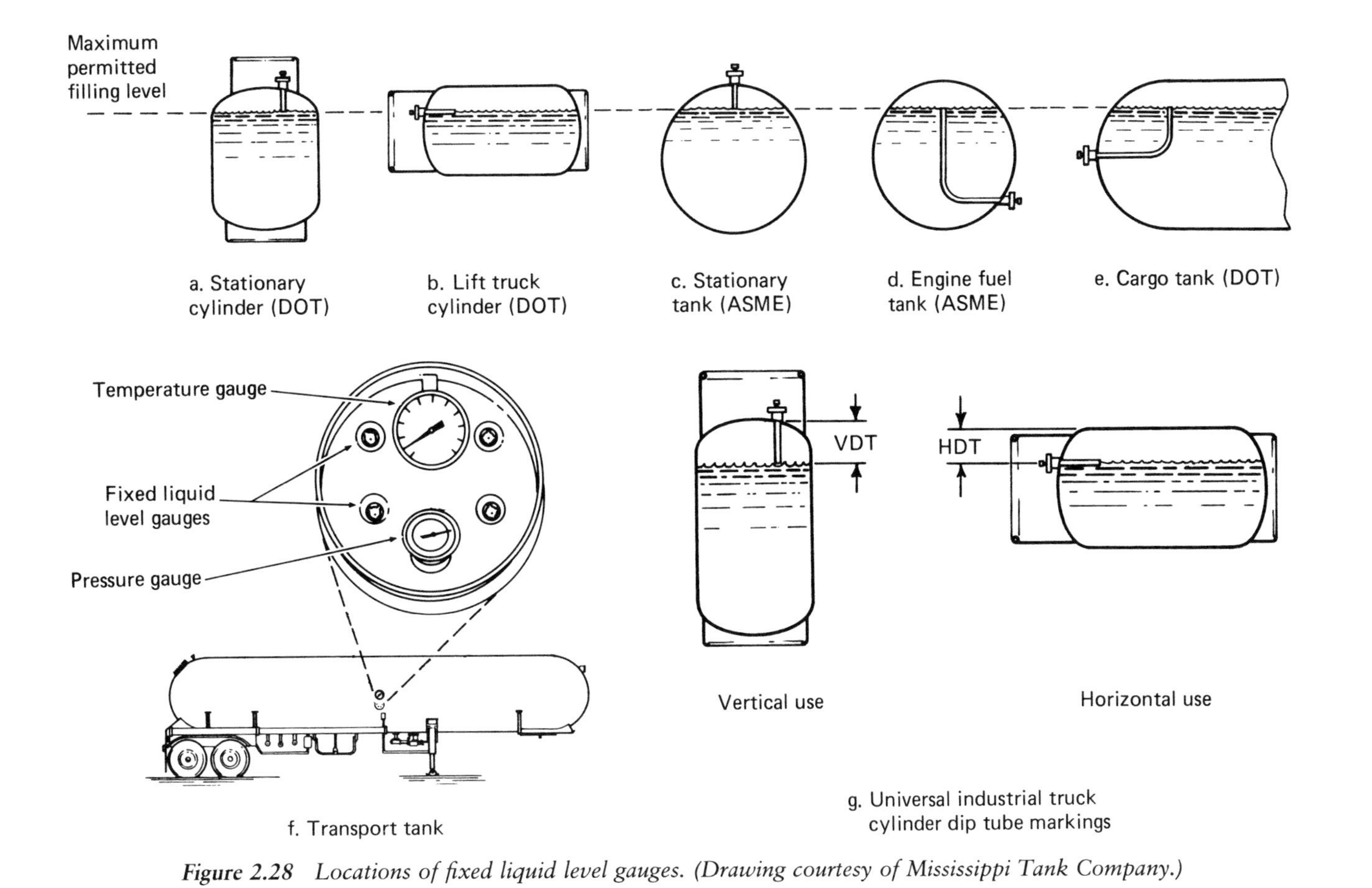

Figure 2.28 Locations of fixed liquid level gauges. (Drawing courtesy of Mississippi Tank Company.)

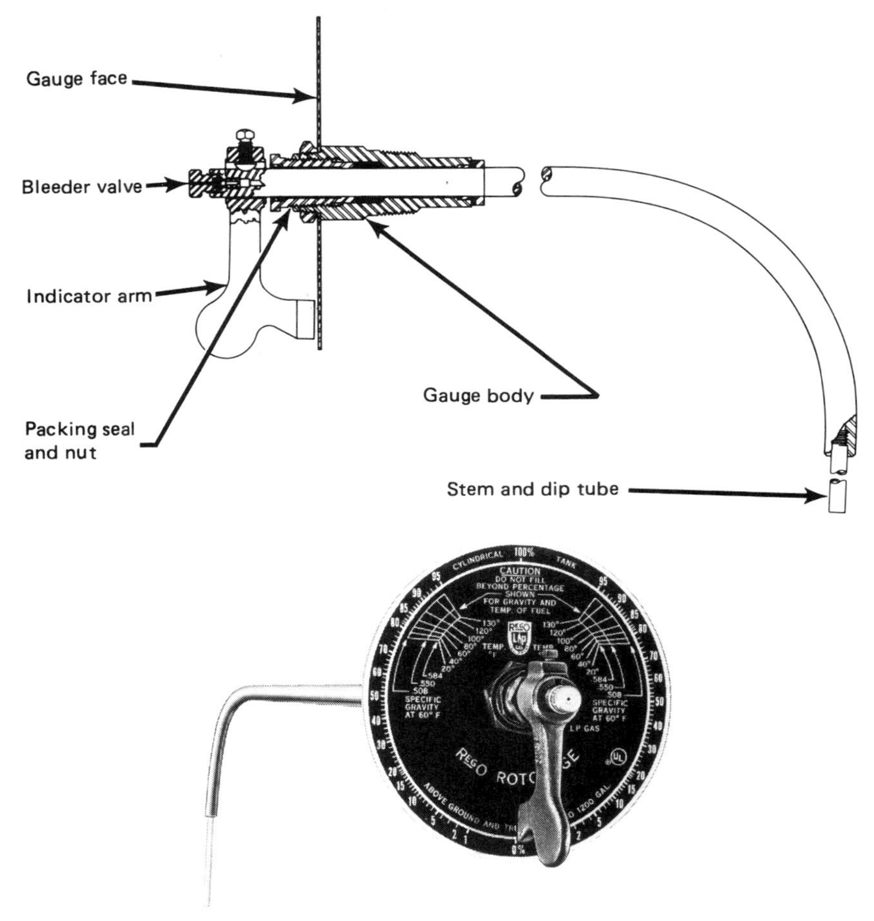

Figure 2.29 *Rotary type of variable liquid level gauge. (Photo and drawing courtesy of Engineered Controls, Inc.)*

Exception No. 3: Liquid level gauging devices in accordance with 2-3.4.

Exception No. 4: Pressure gauges in accordance with 2-3.5.

This provides a complete list in (a) through (e) of all appurtenances that may by installed in container openings. Paragraph (d) was modified by the addition of "actuated liquid withdrawal excess-flow valve" to recognize this new term here for consistency with Table 2-3.3.2 (a). The exceptions, which apply to all of 2-3.6, were added in the 1992 edition to replace part of 2-3.6 that stated, "other than those equipped as provided in 2-3.2, 2-3.4, and 2-3.5." While clear, the statement required the user of the standard to refer to those paragraphs. The use of the exceptions make it easier on the user.

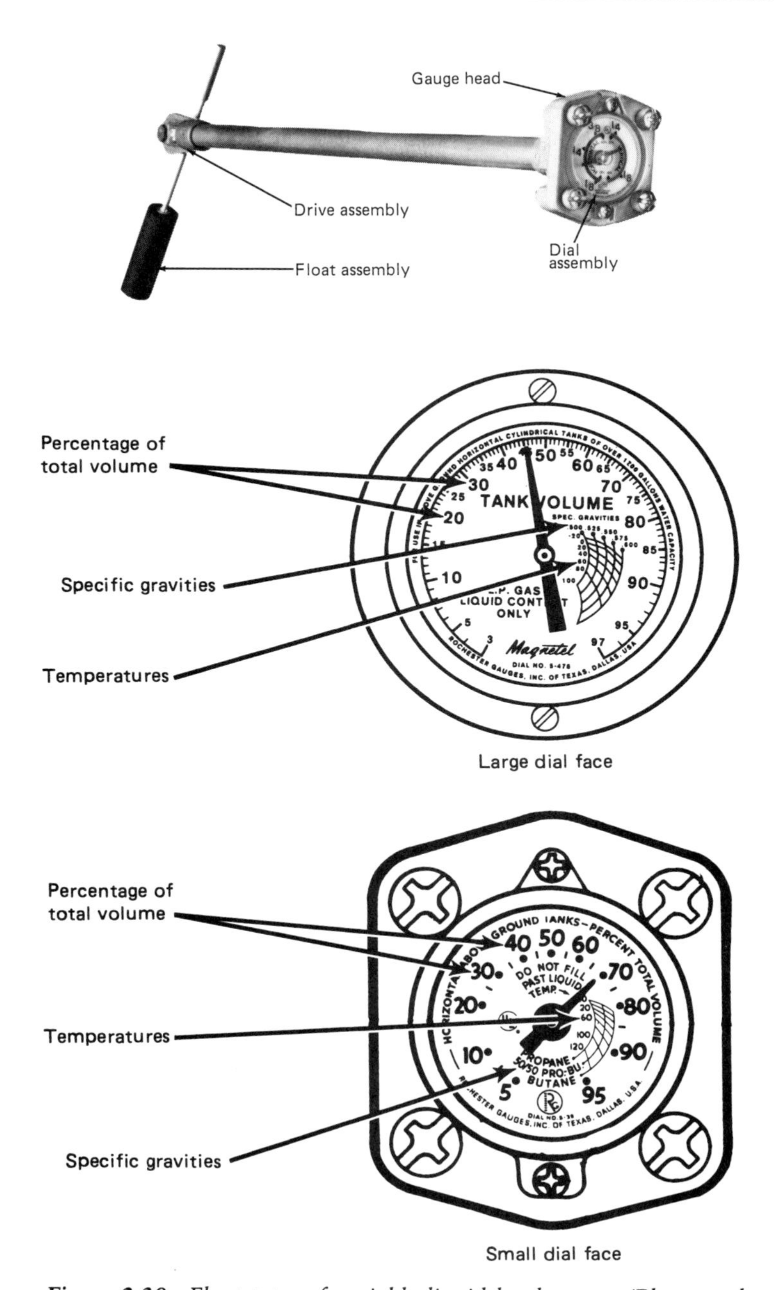

Figure 2.30 *Float type of variable liquid level gauge. (Photo and drawings courtesy of National Propane Gas Association.)*

2-3.7 Container Appurtenance Protection. Container appurtenances other than pressure relief devices shall be installed and protected as follows:

In the 1992 edition, this section, formerly numbered 3-2.4.6, was relocated from Chapter 3. This was done because the section contains equipment requirements, which belong in Chapter 2.

(a) All container openings except those used for pressure relief devices (*see 2-3.2*), liquid level gauging devices (*see 2-3.4*), pressure gauges (*see 2-3.5*), those equipped with double check valves as allowed in Table 2-3.3.2(a), and plugged openings shall be equipped with internal valves [*see 2-3.3.3(d)*] or with positive shutoff valves and either excess-flow or backflow check valves (*also see 2-3.3 for specific application*) as follows:

An excess-flow valve must never be installed in a pressure relief device connection because the high-pressure condition that causes the pressure operation of this safety device would result in a very high flow through the pressure relief device and the excess-flow valve, possibly closing the excess-flow valve and defeating the safety device.

The exception of container openings used for level gauging devices and pressure gauges recognizes that these openings are restricted to a size where it is not a significant hazard and an excess-flow or backflow check valve is not only unnecessary but, if provided, could close and negate the gauging device or pressure gauge.

On all other connections, either an excess-flow valve or a double check valve must be used.

1. On ASME containers excess-flow or backflow check valves shall be located between the LP-Gas in the container and the shutoff valves, either inside the container, or at a point immediately outside where the line enters or leaves the container. If outside, installation shall be made so that any undue strain beyond the excess-flow or backflow check valve will not cause breakage between the container and the valve. All connections, including couplings, nozzles, flanges, standpipes and manways, that are listed in the ASME Manufacturers' Data Report for the container shall be considered part of the container.
2. If an excess-flow valve is required on DOT cylinders, it shall be permitted to be located at the outlet of the cylinder shutoff valve.
3. Shutoff valves shall be located as close to the container as practical. The valves shall be readily accessible for operation and maintenance under normal and emergency conditions, either because of location or by means of permanently installed special provisions. Valves installed in an unobstructed location not more than 6 ft (1.8 m) above ground level shall be considered accessible. Special provisions include, but are not limited to, stairs, ladders, platforms, remote operators, or extension handles.

An excess-flow valve is used where there might be flow in either direction through a given connection. A backflow check valve can be used where the flow would be only into the container and where flow in the opposite direction must be prevented.

It is preferable to have an excess-flow valve or backflow check valve inside the container so that if there is a breakage of the piping outside of the container the safety device will not be adversely affected. Where necessary, however, they can be outside but should be upstream of the manual shutoff valve.

It should be recognized that an excess-flow valve will operate only if the flow is in excess of the valve's rated flow. If a partial break in a line occurs that will not allow a flow up to the rated capacity, the excess-flow valve will not operate. Many operators believe that an excess-flow valve provides 100 percent protection in case of a pipe break, but this is true only if the break is sufficiently sized to permit enough flow to operate the excess-flow valve. While excess-flow valves do not provide 100 percent safety, they are used because there is nothing else available.

It is also important to size the piping downstream of the excess-flow valve so that it does not restrict the flow and prevent the excess-flow valve from working. In many cases it is necessary to put in additional excess-flow valves in the downstream line to ensure a shutoff with a break some distance from the container, or where there is a reduction in the piping size. It is also important that the operator, in opening a line or container valve, open it slowly, so that the excess-flow valve will not "slug" in to shut off the flow when it is not supposed to. In some cases where this has occurred, the operator has removed the working portion of the excess-flow valve which, of course, negates its function as a safety device.

4. The connections, or line, leading to or from any individual opening shall have greater capacity than the rated flow of the excess-flow valve protecting the opening.

(b) Valves, regulators, gauges, and other container appurtenances shall be protected against physical damage.

(c) Valves that are part of the assembly of portable multicontainer systems shall be arranged so that replacement of containers can be made without shutting off the flow of gas in the system. This provision shall not be construed as requiring an automatic changeover device.

(d) Connections to containers installed underground shall be located within a substantial dome, housing, or manhole and shall have access thereto protected by a substantial cover. Underground systems shall be installed so that all terminals for connecting hose and any opening through which there can be a flow from pressure relief devices or pressure regulator vents are located above the normal maximum water table. Terminals for connecting hoses, openings for flow from pressure relief devices, and the interior of domes, housing, and manholes shall be kept clean of debris. Such manholes or housings shall be provided with ventilated louvers or their equivalent. The area of such openings shall equal or exceed the combined discharge areas of the pressure relief devices and other vent lines that discharge into the manhole or housing.

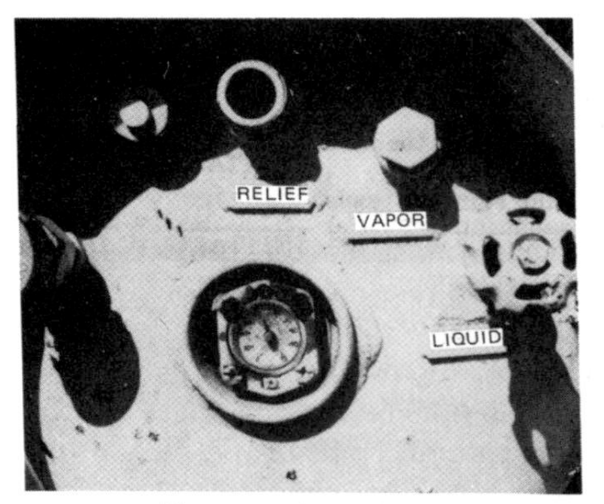

a. Die-stamped on cylinder

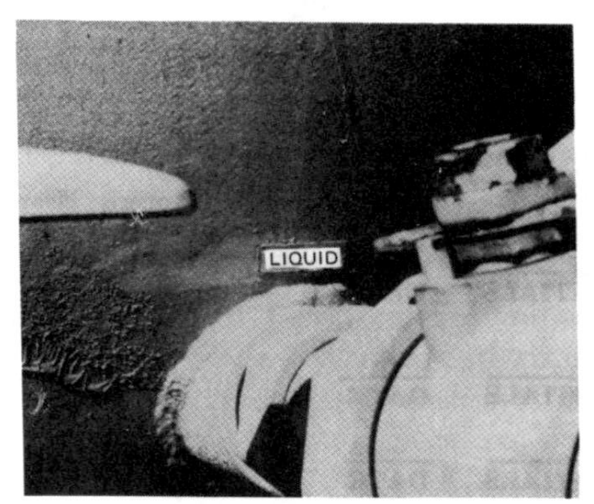

b. Labeled on tank

c. Tagged to service valve

Figure 2.31 *Typical container inlet and outlet connection markings. (Photos courtesy of the National Propane Gas Association.)*

The housing around appurtenances on underground containers is subject to accumulations of water and debris and should be checked often.

(e) Container inlet and outlet connections on containers of more than 2,000 gal (7.6 m^3) water capacity shall be labeled to designate whether they communicate with the vapor or liquid space. Labels shall be permitted to be on valves.

Exception No. 1: Connections for pressure relief devices.

Exception No. 2: Connections for liquid level gauging devices.

Exception No. 3: Connections for pressure gauges on containers of 2,000 gal (7.6 m^3) water capacity or more.

NOTE: See 3-8.2.5(e) and 8-2.2(g) for requirements for labeling smaller containers used for vehicular installations.

(f) Every storage container of more than 2,000 gal (7.6 m^3) water capacity shall be provided with a pressure gauge.

2-4 Piping (Including Hose), Fittings, and Valves.

The term "piping," as used in NFPA 58, includes pipe, tubing, hose, and the valves and fittings used in the piping system. Provisions for the basic design of this equipment are given in this section for use by manufacturers in producing piping system components. Guidelines both for testing laboratories to list this equipment for possible acceptance by authorities having jurisdiction and for users to select proper materials in making a piping installation are also included.

2-4.1 General.

2-4.1.1 This section includes basic design provisions and material specifications for pipe, tubing, pipe and tubing fittings, valves (including hydrostatic relief valves), hose, hose connections, and flexible connectors used to connect container appurtenances with the balance of the LP-Gas system in accordance with the installation provisions of Chapters 3, 8, and 9.

Pressure ratings of fittings and valves are based on the concept that the maximum normal operating discharge pressure of a liquid pump or vapor compressor be designed for 350 psi (2413 kPa), while other piping above 125 psi (862 kPa) be designed for 250 psi (1724 kPa). Note that the metallic pipe used for service above 125 psi (862 kPa) will withstand 350 psi (2413 kPa), and that a pressure limit is not specified in the standard.

2-4.1.2 Piping, pipe and tubing fittings, and valves used to supply utilization equipment within the scope of NFPA 54, *National Fuel Gas Code*, shall comply with that code.

Most LP-Gas piping systems from the outlet of the first-stage regulator are subject to the provisions of NFPA 54, *National Fuel Gas Code*. Therefore, reference should be made to NFPA 54, first to determine whether the particular installation (primarily one of a permanent piping system) is subject to that standard, and second, to Part 2 of that code, "Gas Piping System Design, Materials, and Components."

In general, when an LP-Gas system serves a building, piping between the container and the first-stage regulator is in the scope of NFPA 58 and piping downstream of the first-stage regulator is in the scope of NFPA 54. LP-Gas piping for agricultural applications in buildings such as brooders, dehydrators, dryers, and irrigation equipment fall under NFPA 58 even when served by fixed piping, however. This is because farms are rarely served by natural gas mains and this equipment has been specifically excluded by NFPA 54. Other piping systems beyond the first-stage regulator in applications not usually served by natural gas mains are covered under NFPA 58. These include piping systems serving railroad switch heaters and microwave antennae that must be kept free of ice.

2-4.1.3 Pipe and tubing shall comply with 2-4.2 and 2-4.3 or shall be of material that has been investigated and tested to determine that it is safe and suitable for the proposed service and is recommended for that service by the manufacturer, and shall be acceptable to the authority having jurisdiction.

Materials specified for pipe (2-4.2), tubing (2-4.3), and fittings (2-4.4) have been investigated and tested for recommended use. If materials are not referenced they should not be used unless given special approval by the authority having jurisdiction. Paragraph 1-1.4, Alternate Materials, Equipment, and Procedures, provides a basis for such approval. For example, prior to the 1972 edition of NFPA 58 provisions were included for the use of aluminum alloy pipe and tubing. However, due to adverse experience under a variety of conditions, the provisions were so restrictive that their use was eliminated from the standard. Provisions for plastic pipe and tubing were first incorporated in the 1979 edition and revised in the 1992 edition, but they too are quite restrictive. Pipe, tubing, and fittings are limited to polyethylene types meeting certain ASTM standards and have to be recommended for the service by the manufacturer. Joints can be made by heat fusion and by compression-type mechanical fittings meeting ASTM D 2513, or factory-assembled transition fittings may be used. The service is limited to vapor (no liquid) and even then must not exceed 30 psi (208 kPa). The system can only be installed outside and underground. With reference to copper tubing specifications, Type M is not included since experience has indicated that a minimum wall thickness of 0.032 in. (0.81 mm) should be used to avoid kinking.

2-4.1.4 Piping that can contain liquid LP-Gas and that can be isolated by valving and that requires hydrostatic relief valves, as specified under Section 3-2.9, shall have a minimum design pressure of 350 psi (2.41 MPa) or a design pressure that is equivalent to the maximum discharge pressure of any pump or other source feeding the piping system if it is greater than 350 psi (2.41 MPa).

This new requirement recognizes that the set pressure for hydrostatic relief valves is 350 psi for piping systems operating at up to 350 psi, or higher for systems operating at higher pressure (see 3-4.7). The piping is now mandated to withstand this higher pressure in the portions of piping systems where hydrostatic relief valves are mandated.

2-4.2 **Pipe.** Pipe shall be wrought iron or steel (black or galvanized), brass, copper, or polyethylene (*see 3-2.8.7*) and shall comply with the following:

(a) Wrought-iron pipe: ANSI B36.10M, *Welded and Seamless Wrought Steel Pipe.*

(b) Steel pipe: ASTM A53, *Specification for Pipe, Steel, Black and Hot-Dipped, Zinc-Coated Welded and Seamless.*

(c) Steel pipe: ASTM A106, *Specification for Seamless Carbon Steel Pipe for High-Temperature Service.*

(d) Brass pipe: ASTM B43, *Specification for Seamless Red Brass Pipe, Standard Sizes.*

(e) Copper pipe: ASTM B42, *Specification for Seamless Copper Pipe, Standard Sizes.*

(f) Polyethylene pipe: ASTM D2513, *Specification for Thermoplastic Gas Pressure Pipe, Tubing and Fittings.* Pipe shall be recommended by the manufacturer for use with LP-Gas. Polyethylene pipe shall be marked in full compliance with the product marking requirements of ASTM D2513, and shall include the manufacturer's name or trademark, the Standard Dimensional Ratio of the pipe, the size of the pipe, the designation polyethylene (PE), the date manufactured, and the designation ASTM D2513.

In the 1992 edition the requirement for listing or approval of polyethylene pipe, which had been mandated since polyethylene pipe was first recognized by NFPA 58 in 1979, was dropped. This change had been made in a Tentative Interim Amendment to the 1989 edition. The deletion of the requirement for listing recognized that listed polyethylene pipe was not available, that some authorities having jurisdiction were prevented from approving products not listed, and that no other piping materials had this restriction. Listing requirements is valid for new products without a proven safety record and for components that are critical to safety, and are subject to review after extensive experience. The vast majority of polyethylene pipe and tubing in gas service is used for natural gas service (covered in the United States by the pipeline safety regulations of the U.S. Department of Transportation), where listing is not required, and there was no reason to continue the requirement for propane service.

The requirements for polyethylene pipe were expanded in the 1992 edition to include mandatory marking of manufacturer, size, the words polyethylene (PE) and ASTM D 2513, and the date of manufacture on the pipe to agree with ASTM specification D 2513. These were already being marked by most manufacturers.

The service pressure of polyethylene pipe (and tubing) is limited to 30 psi (207 kPa). [*See 3-2.7(b).*]

2-4.3 **Tubing.** Tubing shall be steel, brass, copper, or polyethylene (*see 3-2.8.7*) and shall comply with the following:

(a) Steel tubing: ASTM A539, *Specification for Electric-Resistance-Welded Coiled Steel Tubing for Gas Fuel Oil Lines.*

(b) Brass tubing [*see 3-2.7(d), Exception No. 3*]: ASTM B135, *Specification for Seamless Brass Tube.*

(c) Copper tubing [*see 3-2.7(d), Exception No. 3*]:

1. Type K or L, ASTM B88, *Specification for Seamless Copper Water Tube.*
2. ASTM B280, *Specification for Seamless Copper Tube for Air Conditioning and Refrigeration Field Service.*

(d) Polyethylene tubing: ASTM D2513, *Specification for Thermoplastic Gas Pressure Pipe, Tubing and Fittings.* Tubing shall be recommended by the manufacturer for use with LP-Gas. Polyethylene tubing shall be marked in full compliance with the product marking requirements of ASTM D2513, and shall include the manufacturer's name or trademark, the Standard Dimensional Ratio of the pipe, the size of the pipe, the designation polyethylene (PE), the date manufactured, and the designation ASTM D2513.

See commentary following 2-4.2(f).

2-4.4 **Fittings for Pipe and Tubing.** Fittings shall be steel, brass, copper, malleable iron, ductile (nodular) iron, or plastic and shall comply with paragraphs (a) through (c). Cast- iron pipe fittings (ells, tees, crosses, couplings, unions, flanges, and plugs) shall not be used. Thermoplastic fittings fabricated from materials listed in ASTM D2513, *Specification for Thermoplastic Gas Pressure Pipe, Tubing and Fittings,* used to join polyethylene pipe, shall comply with ASTM D2513, and shall be recommended for LP-Gas use by the manufacturer.

This section provides the materials requirements for pipe and tubing fittings. The requirements for metallic fittings have not been revised for many years. In both the 1992 and 1995 editions, the Committee has expanded the types of fittings that are permitted for joining polyethylene pipe and tubing. Prior to the 1992 edition, only heat fusion methods were permitted for joining polyethylene. Fusion joints are made by inserting the ends of polyethylene tubing sections into a fitting and heating the fitting using a special electrically heated tool. This method is used widely for polyethylene tubing in natural gas service where the vast majority of this piping material is used. An installer using this equipment must be trained and certified in its use, and remain proficient in its use in order to make leak free joints. Polyethylene piping materials have found less application in propane service than natural gas service, due to the expense of purchasing the special tools required for joining polyethylene and the difficulty of maintaining that employees be qualified in its use.

The revisions include:

- Fittings to be made of nonmetallic materials, other than polyethylene, and that rely on ASTM D2513 for selection of materials
- Use of mechanical fittings to join polyethylene piping materials
- Use of "risers," which provide the transition from polyethylene underground to metallic pipe aboveground in one fitting. Risers must be factory assembled or designed for field assembly.

These changes recognize the good experience with materials other than polyethylene (and fittings other than heat fusion type) in natural gas service, where it is widely used for underground service between gas mains and service meters.

(a) Pipe joints in wrought iron, steel, brass, or copper pipe shall be permitted to be screwed, welded, or brazed.

1. Fittings used at pressures higher than container pressure, such as on the discharge of liquid transfer pumps, shall be suitable for a working pressure of at least 350 psi (2.4 MPa).
2. Fittings used with liquid LP-Gas, or with vapor LP-Gas at operating pressures over 125 psi (0.9 MPa), shall be suitable for a working pressure of 250 psi (1.7 MPa).

Exception: Fittings used at higher pressure as specified in 2-4.4(a)1.

3. Fittings for use with vapor LP-Gas at pressures not exceeding 125 psi (0.9 MPa) shall be suitable for a working pressure of 125 psi (0.9 MPa).
4. Brazing filler material shall have a melting point exceeding 1,000°F (538°C).

Brazing is an acceptable method of joining fittings to pipe and tubing (although it is used mainly for tubing) when the filler metal has a melting point in excess of 1,000°F (538°C). Nearly all brazes meet this requirement, as a braze is defined as a metal with a melting point above 850°F (454°C). Any filler metal with a lower melting point is defined as solder. This prohibits soldering as a method of joining fittings to pipe and tubing. The 1,000°F (538°C) minimum temperature is intended to ensure the integrity of a gas piping system when exposed to fire.

Although not specified in this standard, it should be recognized that NFPA 54, *National Fuel Gas Code,* states that brazing alloys must not contain more than 0.05 percent phosphorus because this can lead to a deterioration of the joint. The editor is advised that all brazing materials currently do not exceed this phosphorous level.

(b) Tubing joints in steel, brass, or copper tubing shall be flared, brazed, or made up with approved gas tubing fittings.

It should be recognized that this does not specify that flared fittings be approved. Flared fittings are fabricated when installed in the field and do not need to be specially approved.

1. Fittings used at pressures higher than container pressure, such as on the discharge of liquid transfer pumps, shall be suitable for a working pressure of at least 350 psi (2.4 MPa).
2. Fittings used with liquid LP-Gas, or with vapor LP-Gas at operating pressures over 125 psi (0.9 MPa), shall be suitable for a working pressure of 250 psi (1.7 MPa).

Exception: Fittings used at higher pressure as specified in 2-4.4(a)1.

3. Fittings for use with vapor LP-Gas at pressures not exceeding 125 psi (0.9 MPa) shall be suitable for a working pressure of 125 psi (0.9 MPa).
4. Brazing filler material shall have a melting point exceeding 1,000°F (538°C).

See commentary following 2-4.4(a)(4).

(c) Joints in polyethylene pipe and polyethylene tubing shall be made by heat fusion, by compression-type mechanical fittings, or by factory-assembled transition fittings. Heat fusion and factory-assembled transition fittings shall be permitted to be used to make joints in all sizes of polyethylene pipe being used. Mechanical compression-type fittings shall not be used on any polyethylene pipe above 2 in. IPS size. All fittings used to join polyethylene pipe or polyethylene tubing shall be tested and recommended by the manufacturer for use with polyethylene (PE) pipe and shall be installed according to the manufacturer's written procedure. For heat fusion, these instructions shall be specific to the type and grade of polyethylene being joined. Polyethylene pipe shall not be joined by a threaded or miter joint.

1. Polyethylene fusion fittings shall conform to ASTM D2683, *Specification for Socket-type Polyethylene (PE) Fittings for Outside Diameter Controlled Polyethylene Pipe;* or ASTM D3261, *Specification for Butt Heat Fusion Polyethylene (PE) Plastic Pipe and Tubing;* or ASTM F1055, *Specification for Electrofusion Type Polyethylene Fittings for Outside Diameter Controlled Polyethylene Pipe and Tubing,* and shall be recommended by the manufacturer for use with LP-Gas.
2. Mechanical fittings shall comply with Category 1 of ASTM D2513 for mechanical joints and shall be tested and shown to be acceptable for use with polyethylene pipe and polyethylene tubing.

This requirement was changed in the 1992 edition to expand the permitted methods of joining polyethylene tubing by:

(a) Adding mechanical fittings (on tubing up to 2 in.) and transition fittings to heat fusion.

(b) Adding fittings meeting ASTM F 1055 to permit electrofusion-type fittings in addition to socket and butt weld fittings.

(c) Requiring that fittings be recommended by the manufacturer for use with LP-Gas, rather than be listed or approved.

The heat fusion method requires special equipment and training to execute properly. This restricted the use of polyethylene in LP-Gas service, as many propane installers did not purchase the heat fusion equipment for the relatively few applications for polyethylene tubing in propane service. By authorizing the use of mechanical fittings for polyethylene tubing, use of polyethylene tubing in the services where it is permitted under other provisions of the standard (vapor service, outside, and underground) have been expanded. Note that mechanical fittings must comply with Category 1 of ASTM D2513, be tested and recommended for use with LP-Gas by the fitting manufacturer, and be installed in accordance with the fitting manufacturer's instructions.

Also added in the 1992 edition is the requirement that all fitting installers be trained in the installation of the fittings they install and that the training be documented. This is consistent with changes to 1-6.1 also made in the 1992 edition to require that, effective January 1, 1993, all employees carry written certification of their job qualifications. (*See commentary on 1-5.*)

Formal Interpretation 92-1
Reference: 2-4.4(c)(2)

Question: Was it the intention of the Technical Committee on Liquefied Petroleum Gases, when they adopted 2-4.4(c)(2) in the 1992 edition of NFPA 58, to restrict the choice of any, or all, of the materials that might be utilized in the several components that comprise the total assembly of mechanical joints to those specifically "listed" or mentioned in the ASTM Standard D2513-90.

Answer: No. It was not the intent of the Committee in 2-4.4(c)(2) to specify materials of construction. Materials of construction of mechanical fittings are covered in 2-4.4 where it was the Committee's intent to limit fittings to be constructed of materials listed in ASTM D2513, except for gasket materials, which are covered in 2-4.4(c)(2)(a).

In 1-2.4, Alternate Materials and Provisions, the Committee provides a method of use of alternate materials when supported by sufficient evidence acceptable to the authority having jurisdiction.

Issue Edition: 1992
Reference: 2-4.4(c)(2)
Issue Date: January 15, 1993
Effective Date: February 3, 1993 ■

a. Compression-type mechanical fittings shall include a rigid internal tubular stiffener, other than a split tubular stiffener to support the pipe. Gasket material in the fitting shall be resistant to the action of LP-Gas and shall be compatible with the polyethylene pipe (PE) material.

b. Fittings shall be installed according to the procedure provided by the manufacturer.

3.* Anodeless risers shall comply with the following:

A-2-4.4(c)3 The *Code of Federal Regulations*, Title 49, Part 192.281(e) states:

Mechanical joints. Each compression-type mechanical joint on plastic pipe must comply with the following:

(1) The gasket material in the coupling must be compatible with the plastic.

(2) A rigid internal tubing stiffener, other than a split tubular stiffener, must be used in conjunction with the coupling.

Part 192.283(b) states:

(b) Mechanical joints. Before any written procedure established under 192.273(b) is used for plastic making mechanical plastic pipe joints that are designed to withstand tensile forces, the procedure must be qualified by subjecting 5 specimen joints made according to the procedure to the following tensile test:

(1) Use an apparatus for the test as specified in ASTM D638.77a (except for conditioning).

(2) The specimen must be of such length that the distance between the grips of the apparatus and the end of the stiffener does not affect the joint strength.

(3) The speed of testing is 5.0 mm (0.2 in.) per min, plus or minus 25 percent.

(4) Pipe specimens less than 102 mm (4 in.) in diameter are qualified if the pipe yields to an elongation less than 25 percent or failure initiates outside the joint area.

(5) Pipe specimens 102 mm (4 in.) and larger in diameter shall be pulled until the pipe is subjected to a tensile stress equal to or greater than the maximum thermal stress that would be produced by a temperature change of 55°C (100°F) or until the pipe is pulled from the fitting. If the pipe pulls from the fitting, the lowest value of the five test results or the manufacturer's rating, whichever is lower, must be used in the design calculations for stress.

(6) Each specimen that fails at the grips must be retested using new pipe.

(7) Results obtained pertain only to the outside diameter, and material of the pipe tested, except that testing of a heavier wall pipe may be used to qualify pipe of the same material but with a lesser wall thickness.

This new provision was added in the 1995 edition to permit the use of anodeless risers, which are fittings that are used to connect polyethylene underground to metallic piping aboveground. Risers have been in use in natural gas service for many years.

The term "anodeless," used in conjunction with riser, indicates that the riser is designed to retard corrosion without the use of a sacrificial anode. Sacrificial anodes are used with underground propane tanks.

Two types of risers are permitted, factory-assembled and field-assembled (kit type and service head adapter type):

- A factory-assembled riser requires only connection to the polyethylene pipe or tubing below grade and metallic piping above grade.
- A field-assembled riser (kit) is a metallic casing incorporating the transition from the polyethylene gas pipe or tubing to the steel thread allowing this transition to be made in the field. It consists of a transition fitting and a prefabricated, prebent steel riser casing. The polyethylene pipe or tubing is inserted up through the casing until it exits the upper end. The transition fitting is installed on the top of the casing, completing the connection from polyethylene to steel. The transition fitting must be installed aboveground to minimize corrosion to the steel pipe threads. The polyethylene pipe or tubing is protected by the metal pipe and is sealed at the lower end of the casing to prevent water from entering the riser. This is an exception to the rule that polyethylene be used only belowground.
- A field-assembled riser can also be field fabricated using a service head adapter and steel pipe. The pipe is threaded at the top and polyethylene pipe or tubing is inserted in it so that it protrudes from the top. The service head adapter is then installed to the pipe thread and seals the polyethylene. As with the field-assembled riser, the service head adapter must be installed aboveground to minimize corrosion. This can be used with existing pipe to line it with polyethylene pipe, and is used for this purpose extensively in natural gas service. The riser casing (pipe or tubing) must meet the materials requirements of 2-4.2.

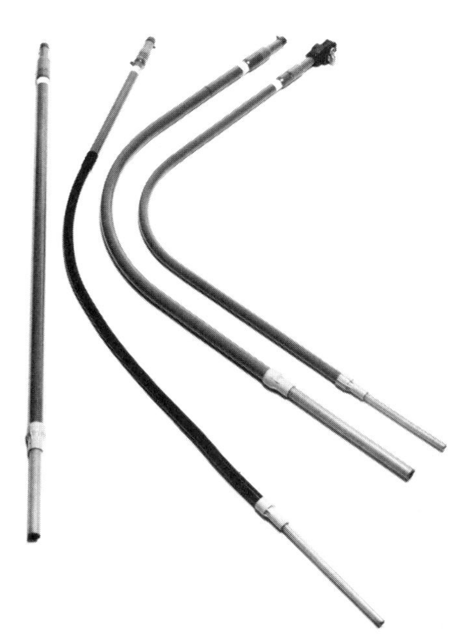

Figure 2.32 *Factory assembled risers. (Photo courtesy of R. W. Lyall & Company, Inc.)*

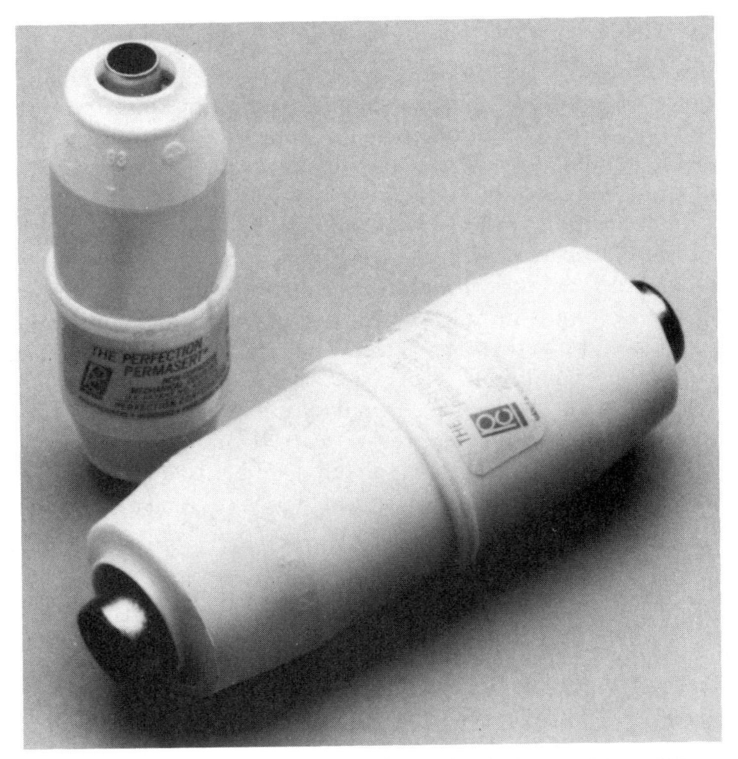

***Figure** 2.33 Mechanical fitting for polyethylene tubing. (Photo courtesy of Perfection Corporation.)*

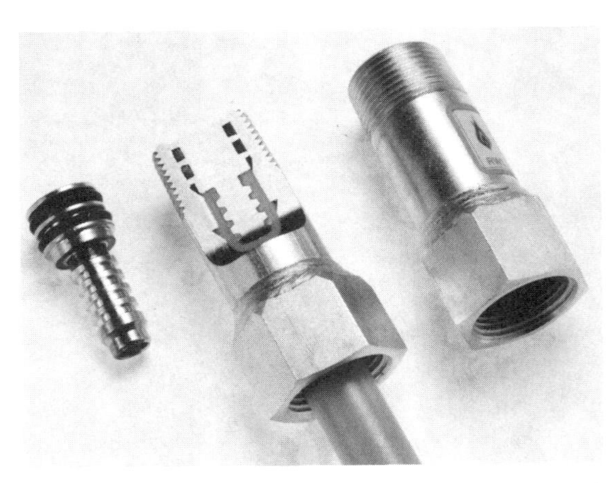

***Figure** 2.34 Cutaway of riser showing internal stiffener. (Photo courtesy of R. W. Lyall & Company, Inc.)*

a. Factory-assembled anodeless risers shall be recommended for LP-Gas use and shall be leak tested by the manufacturer in accordance with written procedures.

b. Field-assembled anodeless risers with service head adapters shall be recommended for LP-Gas use by the manufacturer and shall be design certified to meet the requirements of Category 1 of ASTM D2513, and U.S. Department of Transportation, *Code of Federal Regulations*, Title 49, Part 192.281(e), and the requirements of 3-2.8.7. The manufacturer shall provide the user qualified installation instructions as prescribed by U.S. Department of Transportation, *Code of Federal Regulations*, Title 49, Part 192.283(b).

4.* All persons installing polyethylene piping shall be trained in the applicable joining procedure. The training shall be documented.

A-2-4.4(c)4 Persons joining PE pipe should be trained under the applicable joining procedure established by the manufacturer including the following:

(a) Appropriate training in the use of joining procedures.

(b) Making a specimen joint from pipe sections joined according to the procedures.

(c) Visually examining these joints during and after assembly.

2-4.5 Valves, Other than Container Valves.

2-4.5.1 Pressure-containing metal parts of valves (except appliance valves), including manual positive shutoff valves, excess-flow check valves, backflow check valves, emergency shutoff valves (*see 2-4.5.4*), and remotely controlled valves (either manually or automatically operated), used in piping systems shall be of steel, ductile (nodular) iron, malleable iron, or brass. Ductile iron shall meet the requirements of ASTM A395, *Specification for Ferritic Ductile Iron Pressure-Related Castings for Use at Elevated Temperatures*, or equivalent and malleable iron shall meet the requirements of ASTM A47, *Specification for Ferritic Malleable Iron Castings*, or equivalent. All materials used, including valve seat discs, packing, seals, and diaphragms, shall be resistant to the action of LP-Gas under service conditions.

2-4.5.2 Valves shall be suitable for the appropriate working pressure, as follows:

(a) Valves used at pressures higher than container pressure, such as on the discharge of liquid transfer pumps, shall be suitable for a working pressure of at least 350 psi (2.4 MPa) [400 psi (2.8 MPa) WOG valves comply with this provision].

(b) Valves to be used with liquid LP-Gas, or with vapor LP-Gas at pressures in excess of 125 psi (0.9 MPa), but not to exceed 250 psi (1.7 MPa), shall be suitable for a working pressure of at least 250 psi (1.7 MPa).

Exception: Valves used at higher pressure as specified in 2-4.5.2(a).

(c) Valves (except appliance valves) to be used with vapor LP-Gas at pressures not to exceed 125 psi (0.9 MPa) shall be suitable for a working pressure of at least 125 psi (0.9 MPa).

2-4.5.3 Manual shutoff valves, emergency shutoff valves (*see 2-4.5.4*), excess-flow check valves, and backflow check valves used in piping systems shall comply with the provisions for container valves. [*See 2-3.3.3(a), (b), and (c).*]

2-4.5.4 Emergency shutoff valves shall be approved and incorporate all of the following means of closing (*see 3-2.8.10, 3-3.3.8, and 3-3.3.9*):

(a) Automatic shutoff through thermal (fire) actuation. Where fusible elements are used they shall have a melting point not exceeding 250°F (121°C).

(b) Manual shutoff from a remote location.

(c) Manual shutoff at the installed location.

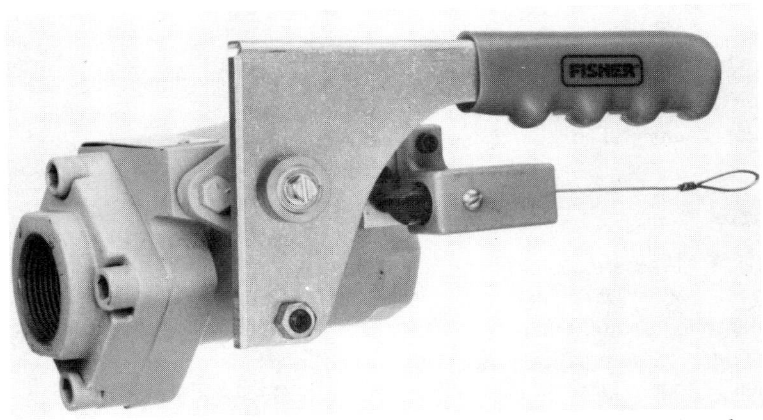

Figure 2.35 *Emergency shutoff valve. (Photo courtesy of Fisher Controls Company.)*

This provision sets forth the requirements for the emergency shutoff valve, a key valve in the protection of liquid transfer operations where 4,000 gal (1.5 m^3) or more of LP-Gas can be stored in one or more containers connected to a common liquid transfer line. The installation of these valves is established in 3-2.8.9. The actuating means for remote shutoff is not specified, but electrical, mechanical, and pneumatic systems are used. Many systems use a pneumatic plastic tubing system, in which the tubing itself acts as a fusible element releasing the pressure holding the valve open. With respect to the feature of manual shutoff at the installed location, it is recommended that this valve be operated occasionally. Also, the system should be tested periodically to determine that it will function properly. See commentary following 3-2.8.10.

Figure 2.36 Emergency shutoff valve/tank car unloading adapter combination. (Photo courtesy of Engineered Controls, Inc.)

2-4.6 Hose, Quick Connectors, Hose Connections, and Flexible Connectors.

2-4.6.1 Hose, hose connections, and flexible connectors (*see definition*) shall be fabricated of materials resistant to the action of LP-Gas both as liquid and vapor. If wire braid is used for reinforcement, it shall be of corrosion-resistant material such as stainless steel.

Note that carbon steel may not be used in hose or hose connections. This prohibition is based on experience with hoses that weakened as the carbon steel braid corroded.

2-4.6.2 Hose and quick connectors shall be approved.

2-4.6.3 Hose, hose connections, and flexible connectors used for conveying LP-Gas liquid or vapor at pressures in excess of 5 psi (34 kPa), and as provided in Section 3-4 regardless of the pressure, shall comply with the following:

Exception: Hoses at a pressure of 5 psi (34 kPa) or less used in agricultural buildings not normally occupied by the public.

(a) Hose shall be designed for a working pressure of 350 psi (240 MPa) with a safety factor of 5 to 1 and be continuously marked "LP-GAS," "PROPANE," "350 PSI WORKING PRESSURE," and the manufacturer's name or trademark.

It is important to note that these provisions for hose, hose connections, and flexible connectors apply only to those involved with pressures in excess of 5 psi (34.5 kPa). Requiring continuous marking, which is current industry practice, permits identification of short connectors. The term "propane" is added to carry out the intent for industry identification internationally. The current provision is consistent with standards of the Rubber Manufacturers Association.

Exception: Hoses at a pressure of 5 psi (34 kPa) or less used in agricultural buildings not normally occupied by the public.

(b) Hose assemblies, after the application of connections, shall have a design capability of withstanding a pressure of not less than 700 psi (4.8 MPa). If a test is performed, such assemblies shall not be leak tested at pressures higher than the working pressure [350 psi (2.4 MPa) minimum] of the hose.

The words "design capability" have been carefully chosen to convey the Committee's intent that hose assemblies be capable of withstanding the anticipated pressure, rather than be tested. While it might be desirable to verify that connection fittings have been installed properly, past experience with hydrostatic testing of hose at this pressure has resulted in injury to the hose. An assembly procedure can be verified to determine that the design capability has been met, and production samples can be tested to ensure that the hoses meet the requirement.

Exception: Hoses at a pressure of 5 psi (34 kPa) or less used in agricultural buildings not normally occupied by the public.

2-4.6.4 Hoses or flexible connectors used to supply LP-Gas to utilization equipment or appliances shall be installed in accordance with the provisions of 3-2.8.9 and 3-2.8.11.

2-4.7 Hydrostatic Relief Valves. Hydrostatic relief valves designed to relieve the hydrostatic pressure that might develop in sections of liquid piping between closed shutoff valves shall have pressure settings not less than 400 psi (2.8 MPa) or more than 500 psi (3.5 MPa) unless installed in systems designed to operate above 350 psi (2.4 MPa). Hydrostatic relief valves for use in systems designed to operate above 350 psi (2.4 MPa) shall have settings not less than 110 percent or more than 125 percent of the system design pressure.

The number of types of hydrostatic relief valves has been reduced and the possibility of using the wrong valve eliminated through a rational consolidation of pressure settings. Small-diameter pipe and tubing usually used in smaller systems can easily handle 500 psi (3.5 MPa) without danger. In large industrial or commercial systems where higher pressures might be used, the specified 110 to 125 percent of design pressure is reasonable. Liquid systems connected to DOT containers are rare and require no special consideration.

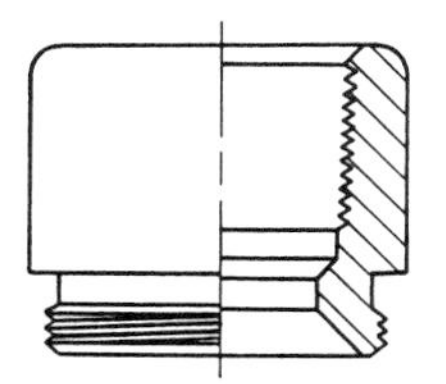

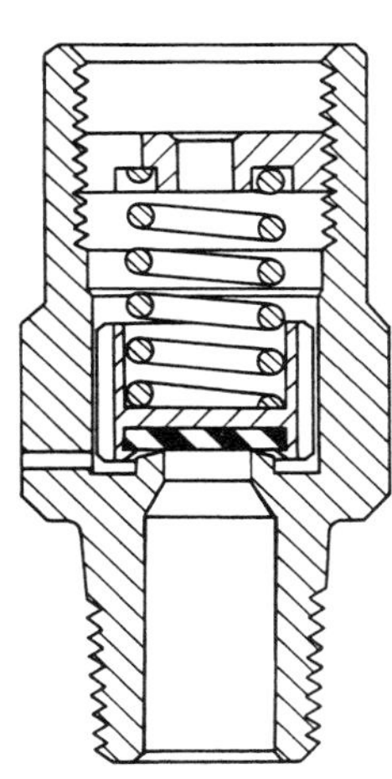

Figure 2.37 Hydrostatic relief valve. (Photo and drawings courtesy of Engineered Controls, Inc.)

2-5 Equipment.

2-5.1 General.

2-5.1.1 This section includes fabrication and performance provisions for the pressure containing metal parts of LP-Gas equipment such as pumps, compressors, vaporizers, strainers, meters, sight flow glasses, and regulators. Containers shall not be subject to the provisions of this section.

2-5.1.2 Equipment shall be suitable for the appropriate working pressure as follows:

(a) Equipment to be used at pressures higher than container pressure, such as on the discharge of a liquid pump, shall be suitable for a working pressure of at least 350 psi (2.4 MPa). If pressures above 350 psi (2.4 MPa) are necessary, the pump and all equipment under pressure from the pump shall be suitable for the pump discharge pressure.

(b) Equipment to be used with liquid LP-Gas, or vapor LP-Gas at pressures over 125 psi (0.9 MPa) but not to exceed 250 psi (1.7 MPa), shall be suitable for a working pressure of at least 250 psi (1.7 MPa).

(c) Equipment to be used with vapor LP-Gas at pressures over 20 psi (138 kPa), but not to exceed 125 psi (0.9 MPa), shall be suitable for a working pressure of at least 125 psi (0.9 MPa).

(d) Equipment to be used with vapor LP-Gas at pressures not to exceed 20 psi (138 kPa) shall be suitable for a working pressure adequate for the service in which it is to be used.

2-5.1.3 Equipment shall be fabricated of materials suitable for LP-Gas service and resistant to the action of LP-Gas under service conditions. The following shall also apply:

(a) Pressure-containing metal parts shall be of steel, ductile (nodular) iron (ASTM A395 or A536 Grade 60-40-18 or 65-45-12), malleable iron (ASTM A47), higher strength gray iron (ASTM A48, Class 40B), brass, or the equivalent.

(b) Cast iron shall not be used for strainers or flow indicators that shall comply with provisions for materials for construction of valves (*see 2-4.5.1*).

(c) Aluminum shall be permitted to be used for approved meters.

(d) Aluminum or zinc shall be permitted to be used for approved regulators. Zinc used for regulators shall comply with ASTM B86, *Specification for Zinc-Alloy Die Casting.*

(e) Nonmetallic materials shall not be used for upper and lower casings of regulators.

It is imperative that the materials of construction of equipment components be suitable for LP-Gas. Equipment should not be purchased until this has been ascertained from the manufacturer.

Restrictions on the use of cast iron reflect its tendency to crack under low temperature conditions and when heated and suddenly cooled under fire control conditions. Behavior when subjected to fire is also reflected in the melting points of metals. The use of metals that melt at less than 1,000°F (538.8°C) under the service pressure stresses—e.g., aluminum and zinc—are restricted to situations in which their failure will not constitute an undue hazard. Subparagraph 2-5.1.3(e) was added in the 1986 edition to reflect growing interest in the use of plastic materials and provides restrictions on their use.

2-5.1.4 Engines used to drive portable pumps and compressors shall be equipped with exhaust system spark arresters and shielded ignition systems.

This requirement, which applies to all LP-Gas equipment, was added in the 1992 edition (and replaces a similar requirement that was located in 2-5.3

covering compressors) following requests for interpretation of paragraph 3-6.5.7 in the 1989 edition, which stated: "Engines used to drive portable compressors shall be equipped with exhaust system spark arrestors and shielded ignition systems." (This paragraph has been relocated to 8-5.7 in 1992 and revised.)

Engines are used to drive portable pumps as well as portable compressors, and there was no guidance in the standard on requirements for these engines. The fact that these devices are portable indicates that they are going to be operated by an internal combustion engine of some type, in all probability. This paragraph requires the same level of protection for internal combustion engines driving portable pumps as that for compressors, and is located so as to apply to all LP-Gas equipment. The absence of coverage of portable pumps had been improperly interpreted by some to imply that engines driving portable pumps were unsafe, and Part D of Table 3-7.2.2, which requires an electrical area classification of Class 1, Division 2, for outdoor pumps, was usually cited. It must be noted that:

(a) Table 3-7.2.2 applies only to fixed electrical equipment and wiring, and

(b) Internal combustion engines in good working order equipped with the accessories required are not a source of ignition. There have been LP-Gas incidents where cars have driven into, stalled, and coasted out of vapor clouds of LP-Gas that have been accidentally released without causing ignition.[1]

2-5.2 Pumps.

2-5.2.1 Pumps shall be designed for LP-Gas service.

2-5.2.2 The maximum discharge pressure of a liquid pump under normal operating conditions shall be limited to 350 psi (2.4 MPa).

It is important that pumps be designed and built for use with LP-Gas. Nonspecialized pumps may not work satisfactorily and some may be hazardous. A check should be made with the manufacturer to see whether the pump is capable of a discharge pressure higher than 350 psi (2.4 MPa). If so, provisions must be made to limit the pressure to 350 psi (2.4 MPa), or the system must be designed for the higher pressure.

Because they are pumping a liquefied gas, which will vaporize if the pressure drops even slightly, LP-Gas pumps are either positive displacement pumps or special types of centrifugal pumps (regenerative turbine pumps). Ordinary centrifugal pumps will "vapor lock" rather easily and be rendered ineffective.

[1]Isner, MS. "Propane Tank Truck Incident, Eight People Killed." Memphis, Tennessee, December 23, 1988. *Fire Investigation Report.* Fire Investigations Division, National Fire Protection Association, Quincy, MA. Feb. 6, 1990.

Figure 2.38 *Sliding vane positive displacement pump. (Photo courtesy of Blackmer.)*

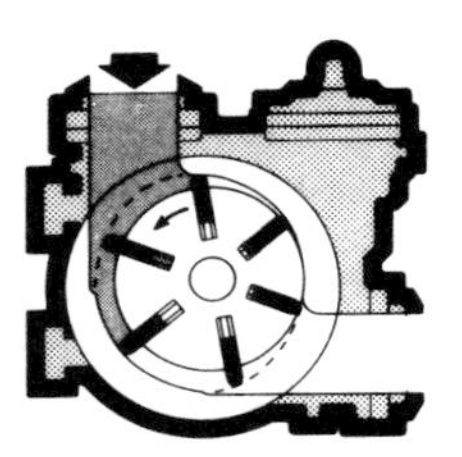

a. Vanes move out, trapping liquid at the pump inlet.

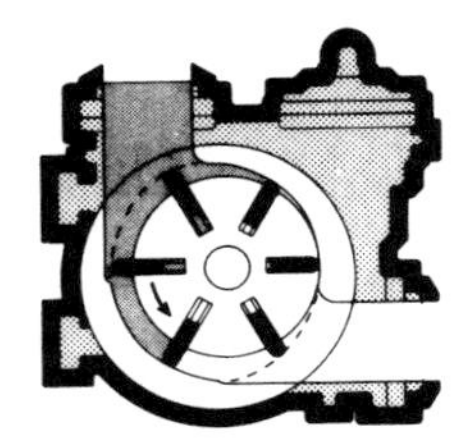

b. Liquid is transferred toward the outlet between the vanes.

c. As the vanes move back into their slots, liquid is discharged through the outlet.

Figure 2.39 *Operation of sliding vane pump. (Drawing courtesy of the National Propane Gas Association.)*

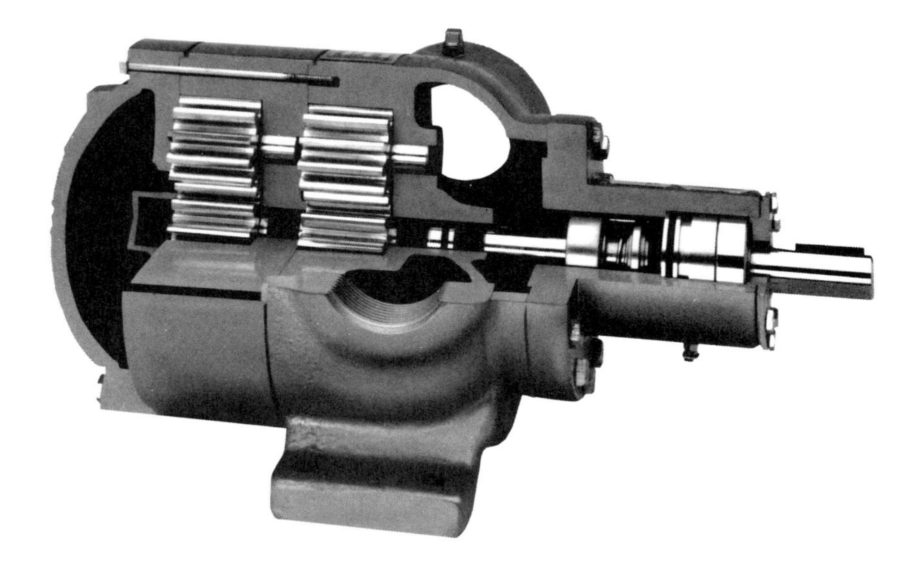

Figure 2.40 *Gear pump. (Photo courtesy of Smith Precision Products Company, Inc.)*

2-5.3 Compressors.

2-5.3.1 Compressors shall be designed for LP-Gas service.

Care should be taken to select compressors properly designed for LP-Gas service. Compressors designed for noncombustible gases may not provide adequate leak protection required for LP-Gases.

2-5.3.2 Means shall be provided to limit the suction pressure to the maximum for which the compressor is designed.

2-5.3.3 Means shall be provided to prevent the entrance of LP-Gas liquid into the compressor suction, either integral with the compressor or installed externally in the suction piping [*see 3-2.13.2(b)*].

Exception: Portable compressors used with temporary connections.

Because all liquids, including liquid LP-Gas, are not compressible, they must not be allowed to enter an operating compressor (if they do, the compressor will probably be damaged and may fail). A number of compressors

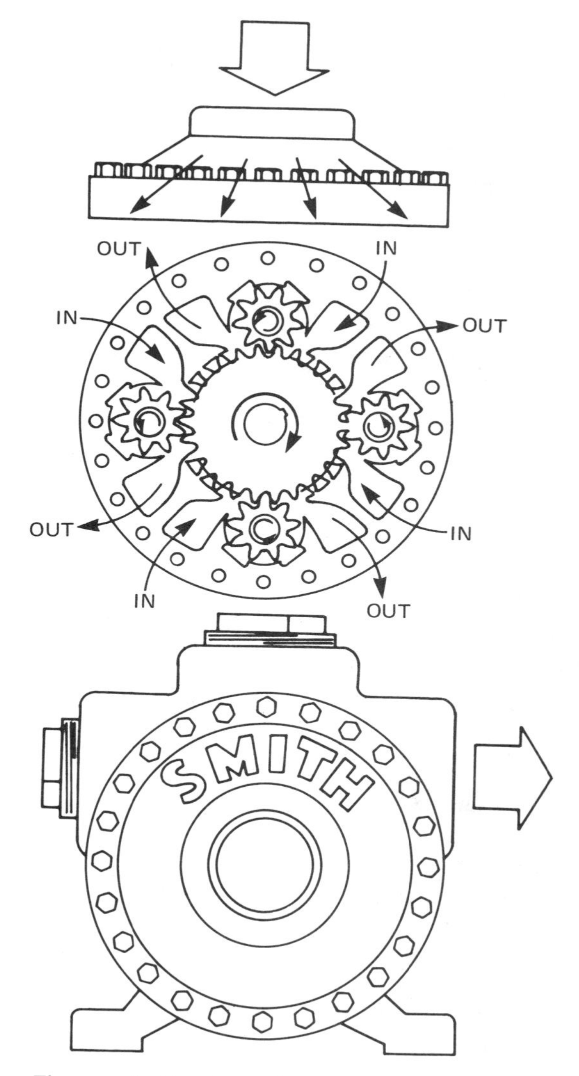

Figure 2.41 Operation of gear pump. (Drawing courtesy of Smith Precision Products Company, Inc.)

manufactured for this service include relief devices built into the cylinder head to prevent destruction of the compressor in case small amounts of liquid are allowed to enter. Continuously taking advantage of this feature, however, will eventually destroy the compressor. Several float-operated devices are designed to prevent liquid from entering the compressor. Some manufacturers use this type of device while others simply use a large receiver as a liquid trap on the suction side of the compressor. Portable compressors are exempted from this type of protection because the temporary connections used are empty at the time they are connected, which usually prevents liquid from entering the compressor. Also, portable units are operated under continuous surveillance by an operator.

Compressors are used to move liquid by pumping vapor from the receiving storage into the supply transport car or truck, thereby increasing the pressure in the transport device and forcing liquid from the transport to the receiver tank. This type of loading or unloading operation is necessary when it is impractical or impossible to use liquid pumps, such as when unloading railroad tank cars. Tank cars have all fittings at the top of the car in the dome. The use of pumps is impractical because of the extended length of piping to the pump suction, which results in pump cavitation and reduced capacity. When compressors are used to unload tank cars, it is common practice to reverse the compressor, after all liquid is removed from the car, to recover the vapor. Normally, this continues until the pressure in the car is about 30 psi (207 kPa).

Figure 2.42 *Portable gasoline engine-driven compressor. (Photo courtesy of Smith Precision Products.)*

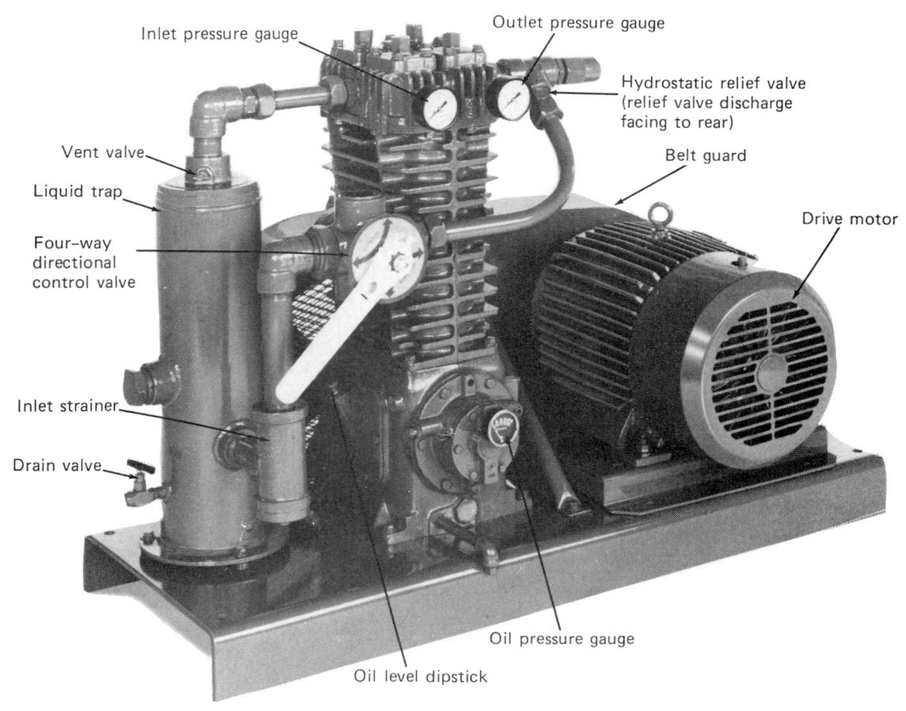

Figure 2.43 *Compressor with protection devices. (Photo courtesy of **Blackmer** Pump Company.)*

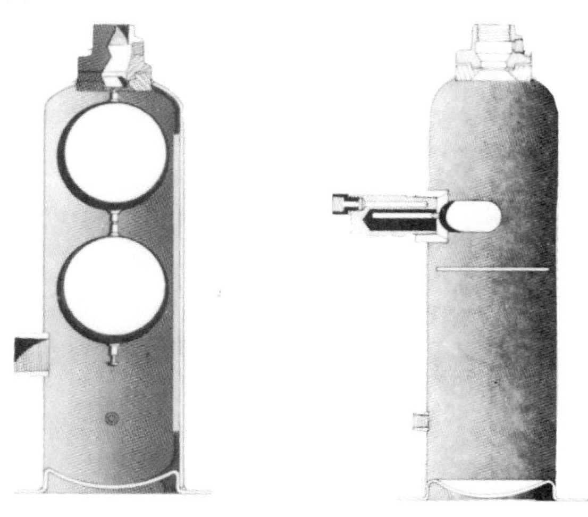

Figure 2.44 *Compressor liquid traps. In part (a), the floats rise with the liquid level and close the valve at the top. In part (b), the float actuates a switch that shuts off the compressor motor. (Drawings courtesy of Corken Pump Company.)*

2-5.4 Vaporizers, Tank Heaters, Vaporizing-Burners, and Gas-Air Mixers.

2-5.4.1 Vaporizers shall be permitted to be of the indirect type (utilizing steam, hot water, or other heating medium), or direct fired. This subsection does not apply to engine fuel vaporizers or to integral vaporizer-burners such as those used with weed burners or tar kettles.

See definitions in Section 1-6 for descriptions of various types of vaporizers.

2-5.4.2 Indirect vaporizers shall comply with the following:

(a) Indirect vaporizers shall be constructed in accordance with the applicable provision of the ASME Code for a design pressure of 250 psi (1.7 MPa) and shall be permanently and legibly marked with:

1. The marking required by the Code.
2. The allowable working pressure and temperature for which designed.
3. The name or symbol of the manufacturer.

Exception: Indirect vaporizers having an inside diameter of 6 in. (152 mm) or less are exempt from the ASME Code and shall not be required to be marked. They shall be constructed for a minimum 250-psi (1.7-MPa) design pressure.

This requires that all indirect vaporizers, which are heat exchangers that transfer heat from a process fluid heated outside the vaporizer to liquid LP-Gas, be built in accordance with the ASME Code to a design pressure of 250 psi (1.7 MPa), which is the same pressure required for most propane containers. The ASME Code has a major exception; vessels with an internal diameter of 6 in. or less. This is reflected in the exception to 2-5.4.2. These small vaporizers must also be designed for an operating pressure of 250 psi (1.7 MPa), but need not be inspected by an ASME-approved inspector or marked with the code marking. It must be noted that some jurisdictions do not allow this ASME exemption, e.g., the State of California requires any vaporizer with an internal volume that would hold more than 1 gal (4 L) of liquid be built, inspected, and marked in accordance with the ASME Code.

Coverage of vaporizers with a diameter of 6 in. or less was relocated to an exception in the 1992 edition and modified to reflect that the coverage duplicated ASME Code requirements already adopted.

In the 1995 edition, the requirement for marking vaporizers with the sum of the outside surface area and the inside heat exchange surface area was deleted because it had no bearing on safety. A heat exchanger is selected to provide a minimum heat exchange. Once this is done, there is no need to provide design information on the exchanger which cannot be field verified.

(b) Indirect vaporizers shall be provided with a suitable automatic means to prevent the passage of liquid through the vaporizer to the vapor discharge piping. This means shall be permitted to be integral with the vaporizer or otherwise provided in the external piping.

The provision for preventing liquid from leaving the vaporizer through vapor discharge piping is an important one. If the vaporizer is sized properly and is functioning properly, liquid will not be present in the vapor outlet. This provision is intended to protect against malfunction where the heat source of the vaporizer failed or the unit was overloaded intentionally or accidentally. The hazard of liquid in the vapor piping is that none of the vapor control gear, such as regulators, control valves, and burners, is designed to handle liquid. Liquid passing through vapor regulators will flash and can result in significantly increased flame size, which can present a fire hazard.

(c) Indirect vaporizers, including atmospheric-type vaporizers using heat from the surrounding air or the ground, and of more than 1 quart (0.9 L) capacity, shall be equipped, at or near the discharge, with a spring-loaded pressure relief valve providing a relieving capacity in accordance with 2-5.4.5. Fusible plug devices shall not be used.

The prohibition of fusible plugs (instead of spring-loaded pressure relief valves) is included because fusible plugs, as well as rupture discs, will cause the entire contents of the protected device to be discharged into the atmosphere in case of overtemperature. Spring-loaded relief valves will minimize the quantity discharged because they close when the excessive pressure has been relieved.

(d) Indirect atmospheric-type vaporizers of less than 1 quart (0.9 L) capacity shall not be required to be equipped with pressure relief valves, but shall be installed in accordance with 3-6.2.7.

2-5.4.3 Direct-fired vaporizers shall comply with the following:

(a) Design and construction shall be in accordance with the applicable requirements of the ASME Code for the working conditions to which the vaporizer will be subjected, and the vaporizer shall be permanently and legibly marked with:

1. The markings required by the Code.
2. The maximum vaporizing capacity in gal per hr.
3. The rated heat input in Btu/h.
4. The name or symbol of the manufacturer.

(b) Direct-fired vaporizers shall be equipped, at or near the discharge, with a spring-loaded pressure relief valve providing a relieving capacity in accordance with 2-5.4.5. The relief valve shall be located so as not to be subject to temperatures in excess of 140°F (60°C). Fusible plug devices shall not be used.

(c) Direct-fired vaporizers shall be provided with suitable automatic means to prevent the passage of liquid from the vaporizer to its vapor discharge piping.

(d) A means for manually turning off the gas to the main burner and pilot shall be provided.

(e) Direct-fired vaporizers shall be equipped with an automatic safety device to shut off the flow of gas to the main burner if the pilot light is extinguished. If the pilot flow exceeds 2,000 Btuh (2 MJ/h), the safety device shall also shut off the flow of gas to the pilot.

(f) Direct-fired vaporizers shall be equipped with a limit control to prevent the heater from raising the product pressure above the design pressure of the vaporizer equipment, and to prevent raising the pressure within the storage container above the pressure specified in the first column of Table 2-2.2.2 that corresponds with the design pressure of the container (or its ASME Code equivalent—*see Notes to Table 2-2.2.2*).

Note that all direct-fired vaporizers must be constructed in accordance with the ASME Code, which includes inspection and marking. The prohibition against fusible plugs and rupture discs as a means to prevent liquid passing into the vapor discharge piping is the same as the provisions for indirect vaporizers. [*See commentary on 2-5.4.2(c).*]

Additional provisions for direct-fired vaporizers include a requirement for the relief valve to be located or protected in such a manner as to prevent excessive temperature, because a direct-fired unit contains a burner that can create temperatures—near the burner itself or in the path of the products of combustion—that would destroy the seats and mechanism of a relief valve that might be subjected to those excessive temperatures. "Marking required by the Code" refers to the ASME Code. Paragraphs 2-5.4.3 (a)(2), (a)(3), (a)(4), and (a)(5) are additional markings not stipulated in the ASME Code. The "outside surface" marking required in (a)(2) is not the surface of the outside of the cabinet or enclosure, but is that surface in contact with LP-Gas on the inside and the atmosphere on the outside, which could add to the vaporization rate of the vaporizer if subjected to an external fire. Referring to Figure 2.43, the outside surface includes all of the heat exchange surface plus the top of the vessel. However, the marking for (a)(2) should be for the top only and (a)(3) should include the bottom and sides.

Subparagraph 2-5.4.3 (e) reiterates the requirement for 100 percent shutoff to a gas burner and is not unique to a direct-fired vaporizer burner.

While 2-5.4.3(f) implies that the limit control should be a pressure-operated device, temperature-sensing devices are commonly used for this purpose. Controlling the temperature of the output of the vaporizer is the only practical method of preventing the heater from raising the product pressure (the heater being the burner of the vaporizer). The vaporizer cannot produce a pressure higher than provided to the liquid inlet of the vaporizer since, if the pressure in the vaporizer were to exceed the inlet pressure, flow of liquid into the vaporizer would stop and, in fact, flow backward.

The reference in 2-5.4.3(f) to raising the pressure within the storage container is descriptive of a type of system seldom used in present-day systems. This was a system in which liquid ran by gravity from a storage tank into a vaporizer and vapor was returned from the vaporizer back into the top of the tank. Vapor was then removed from the tank through a pressure regulator into the distribution system. With this type of system it is theoretically possible, in extremely hot weather, for a vaporizer connected in this manner to raise the pressure in the container. This, of course, would cause the container pressure relief valves to operate, discharging vapor that, because of the existence of the direct-fired vaporizer, would almost certainly become ignited. Again, temperature controls in the vaporizer would prevent this type of malfunction from taking place. Accordingly, a temperature control is the only practical type of control to satisfy this requirement.

Figure 2.45 Direct-fired vaporizer. (Photo courtesy of Pete Freeman.)

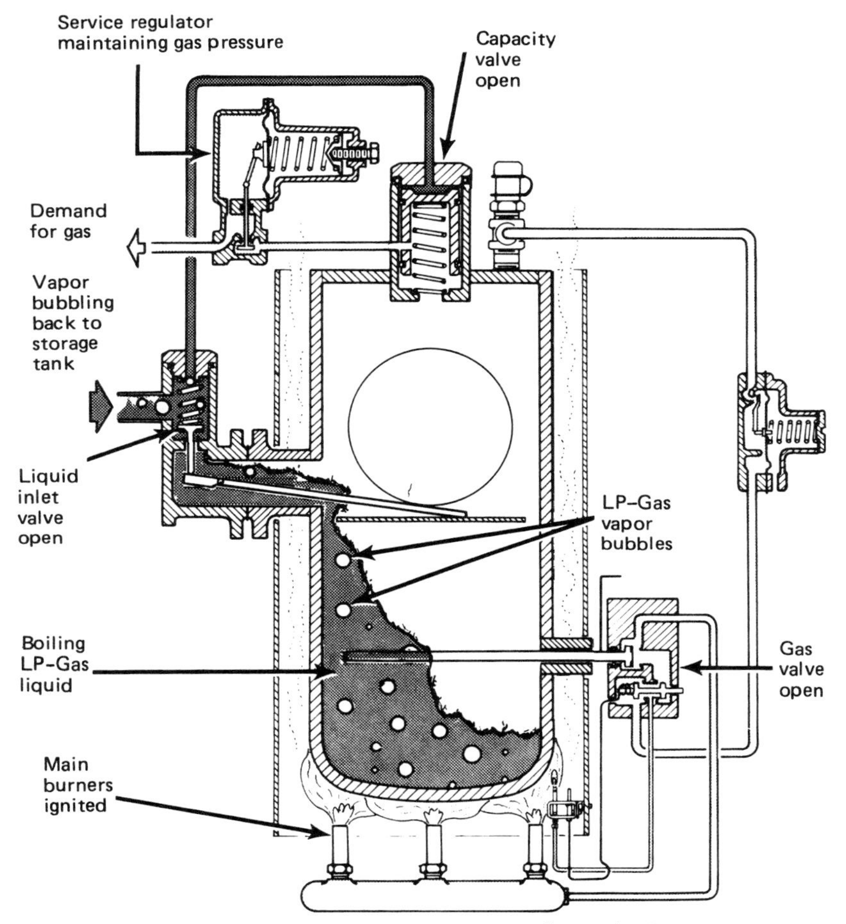

Figure 2.46 Direct-fired vaporizer operating on demand. (Drawing courtesy of National Propane Gas Association.)

2-5.4.4 Waterbath vaporizers shall comply with the following:

(a) The vaporizing chamber, tubing, pipe coils, or other heat exchange surface containing the LP-Gas to be vaporized, hereinafter referred to as "heat exchanger," shall be constructed in accordance with the applicable provisions of the ASME Code for a minimum design pressure of 250 psi (1.7 MPa) and shall be permanently and legibly marked with:

1. The marking required by the Code.
2. The allowable working pressure and temperature for which designed.
3. The name or symbol of the manufacturer.

Exception: Heat exchangers for waterbath vaporizers having an inside diameter of 6 in. (150 mm) or less are exempt from the ASME Code and shall not be required to be marked.

(b) Heat exchangers for waterbath vaporizers shall be provided with a suitable automatic control to prevent the passage of liquid through the heat exchanger to the vapor discharge piping. This control shall be integral with the vaporizer.

(c) Heat exchangers for waterbath vaporizers shall be equipped at or near the discharge with a spring-loaded pressure relief valve providing a relieving capacity in accordance with 2-5.4.5. Fusible plug devices shall not be used.

(d) Waterbath sections of waterbath vaporizers shall be designed to eliminate a pressure buildup above the design pressure.

(e) The immersion heater that provides heat to the waterbath shall be installed so as not to contact the heat exchanger and shall be permitted to be electric or gas-fired.

(f) A control to limit the temperature of the waterbath shall be provided.

(g) Gas-fired immersion heaters shall be equipped with an automatic safety device to shut off the flow of gas to the main burner and pilot in the event of flame failure.

(h) Gas-fired immersion heaters with an input of 400,000 Btuh (422 MJ/h) or more shall be equipped with an electronic flame safeguard and with programming to provide for prepurge prior to ignition, proof of pilot before the main burner valve opens, and full shutdown of the main gas and pilot upon flame failure.

(i) A means shall be provided to shut off the source of heat in case the level of the heat transfer medium falls below the top of the heat exchanger.

The waterbath vaporizer differs from a direct-fired vaporizer in that there is no flame impingement on the vaporizer chamber itself. It is to some extent a cross between a direct-fired vaporizer and an indirect-fired vaporizer. The portion of such a unit that contains and vaporizes the LP-Gas must be built in accordance with the provisions of the ASME Code for a minimum of 250 psi (1.7 MPa) and be marked as stipulated. Again, those less than 6 in. (152 mm) in diameter are exempt from the ASME Code, but must be constructed to a minimum of 250 psi (1.7 MPa). [*See also commentary under 2-5.4.2(a), Exception.*] The same automatic safety controls used with indirect- and direct-fired vaporizers to prevent liquid passing into the vapor discharge piping, etc., must be provided. The immersion heater that provides heat to the waterbath must not contact the LP-Gas heat exchanger surface to avoid overheating, and it must be properly equipped with an automatic shutoff device to prevent overheating of the liquid bath temperature should the level of the bath liquid fall below the top of the heat exchanger.

Subparagraph 2-5.4.4(e) states that the immersion heater that heats the waterbath is permitted to be electric or gas-fired. The intent is not to prevent any other form of heat source, but at this writing the only two that seem practical are electric or gas-fired immersion heaters. There is no provision preventing the use of steam coils or some form of waste heat coils to heat the waterbath. Such forms of heat would be better applied to indirect vaporizers, as covered by 2-5.4.2, rather than adding the unnecessary additional dimension of a second interface between heat source and vaporizing coil. Although this type of operation would be perfectly acceptable, the economics would probably preclude its use. There is no apparent safety advantage over the use of indirect units when the source of heat would be the same as that required by the indirect unit.

2-5.4.5 The minimum rate of discharge in cu ft of air per minute for pressure relief valves for LP-Gas vaporizers, either of the indirect type or direct-fired, shall be determined as follows:

(a) Based upon conservative heat transfer calculations (i.e., assuming that the vaporizing chamber is liquid full), the maximum vapor generating capacity (rate) shall be determined when maximum heat is available. That vapor rate shall be converted to an equivalent air rate.

(b) If the vaporizer is direct fired or if a substantial exterior surface is in contact with the LP-Gas, the sum of the vaporizer surface and the LP-Gas wetted exterior surface shall be permitted to be used in conjunction with Table E-2.2.2 to determine the required relief valve capacity.

These requirements were revised in the 1995 edition to include sizing of pressure relief valves for all vaporizers.

Pressure relief valves (and all other safety devices) are designed for the worst-case condition, which is fire exposure over the entire surface of the vessel. In the event of fire exposure to a vaporizer, the pressure relief valve must be sized to handle vapor produced as a result of heating of the liquid by both the normal heat source and the exposure fire. Note that the reference to Appendix E refers to text extracted from the ASME Code, which is a mandatory reference. The text is located in Appendix E for the convenience of the reader of the standard.

2-5.4.6 Direct gas-fired tank heaters shall be designed exclusively for outdoor aboveground use and so that there is no direct flame impingement upon the container. The following shall also apply:

(a) Tank heaters shall be approved and shall be permanently and legibly marked with:

1. The rated input to the burner in Btuh.
2. The maximum vaporizing capacity in gal per hour.
3. The name or symbol of the manufacturer.

(b) The heater shall be designed so that it can be readily removed for inspection of the entire container.

(c) The fuel gas supply connection to the tank heater shall originate in the vapor space of the container being heated and shall be provided with a manually operated shutoff valve at the heater.

(d) The heater control system shall be equipped with an automatic safety shutoff valve of the manual-reset type arranged to shut off the flow of gas to both the main and pilot burners if the pilot flame is extinguished.

(e) Where installed on a container exceeding 1,000 gal (3.8 m^3) water capacity, the heater control system shall include a valve to automatically shut off the flow of gas to both the main and pilot burners if the container becomes empty of liquid.

(f) Direct gas-fired tank heaters shall be equipped with a limit control to prevent the heater from raising the pressure in the storage container to more than 75 percent of the pressure shown in the first column of Table 2-2.2.2 that corresponds with the design pressure of the container (or its ASME *Boiler and Pressure Vessel Code* equivalent).

Direct-fired tank heaters are no longer widely used. They are heaters that attach to the bottom of the tank to heat the tank shell at the point of application, which then heats the liquid product inside. In order to avoid loss of container strength, however, there can be no flame impingement upon the container itself. Most of these units are constructed so that they can be easily removed for inspection for possible corrosion of the tank at the point of installation. These units have to be carefully engineered so that there will be automatic means to prevent overheating the product in the tank, to shut off the heater in case the tank goes empty of liquid, to shut off the gas flow to the heater in case the pilot light goes out, etc.

2-5.4.7 Vaporizing-burners shall be constructed with a minimum design pressure of 250 psi (1.7 MPa) with a safety factor of 5 and shall comply with the following:

(a) The vaporizing-burner, or the appliance in which it is installed, shall be permanently and legibly marked with:

1. The maximum burner input in Btuh.
2. The name or symbol of the manufacturer.

(b) Vaporizing coils or jackets shall be made of ferrous metals or high-temperature alloys.

(c) The vaporizing section shall be protected by a relief valve, located where it will not be subject to temperatures in excess of 140°F (60°C), and with a pressure setting sufficient to protect the components involved but not lower than 250 psi (1.7 MPa). The relief valve discharge shall be directed upward and away from the component parts of the vaporizing burner. Fusible plug devices shall not be used.

(d) A means shall be provided for manually turning off the gas to the main burner and the pilot.

(e) Vaporizing-burners shall be provided with an automatic safety device to shut off the flow of gas to the main burner and pilot in the event the pilot is extinguished.

(f) Dehydrators and dryers utilizing vaporizing-burners shall be equipped with automatic devices both upstream and downstream of the vaporizing section. These devices shall be installed and connected to shut off in the event of excessive temperature, flame failure, and if applicable, insufficient air flow.

NOTE: See NFPA 61B, *Standard for the Prevention of Fires and Explosions in Grain Elevators and Facilities Handling Bulk Raw Agricultural Commodities*, for ignition and combustion controls applicable to vaporizing-burners associated with grain dryers.

(g) Pressure-regulating and control equipment shall be so located or so protected to prevent its exposure to temperatures above 140°F (60°C), unless designed and recommended for use at a higher temperature by the manufacturer.

(h) Pressure-regulating and control equipment located downstream of the vaporizing section shall be designed to withstand the maximum discharge temperature of hot vapor.

These burners actually are liquid-fed, and the vaporization takes place in the burner. They are used where large quantities of vapor are required. The minimum design pressure must be 250 psi (1.7 MPa), the same as that of the container. The coils and jackets must be made of ferrous metals or high-temperature alloys because of the high temperature created by the flame.

Because liquid is fed to the unit, there must be a hydrostatic relief valve in the unit to protect it when the burner is shut off and the liquid is trapped in the burner and piping. If this were not done, an increase in the temperature of the liquid would create a pressure that could burst the heater or the attached piping. These units are often used for temporary heating during construction or during emergencies and are unattended. As a result, the careful use of automatic equipment is extremely important. The units are also used as agricultural product dehydrators and dryers, and should be properly equipped with automatic devices on both the inlet and the outlet of the vaporizer section of the burner. As these units are generally portable, extreme care must be used in their placement and in ensuring that fuel connections to them are not subjected to mechanical damage.

2-5.4.8 Gas-air mixers shall comply with the following:

(a) Gas-air mixers shall be designed for the air, vapor, and mixture pressures to which they are subjected. Piping materials shall comply with applicable portions of this standard.

(b) Gas-air mixers shall be designed so as to prevent the formation of a combustible mixture. Gas-air mixers that are capable of producing combustible mixtures shall be equipped with safety interlocks on both the LP-Gas and air supply lines to shut down the system if combustible limits are approached.

(c) In addition to the interlocks provided for in 2-5.4.8(b), a method shall be provided to prevent air from accidentally entering gas distribution lines without LP-Gas being present. Check valves shall be installed in the air and LP-Gas supply lines close to the mixer to minimize the possibility of backflow of gas into the air supply lines or of air into the LP-Gas system. Gas-mixing control valves in the LP-Gas and air supply lines that are arranged to fail closed when actuated by safety interlock trip devices shall be considered as acceptable shutdown devices.

(d) Where it is possible for condensation to take place between the vaporizer and the gas-air mixer, an interlock shall be provided to prevent LP-Gas liquid from entering the gas-air mixer.

(e) Gas-air mixers that utilize the kinetic energy of the LP-Gas vapor to entrain air from the atmosphere, and are so designed that maximum air entrained is less than 85 percent of the mixture, need not include the interlocks specified in 2-5.4.8(b), (c), and (d), but shall be equipped with a check valve at the air intake to prevent the escape of gas to atmosphere when shut down. Gas-air mixers of this type receiving air from a blower, compressor, or any source of air other than directly from the atmosphere shall include a method of preventing air without LP-Gas, or mixtures of air and LP-Gas within the flammable range, from entering the gas distribution system accidentally.

Figure 2.47 *Venturi-type gas-air mixer with five (5) venturis. This type of gas mixer utilizes the kinetic energy of LP-Gas vapor to entrain air from the atmosphere to provide a desired LP-Gas–air mixture. Each venturi includes a solenoid valve on the gas inlet side of the venturi. A controller monitors the outlet manifold pressure of the mixer and energizes each individual solenoid valve to provide a constant flow to the process at the required pressure. The air inlet of the venturi includes a backflow check valve to prevent the escape of LP-Gas to the atmosphere. In addition, the controller is interlocked with a low vapor inlet temperature switch and high and low pressure switches on the gas inlet at gas-air outlet manifolds. (Photo courtesy of Allgas-Interex, division of Allgas Industries.)*

Figure 2.48 *High-pressure orifice gas-air mixer. A high-pressure [greater than 9 psi (62 kPa)] variable orifice type of gas-air mixer with inlet air and gas control trains located in a room complying with Chapter 7. Both the air and gas control trains include pressure regulators, check valves, shutoff valves, safety shutoff valves, temperature and pressure indicators, and high-low pressure switches. Process air for the mixer may be furnished from an air compressor in a separate room or from the plant air system. Air and gas pressures are both regulated to equal pressures on the inlet to the mixer. The variable orifice in the mixing valve will open and close depending on flow to provide a constant Btu valve. The propane-air is then distributed within the plant process piping. (Photo courtesy of Allgas-Interex, division of Allgas Industries.)*

The flammable limits for propane in air are approximately 2.15 percent to 9.6 percent. In a venturi-type blender or mixer where the kinetic energy of the LP-Gas vapor is used to entrain air from the atmosphere, the physical geometry of the venturi design is such that the quantity of air that can be drawn in cannot exceed 85 percent of the mixed gas formed. Thus, it is impossible to have a flammable LP-Gas–air mixture, and the interlock provisions of 2-5.4.8(b) can be eliminated. A check valve at each air inlet venturi will prevent any LP-Gas from escaping to the atmosphere.

To prevent malfunction, most venturi mixers have high and low vapor pressure interlocks, high and low mixed LP-Gas–air pressure interlocks, and a low incoming vapor temperature interlock. The latter interlock prevents any LP-Gas liquid from entering the gas-air mixer [*see 2-5.4.8(d)*].

To prevent any possible vapor condensation problems, it is advisable that the mixer be located as close as possible to the vaporizer, and that the mixer vapor inlet connection be sloped back toward the vaporizer vapor outlet connection. If this is not practical, a drip leg with heater should be installed. The piping can be heated to prevent the low temperature by wrapping electric tracer heater cable around such piping. Unless ambient temperatures are very low, insulating the piping may be sufficient to prevent too low a temperature of the gas vapor.

Usually venturi mixers employing atmospheric air must have an output mixture pressure of less than 9 psi (62 kPa) in order to produce a viable mixed gas. For mixer pressures greater than 10 psi (69 kPa), air under pressure must be supplied to the venturi mixer from an external source (i.e., plant air system or air compressor). For this external air supply system, some method must be provided on the venturi mixer to prevent 100 percent air or a flammable LP-Gas–air mixture from being supplied to the user's gas distribution piping. This is usually accomplished through special controls and low air pressure interlocks that will shut the mixer down in the event of malfunction.

Consideration should also be given to the method of installation of the downstream piping for LP-Gas–air from the mixer to the plant utilization point relative to possible recondensation of the LP-Gas–air mix. For most cases when commercial propane is the LP-Gas–air feedstock and the mixed LP-Gas–air mix is at or below 100 psi (690 kPa), there would be no recondensation of propane in the mix until the mixed gas temperature became 20°F (-6.7°C) or colder. But if commercial butane is the LP-Gas feedstock, considerable care must be taken in the mixed gas piping design. At a 10 psi (69 kPa) mixed gas pressure, butane will start to recondense at about 20°F (–6.7°C).

Thus, for higher mixed gas sendout pressures, recondensation of butane will occur at higher and higher temperatures [e.g., at 50 psi (345 kPa) the recondensation temperature is about 60°F (15.6°C)]. Heat for the butane-air mix gas line must thus be provided in the form of heat tracing or insulation to guard against possible butane recondensation.

2-5.5 Strainers. Strainers shall be designed to minimize the possibility of particulate materials clogging lines and damaging pumps, compressors, meters, or regulators. The strainer element shall be accessible for cleaning.

Strainers are especially important when new systems are first started up, as the construction process creates most of the potential for introducing welding slag, metal particles, packing materials, and other contaminants.

2-5.6 Meters.

2-5.6.1 Vapor meters of the tin or brass case type of soldered construction shall not be used at pressures in excess of 1 psi (7 kPa).

2-5.6.2 Vapor meters of the die cast or iron case type shall be permitted to be used at any pressure equal to or less than the working pressure for which they are designed and marked.

2-5.7 Dispensing Devices.

2-5.7.1 Components of dispensing devices, such as meters, vapor separators, valves, and fittings within the dispenser, shall comply with 2-5.1.2(a) and 2-5.1.3.

2-5.7.2 Pumps of dispensers used to transfer LP-Gas shall comply with 2-5.1.2(a), 2-5.1.3, and 2-5.2. Such pumps shall be equipped to permit control of the flow and to minimize the possibility of leakage or accidental discharge. Means shall be provided on the outside of the dispenser to readily shut off the power in the event of fire or accident. This means shall be permitted to be integral with the dispenser or shall be provided externally when the dispenser is installed.

2-5.7.3 Dispensing hose shall comply with 2-4.6.1 through 2-4.6.3.

In the 1995 edition, 2-5.7.3 was revised by deleting requirements for the installation of an excess-flow valve or an emergency shutoff valve. This was done because this is an installation requirement, (rather than an equipment requirement), which should be covered in Chapter 3. The requirement is already in 3-9.3.4.

2-5.8 Regulators.

2-5.8.1 Single-stage regulators shall have a maximum outlet pressure setting of 1.0 psi (7 kPa) and shall be equipped with one of the following [*see 3-2.6.4 for required protection from the elements*]:

Single-stage regulators are permitted in limited applications. See 3-2.6.1, Exception No. 1, for applications where single-stage regulation is permitted. Single-stage pressure regulation is not permitted for fixed piping systems serving $^1/_2$ psi appliance systems.

(a) An integral pressure relief valve on the outlet pressure side having a start-to-discharge pressure setting within the limits specified in the *Standard for LP-Gas Regulators*, UL 144.

(b) An integral overpressure shutoff device that shuts off the flow of LP Gas vapor when the outlet pressure of the regulator reaches the overpressure limits specified in UL 144. Such a device shall not open to permit flow of gas until it has been manually reset.

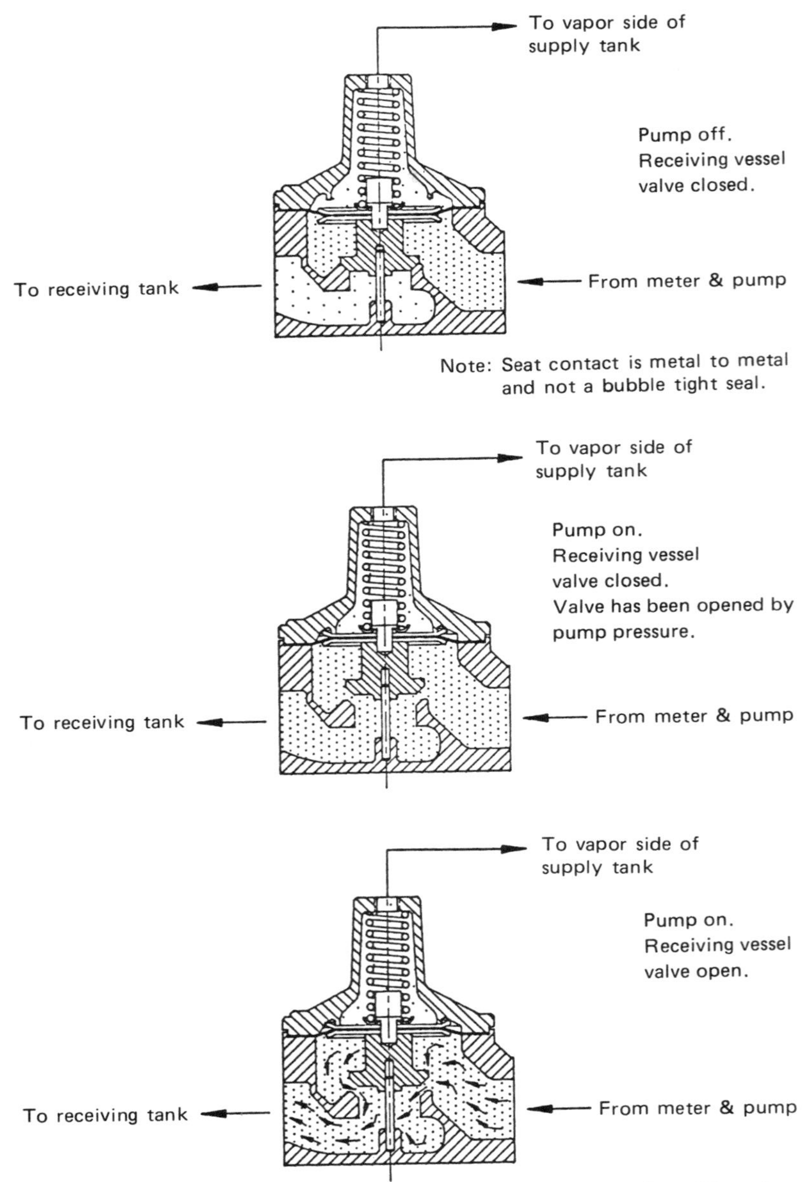

Figure 2.49 *Operation of differential back pressure valve. (Drawings courtesy of Liquid Controls, Inc.)*

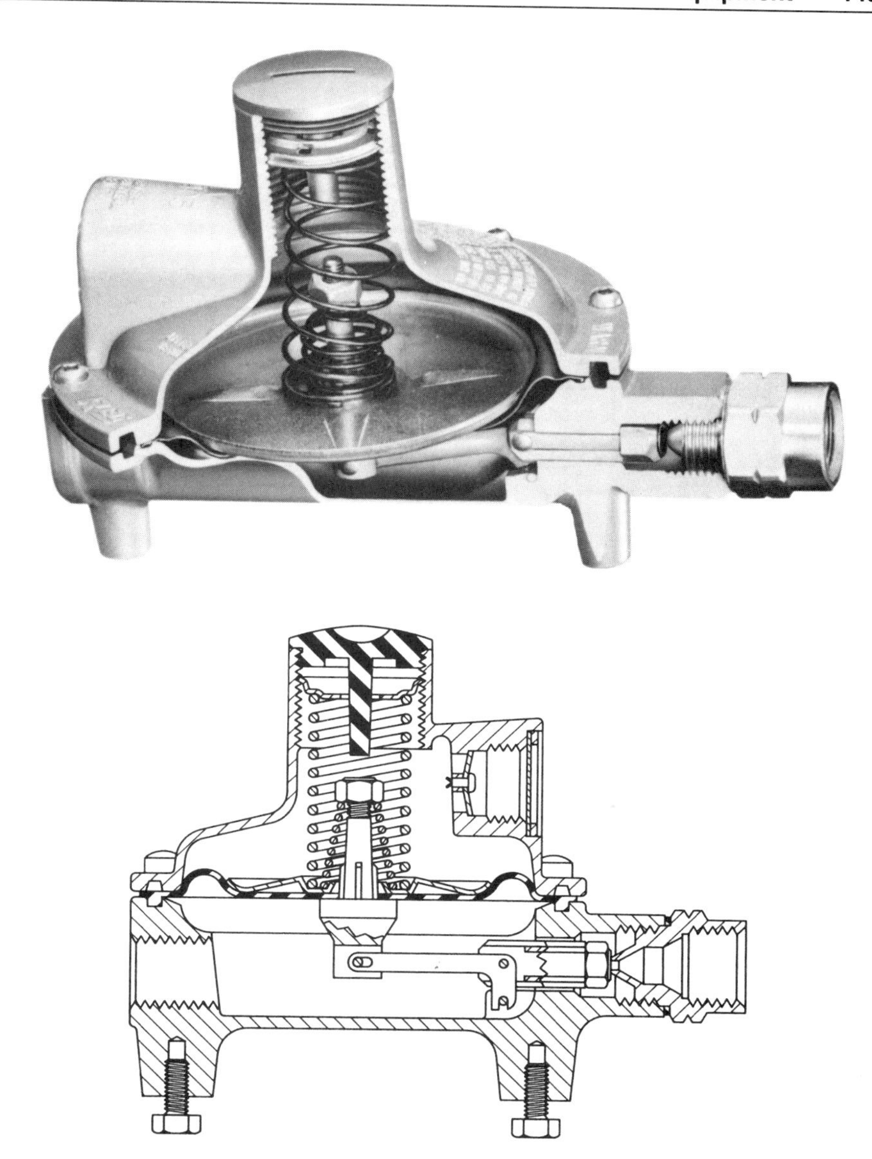

Figure 2.50 Lever-type regulator for vapor service. (Photo and drawing courtesy of Engineered Controls, Inc.)

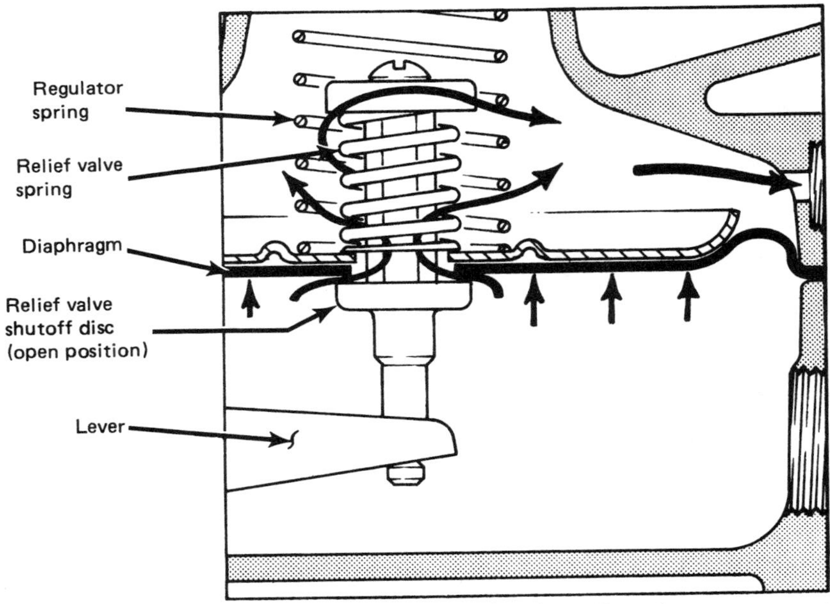

Figure 2.51 *Operation of regulator relief valve. (Drawing courtesy of National Propane Gas Association.)*

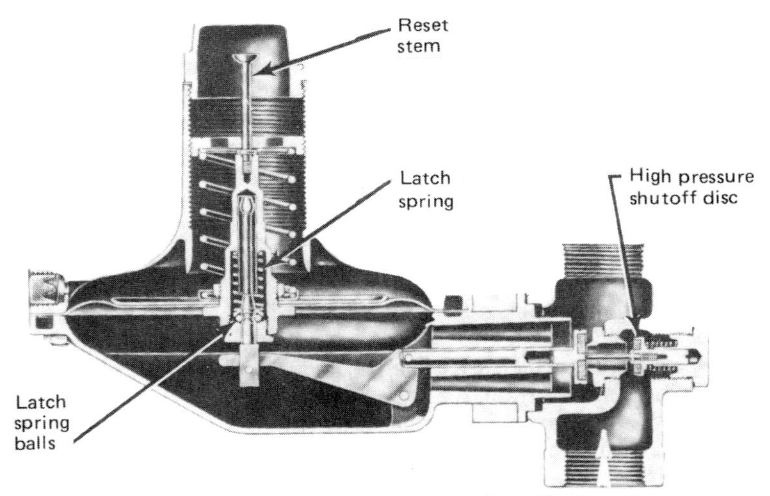

Figure 2.52 *Regulator with high-pressure shutoff. (Drawing courtesy of National Propane Gas Association.)*

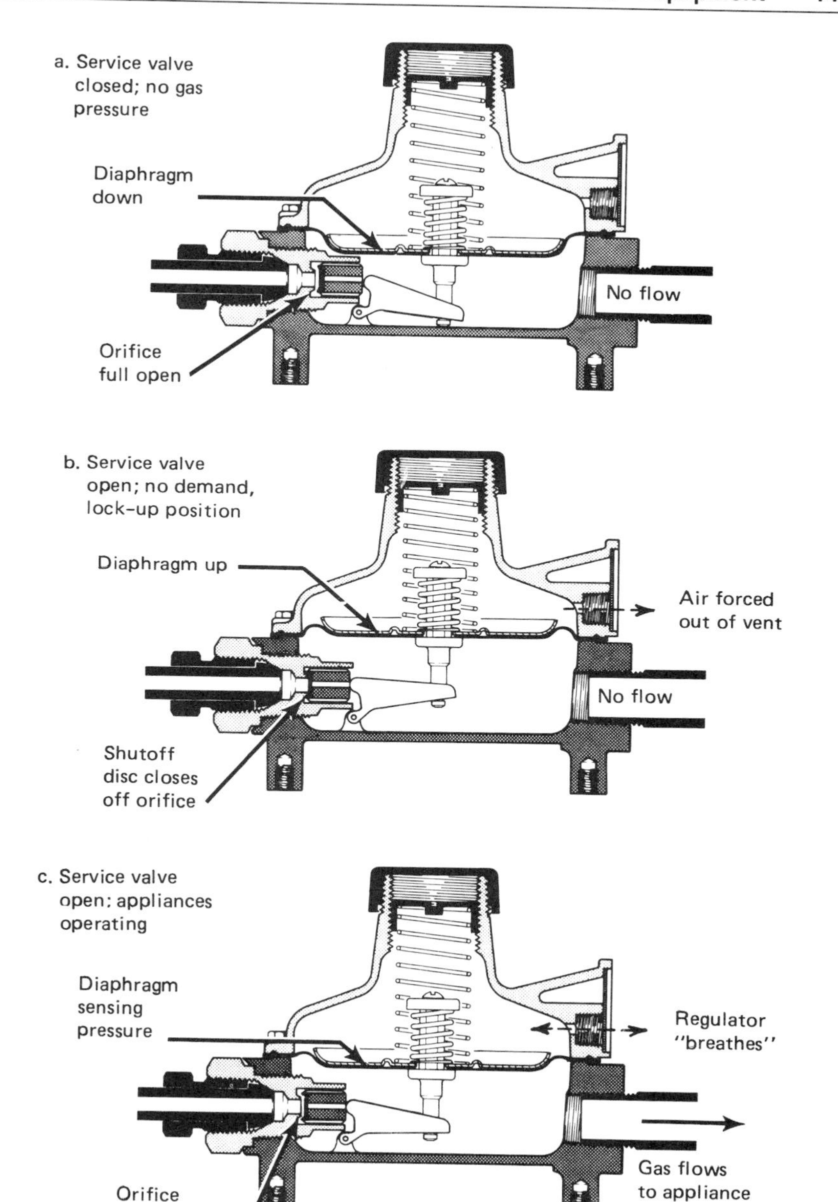

Figure 2.53 *Regulator operation. (Drawing courtesy of National Propane Gas Association.)*

Figure 2.54 Integral Two-Stage Regulator. The second-stage diaphragm cover, with the vent pointing to the left, is seen above the two pressure taps, for reading interstage and final stage pressure. The first-stage regulator is at the threaded inlet with the vent pointing down. Note that this regulator must be installed under a hood as the two vents are not oriented in the same direction. (Photo courtesy of Fisher Controls, Inc.)

This section was rewritten in the 1995 edition to prevent most cases of overpressure of piping systems in buildings. *Overpressure can force flames to protrude beyond an appliance and cause a building fire.* Prevention of overpressure is accomplished by mandating two-stage regulation in most applications in buildings, and mandating the first-stage outlet pressure to be set at 10 psi or less. This change along with changes in UL Standard 144, requiring increased regulator relief valve capacity or an overpressure shutoff device, result in the maximum pressure downstream of the second-stage regulator being limited to 2.0 psi, even with a regulator seat failure (the seat is the resilient sealing component in the regulator). This pressure will not cause appliance controls to fail. Overpressure in piping systems can result from failure of a regulator component, or in some cases, from blockage of a regulator vent. Components that can fail include all the components of the regulator—seats, diaphragms, linkages, etc. While failures of these components occur, they are not common. Blockage of a regulator vent is usually caused by insects or freezing. Requirements for the outdoor installation of regulators mandate that they be "designed, installed, or protected so that their operation will not be affected by the elements" (see 3-2.6.4). This is accomplished by compliance with the requirements of UL 144, which includes a test for blockage of the vent by freezing rain, and requires a vent screen to prevent insect blockage. Unusual weather conditions can cause ice buildup of such thickness that it is beyond the testing mandated by UL 144. It was reported to the editor that one highly unusual storm in Cape May, New Jersey, in March 1984, resulted in ice thickness of about one inch. The changes to this section require a relief valve in both first- and second-stage regulators. The change is designed to eliminate virtually all potential overpressure conditions.

This change is a result of a project of the National Fuel Gas Code Committee to prevent overpressure of all fuel gas piping systems in buildings. The split of the two standards for propane supply to building piping systems places the limit of coverage at the discharge of the first-stage regulator. The National Fuel Gas Code refers back to NFPA 58 for second-stage regulators for propane systems, therefore the requirements are placed here in NFPA 58. Work continues in the National Fuel Gas Code Committee to address prevention of overpressure in natural gas piping systems in buildings.

The changes prevent overpressure in buildings by venting propane through the regulator vent to the outdoors near the building, and permit as an alternate, the use of an overpressure shutoff device requiring manual reset. One proposal that was rejected, proposed mandating at least an automatic shutoff device to prevent overpressure, rather than offering it as an option. One factor in the Committee's action was that regulators already utilizing integral pressure relief valves have performed well, and the changes in UL 144, increasing their capacity to prevent overpressurization, could be accomplished quickly by regulator manufacturers. Overpressure shutoff devices do exist and can be used as an alternate to these requirements.

2-5.8.2 Second-stage regulators and integral two-stage regulators shall have a maximum outlet pressure setting of 14 in. W.C. (4.0 kPa) and shall be equipped with one of the following (*see 3-2.6.4 for required protection from the elements*):

(a) An integral pressure relief valve on the outlet pressure side having a start-to-discharge pressure setting within the limits specified in the *Standard for LP-Gas Regulators*, UL 144. This relief device shall limit the outlet pressure of the second-stage regulator to 2.0 psi (14 kPa) when the regulator seat disc is removed and the inlet pressure to the regulator is 10.0 psi (69 kPa) or less as specified in the *Standard for LP-Gas Regulators*, UL 144.

(b) An integral overpressure shutoff device that shuts off the flow of LP-Gas vapor when the outlet pressure of the regulator reaches the overpressure limits specified in UL 144. Such a device shall not open to permit flow of gas until it has been manually reset.

Exception: Regulators with a rated capacity of more than 500,000 Btu/hr (147 kw/hr) shall be permitted to have a separate overpressure protection device complying with paragraphs 2.9.2 through 2.9.8 of the National Fuel Gas Code, NFPA 54 (ANSI Z223.1). The overpressure protection device shall limit the outlet pressure of the regulator to 2.0 psi (14 kPa) when the regulator seat disc is removed and the inlet pressure to the regulator is 10 psi (69 kPa) or less.

Note that the maximum discharge pressure from a second-stage regulator is limited to a 16 in. water column. It is also anticipated that the integral pressure relief valve option will be used by regulator manufacturers.

Regulators with capacities over 500,000 Btu/hr are exempt from the required internal overpressure protection provisions, as they can use a separate overprotection device as covered by the *National Fuel Gas Code* (1992 edition), which states:

2.9.2 Devices. Any of the following pressure relieving or pressure limiting devices shall be permitted to be used:

(a) Spring-loaded relief device

(b) Pilot-loaded back pressure regulator used as a relief valve so designed that failure of the pilot system or external control piping will cause the regulator relief valve to open

(c) A monitoring regulator installed in series with the service or line pressure regulator

(d) A series regulator installed upstream from the service or line regulator and set to continuously limit the pressure on the inlet of the service or line regulator to the maximum working pressure of the downstream piping system

(e) An automatic shutoff device installed in series with the service or line pressure regulator and set to shut off when the pressure on the downstream piping system reaches the maximum working pressure or some other predetermined pressure less than the maximum working pressure. This device shall be designed so that it will remain closed until manually reset.

(f) A liquid-seal relief device that can be set to open accurately and consistently at the desired pressure

These devices may be installed as an integral part of the service or line pressure regulator or as separate units. If separate pressure relieving or pressure limiting devices are installed, they shall comply with 2.9.3 through 2.9.8.

2.9.3 Construction and Installation. All pressure relief or pressure limiting devices shall:

(a) Be constructed of materials so that the operation of the device will not be impaired by corrosion of external parts by the atmosphere or of internal parts by the gas

(b) Be designed and installed so they can be operated to determine if the valve is free. The devices shall also be designed and installed so they can be tested to determine the pressure at which they will operate, and be examined for leakage when in the closed position.

2.9.4 External Control Piping. External control piping shall be protected from falling objects, excavations, or other causes of damage and shall be designed and installed so that damage to any control piping shall not render both the regulator and the overpressure protective device inoperative.

2.9.5 Setting. Each pressure limiting or pressure relieving device shall be set so that the pressure shall not exceed a safe level beyond the maximum allowable working pressure for the piping and appliances connected.

2.9.6 Unauthorized Operation. Precautions shall be taken to prevent unauthorized operation of any shutoff valve that will make a pressure relief valve or pressure limiting device inoperative. Acceptable methods for complying with this provision are:

(a) Lock the valve in the open position. Instruct authorized personnel of the importance of leaving the shutoff valve open and of being present while the shutoff valve is closed so that it can be locked in the open position before leaving the premises.

(b) Install duplicate relief valves, each having adequate capacity to protect the system, and arrange the isolating valves or three-way valve so that only one safety device can be rendered inoperative at a time.

2.9.7 Vents. The discharge stacks, vents, or outlet parts of all pressure relief and pressure limiting devices shall be located so that gas is safely discharged into the outside atmosphere. Discharge stacks or vents shall be designed to prevent the entry of water, insects, or other foreign material that could cause blockage. The discharge stack or vent line shall be at least the same size as the outlet of the pressure relieving device.

2.9.8 Size of Fittings, Pipe, and Openings. The openings, pipe, and fittings, located between the system to be protected and the pressure relieving device, shall be sized to prevent hammering of the valve and to prevent impairment of relief capacity.

2-5.8.3 Integral two-stage regulators except automatic changeover regulators shall be provided with a means to determine the outlet pressure of the high pressure regulator portion of the integral two-stage regulator.

Integral two-stage regulators, sometimes called "piggyback" regulators, consist of first- and second-stage regulators in the same body and are sold and installed as one unit. The means to determine outlet pressure of the first-stage regulator portion is a plugged opening, with a restrictive orifice for a pressure gauge, or a permanently installed pressure-indicating device.

2-5.8.4 Integral two-stage regulators shall not incorporate an integral pressure relief valve in the high pressure regulator portion of the unit.

2-5.8.5 First-stage regulators shall incorporate an integral pressure relief valve having a start-to-discharge setting within the limits specified in the *Standard for LP-Gas Regulators*, UL 144.

Exception: First-stage regulators with a rated capacity of more than 500,000 Btu/hr (147 kw/hr) shall be permitted to have a separate pressure relief valve.

See commentary following 2-5.8.2.

2-5.8.6 High pressure regulators with a rated capacity of more than 500,000 Btu/hr (147 kw/hr) where permitted to be used in two-stage systems shall incorporate an integral relief valve or shall have a separate relief valve.

See commentary following 2-5.8.2.

2-5.8.7 First-stage regulators shall have an outlet pressure setting up to 10.0 psi (69 kPa) in accordance with the *Standard for LP-Gas Regulators*, UL 144.

2-5.8.8 Regulators shall be designed so as to drain all condensate from the regulator spring case when the vent is directed vertically down.

2-5.9 Sight Flow Glasses. Flow indicators, either of the simple observation type or those combined with a backflow check valve, shall be permitted to be used in applications in which the observation of liquid flow through the piping is desirable or necessary.

Figure 2.55 *Sight flow indicator with backflow check valve. (Photo courtesy of National Propane Gas Association.)*

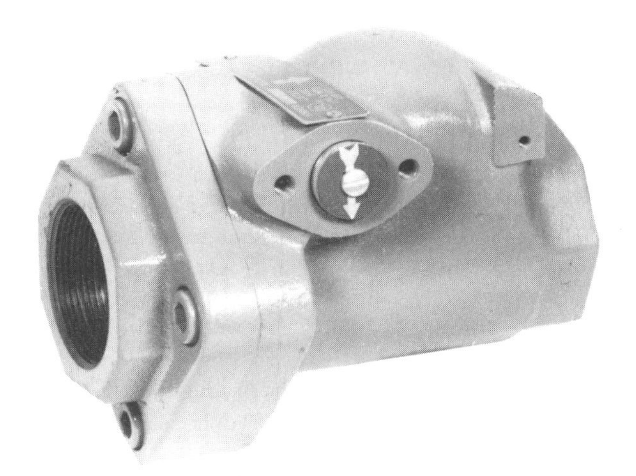

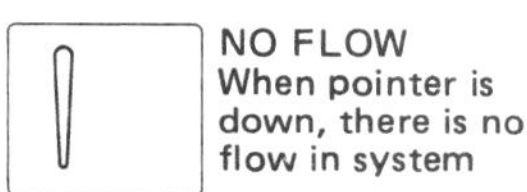

ERRATIC FLOW
Unstable position indicates vapor or cavitation

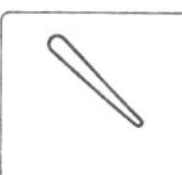

FULL FLOW
When pointer is in position shown, there is flow in the system

Figure 2.56 *Flow indicator with backflow check valve. (Photo courtesy of Fisher Controls.)*

2-6 Appliances.

2-6.1 Approved Appliances.

2-6.1.1 This section includes basic construction and performance provisions for LP-Gas consuming appliances.

In the United States, Canada, and other economically developed countries, it has become customary to subject the more common LP-Gas and natural gas appliances to coverage by specific national consensus or governmental standards. In many instances, private or governmental listing agencies have been active in achieving a level of safety for these appliances by evaluating them against the standards or through the exercise of their judgment. Because of this activity, it has not only been unnecessary to provide detailed technical coverage in NFPA 58, but such coverage could create difficult correlation problems. Therefore, NFPA 58 coverage has been limited to the following situations:

(1) Equipment not addressed in separate standards. This may be due either to a small number or limited use of appliances or to the use being in an environment under the control of specialists. Neither situation warrants the level of activity necessary to develop a separate standard.

(2) Coverage needed as an interim measure. It takes considerable time to develop a separate standard, and frequently such a standard cannot be promulgated unless the use of the appliance is at least permitted by NFPA 58 and basic technical features are included in it.

In such instances, while NFPA 58 serves a useful function, the Technical Committee must be careful to amend the standard as necessary when a separate, specific standard is developed. Such amendment could include deletion of provisions in NFPA 58. In general, a specific appliance standard should govern in recognition of its more complete and intensive coverage, provided the standard has suitable status.

It is reemphasized that the requirement for approval in 2-6.1.2 does not include listing. The decision is that of the authority having jurisdiction. Such authorities have, however, demonstrated a pronounced disposition to accept, or even require, listed appliances and equipment whenever they are available.

It should also be noted that many of the appliances addressed in Section 2-6 are actually installed under circumstances that make them subject to NFPA 54, *National Fuel Gas Code* (ANSI Z223.1), rather than to NFPA 58, in accordance with 1-2.3.1(f.). [*See commentary on 1-2.3.1(f.).*]

2-6.1.2 New residential, commercial, and industrial LP-Gas consuming appliances shall be approved.

The manufacture of most residential and commercial LP-Gas appliances is covered by specific national consensus standards. The largest body of these are the ANSI Z21 series of standards promulgated by the ANSI-accredited Z21 Committee on Performance and Installation of Gas-Burning Appliances and Related Accessories, and the ANSI Z83 series of standards promulgated by the ANSI-accredited Z83 Committee on Industrial Gas Equipment Installation and Utilization, also sponsored by International Approval Services (formed by a merger of the American Gas Association Laboratories and the Canadian Gas Association.). These standards are in the process of being harmonized with their Canadian counterparts. These appliances are also certified ("listed," in NFPA 58 terminology.) by the International Approval Services Laboratories (formed by the merger of the American Gas Association Laboratories and Canadian Gas Association .) and other laboratories in accordance with the relevant standards. An analogous situation exists in Canada under the auspices of the Canadian Standards Association and the Underwriters Laboratories of Canada (not affiliated with Underwriters Laboratories Inc.), and other certification.

Many industrial appliances are also covered by specific national consensus standards. Included in these are the ANSI Z83 series of standards promulgated by the ANSI-accredited Z83 Committee on Industrial Gas Equipment Installation and Utilization, sponsored by the American Gas Association. These are also certified (listed) by the AGA Laboratories. Similar conditions exist in Canada.

A list of current ANSI Z21 and Z83 standards is included at the end of this chapter.

In addition to the preceding, many LP-Gas appliances and equipment are listed by Underwriters Laboratories Inc. (United States and Canada.). In many instances, these listings use national consensus standards adopted by ANSI under its canvass method. In the industrial area, the listing activity (called "approval.") of the Factory Mutual Engineering Corporation is notable with respect to industrial heat-processing equipment.

Exception: For an appliance, class of appliance, or appliance accessory for which no applicable standard has been developed, approval of the authority having jurisdiction shall be permitted to be required before installation is made.

The intent here is to encourage such approval. In many instances where NFPA 58 is adopted as a regulatory instrument and often as a contractual matter, such approval is mandatory. For example, it is common for a purchaser to specify that an appliance be approved by its insurance carrier. However, there are situations where such approval is not feasible—e.g., the authority does not have the needed expertise—and this prevents the standard from requiring such approval.

2-6.2 Provisions for Appliances.

2-6.2.1 Any appliance originally manufactured for operation with a gaseous fuel other than LP-Gas and in good condition shall be permitted to be used with LP-Gas provided it is properly converted, adapted, and tested for performance with LP-Gas before being placed into use.

The conversion of, for example, a natural gas appliance to LP-Gas is not a complex operation. If the original appliance was listed or approved, however, the conversion may void the original listing or approval. Prudence is indicated in these situations.

2-6.2.2 Unattended heaters used inside buildings for animal or poultry production or care shall be equipped with approved automatic devices to shut off the flow of gas to the main burners, and pilots if used, in the event of flame extinguishment or combustion failure.

Exception: Heaters as provided in 3-5.1.3.

This provision is an example of the use of NFPA 58 to fill a gap created by the absence of a specific standard (*see commentary on 2-6.1.1.*). It is noted that this provision addresses only one hazard of such a heater. Although it is a major hazard, others need to be considered. (*See commentary on 3-5.1.3.*)

2-6.2.3 Appliances using vaporizing-burners shall comply with 2-5.4.7.

2-6.2.4 Appliances used in mobile homes and recreational vehicles shall be approved for such service.

This provision is a carry-over from earlier editions that applied to mobile homes and recreational vehicles. At present, recreational vehicles are covered by NFPA 501C, and this requirement is consistent with the requirements of NFPA 501C. Manufactured homes sold in the United States are covered under federal regulations issued by the Department of Housing and Urban Development for new manufactured homes only. This provision is applicable to appliances installed in manufactured homes after original manufacture and sale.

2-6.2.5 LP-Gas appliances used on commercial vehicles (*see Section 3-8*) shall be approved for the service (*see 2-6.1*) and shall comply with the following:

(a) Gas-fired heating appliances and water heaters shall be equipped with automatic devices designed to shut off the flow of gas to the main burner and the pilot in the event the pilot flame is extinguished.

(b) Catalytic heating appliances shall be equipped with an approved automatic device to shut off the flow of gas in the event of combustion failure.

(c) Gas-fired heating appliances and water heaters to be used in vehicles intended for human occupancy shall make provisions for complete separation of the combustion system and the living space. If this separation is not integral with the appliance, it shall be provided otherwise by the method of installation (*see 3-8.4.2*).

References Cited in Commentary

The following publications are available from the National Fire Protection Association, 1 Batterymarch Park, P.O. Box 9101, Quincy, MA 02269-9101.

NFPA 54, *National Fuel Gas Code,* 1988 edition.

NFPA 501C, *Standard on Recreational Vehicles,* 1990 edition.

The following publications are available from the American Gas Association Laboratories, 8501 East Pleasant Valley Road, Cleveland, OH 44131.

ANSI Z21 and Z83 Series Standards:

ANSI Z21.1-90, *Household Cooking Gas Appliances.*

ANSI Z21.5.1-82, *Gas Clothes Dryers, Volume I, Type 1, Clothes Dryers.*

ANSI Z21.5.2-87, *Gas Clothes Dryers, Volume II, Type 2, Clothes Dryers.*

ANSI Z21.10.1-90, *Gas Water Heaters, Volume I, Storage Water Heaters with Inputs of 75,000 Btu per Hour or Less.*

ANSI Z21.10.3-90, *Gas Water Heaters, Volume III, Storage, with Input Ratings Above 75,000 Btu per Hour, and Circulating Instantaneous Water Heaters.*

ANSI Z21.11.1-88, *Gas-Fired Room Heaters, Volume I, Vented Room Heaters.*

ANSI Z21.11.2-89, *Gas-Fired Room Heaters, Volume II, Unvented Room Heaters.*

ANSI Z21.13-87, and Addenda, Z21.13a, *Gas-Fired Low-Pressure Steam and Hot Water Boilers.*

ANSI Z21.19-90, *Refrigerators Using Gas Fuel.*

ANSI Z21.40.1-81, and Addenda, Z21.40.1a, *Gas-Fired Absorption Summer Air Conditioning Appliances.*

ANSI Z21.42-71, *Gas-Fired Illuminating Appliances.*

ANSI Z21.44-88, *Gas-Fired Gravity and Fan Type Direct Vent Wall Furnaces.*

ANSI Z21.47-90, *Gas-Fired Central Furnaces (Except Direct Vent Central Furnaces.).*

ANSI Z21.48-89, *Gas-Fired Gravity and Fan Type Floor Furnaces.*

ANSI Z21.49-89, *Gas-Fired Gravity and Fan Type Vented Wall Furnaces.*

ANSI Z21.50-89, *Vented Decorative Gas Appliances.*

ANSI Z21.56-89, *Gas-Fired Pool Heaters.*

ANSI Z21.57-87, *Recreational Vehicle Cooking Gas Appliances.*

ANSI Z21.58-91, *Outdoor Cooking Gas Appliances.*

ANSI Z21.60-88, *Decorative Gas Appliances for Installation in Vented Fireplaces.*

ANSI Z21.61-83, *Gas-Fired Toilets.*

ANSI Z21.64-88, *Direct Vent Central Furnaces.*

ANSI Z83.3-71, *Standard for Gas Utilization Equipment in Large Boilers.*

ANSI Z83.4-89, *Direct Gas-Fired Make-Up Air Heaters.*

ANSI Z83.6-87, *Gas-Fired Infrared Heaters.*

ANSI Z83.7-87, *Gas-Fired Construction Heaters.*

ANSI Z83.8-90, *Gas Unit Heaters.*

ANSI Z83.9-90, *Gas-Fired Duct Furnaces.*

ANSI Z83.11-89, *Gas Food Service Equipment—Ranges and Unit Broilers.*

ANSI Z83.12-89, *Gas Food Service Equipment—Baking and Roasting Ovens.*

ANSI Z83.13-89, *Gas Food Service Equipment—Deep Fat Fryers.*

ANSI Z83.14-89, *Gas Food Service Equipment—Counter Appliances.*

ANSI Z83.15-89, *Gas Food Service Equipment—Kettles, Steam Cookers and Steam Generators.*

ANSI Z83.16-82, *Unvented Commercial and Industrial Heaters.*

The following publications are available from the American Society for Testing and Materials, 1916 Race Street, Philadelphia, PA 19103.

ASTM D 2513-90, *Standard Specification for Thermoplastic Gas Pressure Pipe, Tubing, and Fittings.*

ASTM F 1055-87, *Standard Specification for Electrofusion Type Polyethylene Fittings for Outside Diameter Controlled Polyethylene Pipe and Tubing.*

The following publications are available from the Compressed Gas Association, Inc., 1235 Jefferson Davis Highway, Arlington, VA 22202.

CGA Pamphlet C-2-87, *Recommendations for the Disposition of Unserviceable Compressed Gas Cylinders.*
CGA Pamphlet C-14-79, *Procedures for Fire Testing of DOT Cylinder Safety Relief Device Systems.*
CGA Standard V-1-87/ANSI B57.1, *Compressed Gas Cylinder Valve Outlet and Inlet Connections.*
CGA Pamphlet S-1.1-89, *Pressure Relief Device Standards.*

The following publication is available from the Underwriters Laboratories Inc., 333 Pfingsten Rd., Northbrook, IL 60062.
UL 132-84, *Safety Relief Valves for Anhydrous Ammonia and LP-Gas.*

3

Installation of LP-Gas Systems

3-1 General.

3-1.1 This chapter applies to the location and field installation of LP-Gas systems utilizing components, subassemblies, container assemblies, and container systems fabricated in accordance with Chapter 2.

The intent of Chapter 3 is to include all information needed by the installer of LP-Gas systems with the exceptions of cargo tank vehicles, which are covered by Chapter 6, and engine fuel systems, which are covered in Chapter 8. Prior to the 1972 edition, this material was scattered throughout the standard. In the 1992 edition, the chapter was modified to clearly cover installation of LP-Gas dispensing systems, some material was relocated to Chapter 3 from Chapter 4, and the section on engine fuel systems, formerly 3-6, was relocated to Chapter 8.

Chapter 3 is the largest chapter in the standard. Its size is indicative of the importance of the installer's role in LP-Gas system safety.

The fire protection systems covered by Section 3-10 generally are not installed by the LP-Gas system installer and, as such, are somewhat of an anomaly in Chapter 3. Section 3-10, however, can be important from a facility location standpoint and, as such, must be considered with Section 3-2.

Exception: Refrigerated containers shall be installed in accordance with Chapter 9.

NOTE: Section 3-2 includes general provisions applicable to most stationary systems. Sections 3-3 to 3-8 extend and modify Section 3-2 for systems installed for specific purposes.

This paragraph was relocated to a note in the 1992 edition because the content is explanatory and such unenforceable material does not belong in a standard.

3-1.2 Installation of systems used in the highway transportation of LP-Gas shall be in accordance with Chapter 6.

3-1.3 LP-Gas systems shall be installed in accordance with this standard and other national standards or regulations that apply. These include:

(a) NFPA 37, *Standard for the Installation and Use of Stationary Combustion Engines and Gas Turbines.*
(b) NFPA 54, *National Fuel Gas Code* (ANSI Z223.1).
(c) NFPA 61B, *Standard for the Prevention of Fires and Explosions in Grain Elevators and Facilities Handling Bulk Raw Agricultural Commodities.*
(d) NFPA 82, *Standard on Incinerators, Waste, and Linen Handling Systems and Equipment.*
(e) NFPA 86, *Standard for Ovens and Furnaces.*
(f) NFPA 96, *Standard for Ventilation Control and Fire Protection of Commercial Cooking Operations.*
(g) NFPA 302, *Fire Protection Standard for Pleasure and Commercial Motor Craft.*
(h) NFPA 501A, *Standard for Fire Safety Criteria for Manufactured Home Installations, Sites, and Communities.*
(i) NFPA 501C, *Standard on Recreational Vehicles.*
(j) U.S. DOT Regulations, 49 CFR 191 and 192, for LP-Gas pipeline systems subject to DOT.

This provision calls attention to the existence of other standards covering specific LP-Gas system applications and to encourage their adoption and use on the same basis as NFPA 58. Unless handled otherwise in enabling legislation, the adoption of 3-1.3 as part of NFPA 58 also adopts the standards cited in 3-1.3(a) through (j). The exception makes it clear that refrigerated containers are covered in Chapter 9, and that the provisions of Chapter 3 should not be applied to them. Refrigerated storage tanks are not pressure vessels and store LP-Gases at atmospheric pressure or slight positive pressure (usually ½ psi max). The requirements in this chapter are intended for pressure vessels.

3-2 General Provisions.

3-2.1 Scope. This section includes criteria for location of containers and liquid transfer systems; the installation of container appurtenances and regulators; piping service limitations; the installation of piping (including flexible connectors and hose); hydrostatic relief valves and equipment (other than vaporizers — *see Section 3-6*); and the testing of piping systems.

3-2.2 Location of Containers.

3-2.2.1 LP-Gas containers shall be located outside of buildings.

Exception No. 1: Portable containers as specifically provided for in Section 3-4.

Exception No. 2: Containers of less than 125 gal (0.5 m^3) water capacity for the purposes of being filled in buildings or structures complying with Chapter 7.

Exception No. 3: Containers on LP-Gas vehicles complying with, and parked or garaged in accordance with, Chapter 6.

Exception No. 4: Containers used with LP-Gas stationary or portable engine fuel systems complying with Chapter 8.

Exception No. 5: Containers used with LP-Gas fueled industrial trucks complying with 8-3.6.

Exception No. 6: Containers on LP-Gas fueled vehicles garaged in accordance with 8-6.

Exception No. 7: Portable containers awaiting use, resale, or exchange when stored in accordance with Chapter 5.

From its inception in 1932, NFPA 58 has taken a basic position that LP-Gas containers should be located outdoors because in the event of leakage of the flammable gas a leak outdoors has a much greater opportunity to dissipate prior to being ignited than an indoor leakage. (*See commentary on Section 3-4 for history.*)

The container holds the LP-Gas in both the liquid and vapor states to an extent dependent upon how much of its contents have been withdrawn in service—a "full" container being approximately 80 percent full of liquid. The pressure inside the container varies with the temperature of the liquid but is substantial—about 130 psi (0.9 MPa) at 70°F (21°C) for propane. (*See Appendix B and texts covering thermodynamic properties of LP-Gases.*) If the LP-Gas escapes in the vapor state, the pressure can result in a substantial leakage rate. If the LP-Gas escapes in the liquid state, it will rapidly vaporize and produce large quantities of vapor. In the case of propane, one volume of liquid will vaporize to about 270 volumes of vapor. Either of these circumstances has the potential to lead to a substantial quantity of a flammable vapor-air mixture which, if ignited in an enclosed space, could result in a combustion explosion.

While considering this, it must also be noted that NFPA 58 requires that all LP-Gas containers be equipped with a pressure relief device. As explained earlier, a principal purpose of this device is to release gas or liquid to prevent container failure caused by overpressurization. This device will operate at normal atmospheric or room temperatures if the container has been overfilled. Therefore, it is not necessary to have a leak as a result of some sort of container or appurtenance failure in order to have vapor or liquid release.

The other major hazard is that of the Boiling Liquid-Expanding Vapor-Explosion (BLEVE). If an LP-Gas container fails in such a way that it suddenly comes apart in two or more pieces at a time when it contains liquid, the results are nearly instantaneous vaporization and dispersal of the contents (and consequently very rapid formation of a flammable mixture), propulsion of the container pieces, and creation of a shock wave. Whether the failure is the result of fire exposure (the usual case among the best known incidents), weakening of the container (e.g., by impact or corrosion), or failure of a pressure relief device to operate in an enclosed space, the results can be similar to a combustion explosion.

Considering either type of hazard, it is apparent that the presence of a room, building, or other enclosure is a key element in both the potential for and severity of the explosion. It follows that the absence of an enclosure is a highly desirable safeguard that should be achieved whenever it is feasible to do so.

In the more than 60 years since the first edition of NFPA 58, however, the convenience of "gas in a bottle" has led society to demand applications where the container must be present inside a structure. These have evolved slowly but more or less at pace with the increasing wealth and technological complexity of the industrialized nations. The NFPA Technical Committee on Liquefied Petroleum Gases has approached these applications with great caution. Where considered acceptable, they have been accompanied by safeguards designed to minimize the chances of an explosion occurring and its severity. These include limiting the size of the container, controlling the presence of ignition sources, relating operation of systems to the presence or absence of persons in the building, and even designing the room or building so as to control its reaction to explosion or fire.

All of these safeguards and more are represented by the seven exceptions to the outdoor location cited in 3-2.2.1 and are addressed more fully in text provisions on the specific exceptions.

It should be noted that while the requirements of NFPA 58 apply in the United States, are the basis for regulations in Canada, and are similar to requirements for LP-Gas in the developed world, in many lesser-developed countries it is common practice for portable LP-Gas containers to be connected directly to appliances within buildings.

3-2.2.2 Containers installed outside of buildings, whether of the portable type replaced on a cylinder exchange basis or permanently installed and refilled at the installation, shall be located with respect to the nearest container, important building, group of buildings, or line of adjoining property that can be built upon, in accordance with Table 3-2.2.2, Table 3-2.2.4, and Table 3-2.2.7(f).

NOTE: See 3-2.3.2, 3-2.3.3, and 3-3.4.3 for minimum distances to buildings as outlined in, or referenced by, these respective sections.

General: The container location provisions of 3-2.2.2 (and Tables 3-2.2.4 and 3-2.2.7(f)), together with those of 3-2.3 regarding location of transfer operations—and, for larger installations, those of Section 3-10—comprise the basic outdoor container siting criteria in the standard. They should be studied together any time a container is to be filled and the point of transfer is not at a container appurtenance, in order to avoid costly mistakes in allocation of land use.

The siting criteria in 3-2.2.2 require that a container be located certain distances from another container, an important building or group of buildings, and a line of adjoining property that may be built upon—and only from these items cited. They reflect the hazard of the container to the items cited, and vice versa.

It is important to recognize that these distances are not sufficient, in and of themselves, to eliminate mutual exposures. A large, sustained liquid leak from a container connection or unignited liquid release during a BLEVE can produce a vapor cloud that can travel much farther than even the largest distance specified [400 ft (122 m)]. The thermal radiation, flying missiles, and shock wave from a BLEVE can also cause damage and injury at distances greater than any of those specified.

The distances specified are intended, then, to buy time and provide space for implementation of emergency activity. The great majority of leaks are of a size such that these distances are adequate to prevent ignition (most ignition sources are in buildings or beyond the property line). If ignition does occur, these distances provide for separation that facilitates the application of water and reduces thermal radiation exposure. For more information on conducting a firesafety analysis, refer to the supplement *Guidelines for Conducting a Firesafety Analysis.*

Important Buildings: Experience has indicated some confusion with respect to the terms "important buildings" and "line of adjoining property that can be built upon." A building can be "important" for a number of reasons. Its replacement value is customarily recognized by all parties to a decision. Its importance by virtue of human occupancy or the value of its contents is also well recognized. Its importance by virtue of the vital role of its production equipment or business records to maintenance of production is not so often recognized.

A frequently overlooked reason for designating building importance is the building's effect upon leak and fire control activities by firefighters and other emergency handling groups. As discussed in other sections, the application of water from hoses is a fundamental fire fighting technique. A poorly sited building having little monetary, structural, or content value could be quite important as an impediment to these control activities.

The bottom line is that the importance or unimportance of a building should be determined by consultation of all parties likely to be involved in the handling of an incident. In most cases, the facility owner, the fire department, and the insurance company have an interest.

In 1957, the NFPA Technical Committee issued Formal Interpretation LPG-5 concerned with "line of adjoining property that may be built upon." This interpretation remains valid and is reproduced on page 166. Also see Figure 3.1 below.

Line of Adjoining Property: Recent questions on the application of "line of adjoining property that may be built upon" have centered on the application of local zoning ordinances. In many areas of the country local zoning laws prohibit building within a specified distance of a lot line. In some areas variances to these setbacks are easily obtained, while in other areas variances are very difficult, if not impossible, to obtain. This should be taken into account by the authority having jurisdiction when reviewing applications.

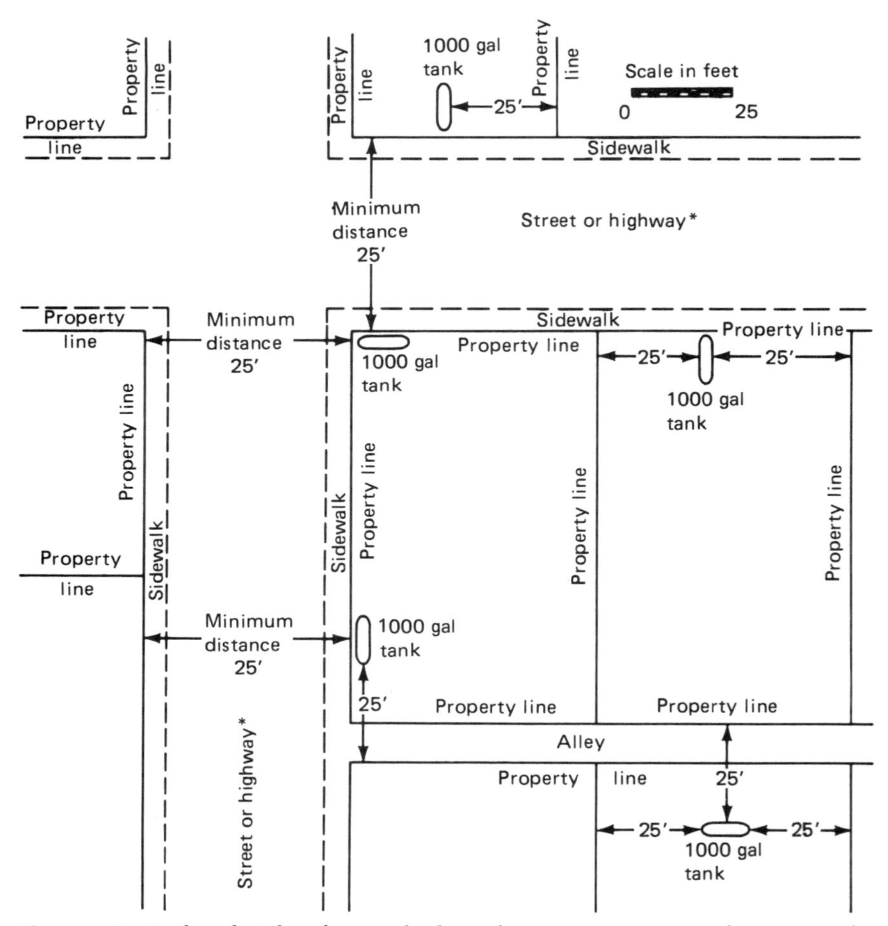

Figure 3.1 Railroad right-of-ways, bodies of water, or property that cannot be built upon should receive the same consideration as streets, highways, or public right-of-ways.

Table 3-2.2.2

Water Capacity Per Container Gallons (m³)	Minimum Distances		
	Mounded or Underground Containers [Note (d)]	Aboveground Containers [Note (f)]	Between Containers [Note (e)]
Less than 125 (0.5) [Note (a)]	10 ft (3 m)	None [Note (b)]	None
125 to 250 (0.5 to 1.0)	10 ft (3 m)	10 ft (3 m)	None
251 to 500 (1.0 + to 1.9)	10 ft (3 m)	10 ft (3 m)	3 ft (1 m)
501 to 2,000 (1.9 + to 7.6)	10 ft (3 m)	25 ft (7.6 m) [Note (c)]	3 ft (1 m)
2,001 to 30,000 (7.6 + to 114)	50 ft (15 m)	50 ft (15 m)	5 ft (1.5 m)
30,001 to 70,000 (114 + to 265)	50 ft (15 m)	75 ft (23 m)	
70,001 to 90,000 (265 + to 341)	50 ft (15 m)	100 ft (30 m)	(¹/₄ of sum of diameters of adjacent containers)
90,001 to 120,000 (341 + to 454)	50 ft (15 m)	125 ft (38 m)	
120,001 to 200,000 (454 to 757)		200 ft (61 m)	
200,001 to 1,000,000 (757 to 3 785)		300 ft (91 m)	
Over 1,000,000 (3,785)		400 ft (122 m)	

Notes to Table 3-2.2.2

Note (a): At a consumer site, if the aggregate water capacity of a multicontainer installation comprised of individual containers having a water capacity of less than 125 gal (0.5 m³) is 501 gal (1.9 + m³) or more, the minimum distance shall comply with the appropriate potion of this table, applying the aggregate capacity rather than the capacity per container. If more than one such installation is made, each installation shall be separated from any other installation by at least 25 ft (7.6 m). Do not apply the MINIMUM DISTANCES BETWEEN CONTAINERS to such installations.

Note (b): The following shall apply to aboveground containers installed alongside of buildings:

1. DOT specification containers shall be located and installed so that the discharge from the container pressure relief device is at least 3 ft (1 m) horizontally away from any building opening that is below the level of such discharge, and shall not be beneath any building unless this space is well ventilated to the outside and is not enclosed for more than 50 percent of its perimeter. The discharge from container pressure relief devices shall be located not less than 5 ft (1.5 m) in any direction away from any exterior source of ignition, openings into direct-vent (sealed combustion system) appliances, or mechanical ventilation air intakes.
2. ASME containers shall be located and installed so that the discharge from the container pressure relief device is at least 5 ft (1.5 m) horizontally away from any building opening that is below the level of such discharge, and not less than 10 ft (3 m) in any direction away from any exterior source of ignition, openings into direct-vent (sealed combustion system) appliances, or mechanical ventilation air intakes.
3. The filling connection and the vent from liquid level gauges on either DOT or ASME containers filled at the point of installation shall be not less than 10 ft (3 m) in any direction away from any exterior source of ignition, openings into direct-vent (sealed combustion system) appliances, or mechanical ventilation air intakes.

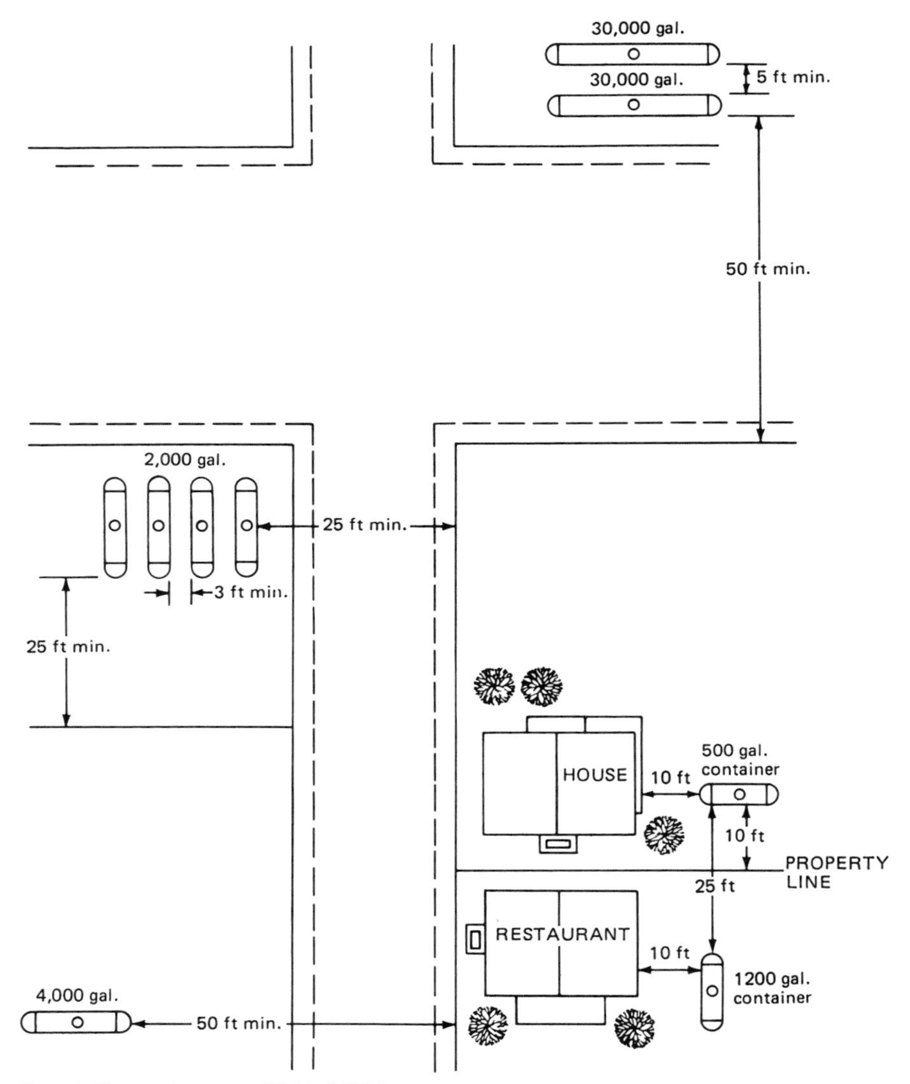

Note: 1. The requirements of Table 3-2.3.3 must also be considered.

2. Installations over 4,000 gal. aggregate must comply with Section 3-10.

***Figure** 3.2 Example of multiple container installations and various size containers. [Illustration of Table 3-2.2.2 Note (c).]*

Note (c): This distance may be reduced to not less than 10 ft (3 m) for a single container of1,200 gal (4.5 m^3) water capacity or less provided such container is at least 25 ft (7.6 m) from any other LP-Gas container of more than 125 gal (0.5 m^3) water capacity.

Note (d): Minimum distances for underground containers shall be measured from the pressure relief device and filling or liquid level gauge vent connection at the container, except that no part of an underground container shall be less than 10 ft (3 m) from a building or line of adjoining property that may be built upon.

Note (e): Where underground multicontainer installations are made of individual containers having a water capacity of 125 gal (0.5 m^3) or more, such containers shall be installed so as to permit access at their ends or sides to facilitate working with cranes or hoists.

Note (f): In applying the distance between buildings and ASME containers of 125 gal (0.5 m^3) or more water capacity, a minimum of 50 percent of this horizontal distance shall also apply to all portions of the building that project more than 5 ft (1.5 m) from the building wall and that are higher than the relief valve discharge outlet. This horizontal distance shall be measured from a point determined by projecting the outside edge of such overhanging structure vertically downward to grade or other level upon which the container is installed. Under no conditions shall distances to the building wall be less than those specified in Table 3-2.2.2.

Exception to Note (f): Not applicable to installations in which overhanging structure is 50 ft (15 m) or more above the relief valve discharge outlet.

Figure 3.3 *Fifty 100-lb (45-kg) LP-Gas capacity [about 29 gal (0.11 m^3) water capacity each] DOT specification containers in two manifolds installed alongside a multistory office building. The installation is equivalent to about a single 1,200-gal (4.5-m^3) container, which would require a 10-ft (3-m) distance from the building per Note (c) to Table 3-2.2.2. (Photo courtesy of Wilbur Walls.)*

Notes to Table 3-2.2.2: In spite of editorial amendments to clarify, Note (a) to Table 3-2.2.2 still is subject to misinterpretation in the field. This provision was added to the standard as Note 1 to Table 1 in the 1969 edition.

Over its entire history, the standard has permitted containers of less than 125 gal (0.5 m^3) water capacity to be installed alongside a building [but some distance away from building openings per Note (b) to Table 3-2.2.2] with no limit to the number of such containers. The presumption was that this number would not exceed 10 or 12 containers because normal industry practice would result in a large single container being used under such service conditions. While this is generally the case today, in the mid-1960s the Technical Committee became aware of several installations of a large number of manifolded containers to take advantage of limited space. An example of one such installation is shown in Figure 3.3.

Note (a) to Table 3-2.2.2 corrects such abuses by placing an aggregate capacity on such multicontainer installations. However, Note (a) is applicable only where the water capacity of each individual container is less than 125 gal (0.5 m^3). It is not applicable where larger containers are involved.

Formal Interpretation
Reference: 3-2.2.2

The following interpretation of the *Standard for the Storage and Handling of Liquefied Petroleum Gases*, NFPA Standard No. 58, has been released by the Interpretation Subcommittee of the NFPA Committee on Gases following concurrence by the entire Committee on Gases.

Question: Does the quoted language in Section B.6(b) of NFPA Standard No. 58, 1957 Edition, when referring to the location of a domestic tank, mean the lot line of the property on which the tank is located or the center line of the street or alley, or is it intended that this language indicate the lot line of an adjoining property that may be built upon?

Answer: It is the Committee's opinion that the "line of adjoining property which may be built upon" refers to the property boundaries of the plot adjacent to the one upon which the tank is located. This is illustrated in Figure 3.1 on page 162, taking into consideration a condition that involves property on the other side of a street, highway, or other right of way. It is the Committee's opinion that the minimum distance limitation is from the tank to the property line where that property line is common to plots of ground of different ownership and would also apply between the tank and the property line on the far side of a street or other public right of way.

Issue Edition: 1957 ■

Note (b) to Table 3-2.2.2 is applicable wherever containers are installed alongside a building [a DOT specification container cannot exceed 1,000 lb (454 kg) water capacity—about 120 gal (0.45 m^3) of water—per DOT regulations]. The criteria are intended to minimize the chances of LP-Gas entering a building, fire exposure to a container during the early stages of a fire originating inside the building, and ignition by some ignition sources commonly found in the immediate vicinity of such installations.

Parts (1) and (2) of Note (b) are concerned with discharge from container pressure relief devices—primarily as a result of overfilling. Both were editorially modified in the 1992 edition for clarity. In both instances noted, the relief device discharge is normally on the container or its shutoff valve (not piped away). A lesser distance [3 ft (1 m)] is permitted for all DOT containers than is permitted for an ASME container [5 ft (1.5 m)] because the start-to-discharge setting for a DOT container is higher than that for an ASME container [nominally 375 psi (2.6 MPa) and 250 psi (1.7 MPa), respectively], resulting in less opportunity for discharge from a DOT container.

For this same reason, the relief device discharge from a DOT container can be beneath certain buildings but the discharge from an ASME container cannot. (These certain buildings, as represented by the "not enclosed for more than 50 percent of their perimeter" qualification, are most often found constructed on the sides of hills or on waterfronts.)

The "building openings" cited are normally doors and windows, which can be either closed or open at any particular time and which, when open, can have airflow through them in either direction. Where the airflow is normally into the building—e.g., a mechanical ventilation air intake—a 5-ft (1.5-m) distance is specified, which could affect the installation of a DOT container in particular.

While direct vent appliances do not represent a pathway for LP-Gas into the building interior, they draw in outside air for combustion. If this air contains LP-Gas in ignitable proportions, the appliance ignitor or burner flame can constitute an ignition source.

Mechanical ventilation system air intakes are an obvious pathway for LP-Gas into a building interior. A mechanical ventilation system air outlet, e.g., a kitchen exhaust fan, is considered a building opening like a door or window.

Part (3) of Note (b) requires a 10-ft (3-m) [rather than 5-ft (1.5-m)] separation because it reflects conditions where LP-Gas vapor and liquid are released intentionally and reasonably often as a result of normal operations. This distance was derived from experience and tests.

The results of one such test are shown in Figure 3.4. Discharge of liquid from such a gauge for as long as 3 minutes is abnormal (normally the gauge discharge valve is closed the instant liquid appears). A lack of wind is conducive to the formation of high concentrations of LP-Gas and calm conditions are seldom found. In any event, in spite of the severity of this test, the concentration of LP-Gas at 10 ft (3 m) did not exceed 20 percent (one-fifth of the lower flammable limit)—or a safety factor of 5.

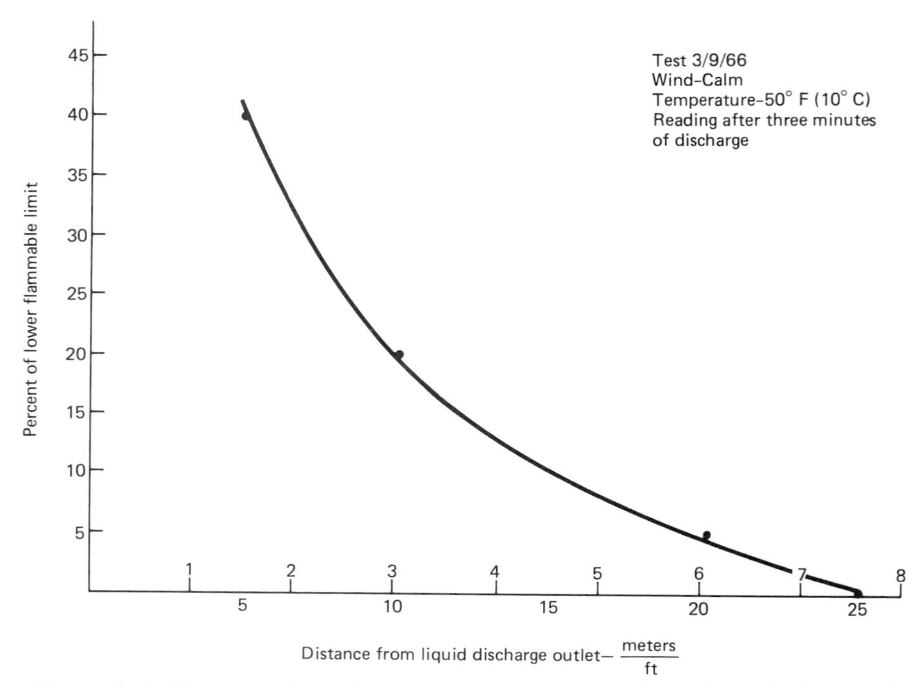

Figure 3.4 Concentration of propane-air mixture resulting from discharge of liquid propane from fixed liquid level gauge on DOT/ICC/CTC cylinder. (Photo courtesy of Wilbur Walls.)

Work on dispersion of flammable gases from relief valves has been conducted by Batelle Laboratories and is reported in the paper "The Effects of Velocity, Temperature, and Molecular Weight on Flammability Limits in Wind-Blown Jets of Hydrocarbon Gases," by Vernon O. Hoenhe and Ross G. Luce. This forms the basis for coverage in API Recommended Practice 521, *Guide for Pressure-Relieving Systems.* These references should be reviewed by those interested in more information in this area.

Note (c) of Table 3-2.2.2 relaxes the 25-ft (7.6-m) requirement applicable to a 1,200 gal (4.5 m^3) or less water capacity container where only one such container is installed. This is a concession to the limited space often found in commercial areas and has not resulted in known serious fires in many years in the Committee's experience. (*See Figure 3.2 for an illustration of this.*)

The lesser distances for underground or mounded containers and the criteria in Note (d) to Table 3-2.2.2 reflect the fact that these containers are not subject to BLEVEs from fire exposure.

Note (f) reflects the exposure to larger overhanging eaves of buildings becoming more common in consideration of aesthetics and energy conservation. A torching container pressure relief device could lead to rapid penetration of fire into the structure.

In the 1989 edition, Table 3-2.2.2 was modified by extending coverage to containers having a capacity of more than 120,000 gal (454 m^3). This was done in conjunction with the change in the scope of the standard to include refrigerated storage, marine and pipeline terminals, and portions of natural gas processing plants, refineries, and petrochemical plants in certain cases (*see 1-2.3.1*).

3-2.2.3 Where storage containers having an aggregate water capacity of more than 4,000 gal (15.1 m^3) are located in heavily populated or congested areas, the siting provisions of 3-2.2.2 and Table 3-2.2.2 shall be permitted to be modified as indicated by the fire safety analysis described in 3-10.2.3.

Historically, the character of LP-Gas applications has been such that they have been concentrated in rural or suburban areas, or in commercial or industrial areas where the hazards are commensurate with other operations and exposure to the public has been limited accordingly. Inevitably, however, exceptions occur, and the versatility of LP-Gas has increased the number of them.

In recent years, for example, the popularity of LP-Gas powered vehicles has resulted in an increase in the number of refueling stations in heavily populated or congested areas. In addition to the use of the containers themselves, filling them means that LP-Gas transports are often located in such areas or may travel through such areas.

Paragraph 3-2.2.3 acknowledges that such circumstances may warrant further attention to the general siting criteria in 3-2.2.2. While the implication is that the distances should be increased in such locations, this is not necessarily the case (in fact, this may be impossible in a heavily populated or congested area). What 3-2.2.3 does require is a firesafety analysis to be conducted in such areas. The character of these areas is such that emergency handling measures are usually more readily available than in more remote areas. A fire safety analysis is much more likely to result in distances no greater than those in Table 3-2.2.2 (or even less), and in leak and fire control provisions exceeding the minimums prescribed, e.g., special protection. In extreme cases, such an analysis may result in a decision that a proposed site is simply unacceptable. (For additional information, see the supplement to this handbook on conducting a firesafety analysis.)

The expression "heavily populated or congested area" is not defined. In most instances it is subject to local determination, and often zoning criteria come into play.

3-2.2.4 Aboveground multicontainer installations comprised of containers having an individual water capacity of 12,000 gal (45 m^3) or more installed for use in a single location shall be limited to the number of containers in one group, and with each group separated from the next group in accordance with the degree of fire protection provided in Table 3-2.2.4.

Table 3-2.2.4

Fire Protection Provided by	Maximum Number of Containers in One Group	Minimum Separation Between Groups—Feet
Hose streams only—see 3-10.2.3	6	50 (15 m)
Fixed monitor nozzles per 3-10.3.5*	6	25 (7.6 m)
Fixed water spray per 3-10.3.4*	9	25 (7.6 m)
Insulation per 3-10.3.1	9	25 (7.6 m)

*In the design of fixed water spray and fixed monitor nozzle systems, the area of container surface to be protected may reflect portion of containers not likely to be subject to fire exposure as determined by good fire protection engineering practices.

This table was added to the standard as a result of installations of greater numbers of larger containers increasing and containers being added to existing installations without adequate consideration of the fire control problems introduced.

3-2.2.5 Underground or mounded containers shall be located outside of any buildings. Buildings shall not be constructed over any underground or mounded containers. Sides of adjacent containers shall be separated in accordance with Table 3-2.2.2 but not less than 3 ft (1 m).

Where containers are installed parallel with ends in line, any number of containers are permitted to be in one group. Where more than one row is installed, the adjacent ends of the tanks in each row shall be separated by not less than 10 ft (3.0 m).

This new requirement was added in the 1995 edition. It recognizes the number of underground tanks in a group—the spacing of such tanks was not previously covered in NFPA 58. The text was based on similar provisions in NFPA 59, *Standard for the Storage and Handling of Liquefied Petroleum Gases at Utility Gas Plants*

3-2.2.6 In the case of buildings of other than wood-frame construction devoted exclusively to gas manufacturing and distribution operations, the distances specified in Table 3-2.2.2 shall be permitted to be reduced provided that containers having a water capacity exceeding 500 gal (1.9 m^3) shall not be located closer than 10 ft (3.0 m) to such gas manufacturing and distributing buildings.

3-2.2.7 The following provisions shall also apply:

The ten items in this provision are siting or operational controls designed primarily to protect the integrity of an installed LP-Gas container. However, some items are applicable to noninstalled containers and may be more appropriate in other parts of the standard.

(a) Containers shall not be stacked one above the other.

This requirement is applicable to both installed and noninstalled containers. Its purpose is to avoid "bonfire" type incidents and, in the case of manifolded installed containers, opportunities for overfilling the lower containers as a result of migration of liquid under the influence of gravity.

Several users of the standard have asked if this paragraph applies to an aboveground container installed over an underground container. While a literal reading of the words would lead the reader to believe that such an installation is prohibited, an understanding of the intent of the requirement usually leaves one in doubt. The intent is to prevent interaction of containers in a fire situation, but it is hard to imagine a scenario where: (a) fire attacks an underground container, or (b) fire of an aboveground container affects an underground container installed beneath it.

The only concern in such an installation would be the location of the relief valve discharge of the underground container (even though operation of a relief valve on an underground container is highly unlikely, except if the container is overfilled). Depending on the installation, it may be prudent to pipe the relief valve discharge of the underground container to a point where it is higher than the top of, or cannot affect an aboveground container installed above. In such cases, care must be taken to ensure that the relief valve discharge piping cannot affect the operation of the relief valve.

(b) Loose or piled combustible material and weeds and long dry grass shall not be permitted within 10 ft (3.0 m) of any container.

This requirement is applicable to both installed and noninstalled containers. It's intent is to prevent a possible grass or brush fire from affecting an LP-Gas container.

(c) Means shall be used to prevent the accumulation or flow of liquids having flash points below 200°F (93.4°C) under adjacent LP-Gas containers such as by dikes, diversion curbs, or grading.

Flash point is a measure of the flammability of liquids. It is a test in which a sample of the liquid is heated until its vapors can be ignited. Gases, such as propane and butane, do not have a flash point as they cannot be contained in an open cup at room temperature. Liquids with flash points below 100°F (37.8°C) are classified as Class I flammable liquids by NFPA 30, *Flammable Liquids Code.* Examples of Class I flammable liquids are gasoline and many alcohols. Liquids with a flash point between 100°F (37.8°C) and 140°F (60.0°C) are classified as Class II combustible liquids. Examples of Class II flammable liquids include diesel fuel and kerosene.

Flash point must not be confused with boiling point. While propane has a boiling point of –44°F (–42°C), it does not have a flash point, and the rules for storing flammable liquids in NFPA 30 do not apply to it.

NOTE: For information on determination of flash points see NFPA 321, *Standard on Basic Classification of Flammable and Combustible Liquids.*

(d) LP-Gas containers shall be located at least 10 ft (3.0 m) from the centerline of the wall of diked areas containing flammable or combustible liquids.

(e) The minimum horizontal separation between aboveground LP-Gas containers and aboveground tanks containing liquids having flash points below 200°F (93.4°C) shall be 20 ft (6 m). No horizontal separation shall be required between aboveground LP-Gas containers and underground tanks containing flammable or combustible liquids installed in accordance with NFPA 30, *Flammable and Combustible Liquids Code.*

Subparagraphs (c), (d), and (e) are applicable to both installed and non-installed containers. The provisions of (c) and (d) are aimed at keeping leaking flammable or combustible liquids from accumulating under or around LP-Gas containers. Subparagraph 3-2.2.7(e) reflects the mutual exposure between tanks or containers themselves, and was derived through a cooperative effort between the NFPA Technical Committees responsible for NFPA 30, *Flammable and Combustible Liquids Code*, and NFPA 58. The separation distances are intended to facilitate cooling and fire extinguishing activities by fire departments. The exception reflects a fairly common farm or residential situation. It is a concession to limited space available, and experience has shown it not to be a problem. The paragraph was revised editorially for clarity in the 1992 edition.

Exception: This provision shall not apply where LP-Gas containers of 125 gal (0.5 m^3) or less water capacity are installed adjacent to fuel oil supply tanks of 660 gal (2.5 m^3) or less capacity.

(f) The minimum separation between LP-Gas containers and oxygen or gaseous hydrogen containers shall be in accordance with Table 3-2.2.7(f).

This requirement and its exception are derived from NFPA standards covering oxygen and gaseous hydrogen container installations. It is included in NFPA 58, along with Table 3-2.2.7(f), as a convenience. The caution stated in the Note to 3-2.2.7(f) is not in the other standards because the problems cited are not pertinent to oxygen or hydrogen.

Exception: Shorter distances shall be permitted where protective structures having a minimum fire resistance rating of 2 hr interrupt the line of sight between uninsulated portions of the oxygen or hydrogen containers and the LP-Gas containers. The location and arrangement of such structures shall minimize the problems cited in the Note to 3-2.2.9.

NOTE: Also, see NFPA 50, *Standard for Bulk Oxygen Systems at Consumer Sites*, and NFPA 51, *Standard for the Design and Installation of Oxygen-Fuel Gas Systems for Welding, Cutting, and Allied Processes*, for oxygen systems and NFPA 50A, *Standard for Gaseous Hydrogen Systems at Consumer Sites*, on gaseous hydrogen systems.

Table 3-2.2.7(f)

LP-Gas Containers Having an Aggregate water capacity of	Separation from Oxygen Containers Having an			Separation from Gaseous Hydrogen Containers Having an		
	Aggregate capacity of 400 CF (11 m^3)* or less	Aggregate capacity of more than 400 CF (11 m^3)* to 20,000 CF (566 m^3),* including unconnected reserves	Aggregate capacity of more than 20,000 CF (566 m^3),* including unconnected reserves	Aggregate capacity of less than 400 CF (11 m^3)*	Aggregate capacity of 400 CF (11 m^3)* to 3000 CF (85 m^3)*	Aggregate capacity of more than 3000 CF (85 m^3)*
1200 gal (4.5 m^3) or less	None	20 ft (6 m)	25 ft (7.6 m)			
Over 1200 gal (4.5 m^3)	None	20 ft (6 m)	50 ft (15 m)			
500 gal (1.9 m^3) or less				None	10 ft (3 m)	25 ft (7.6 m)
Over 500 gal (1.9 m^3)				None	25 ft (7.6 m)	50 ft (15 m)

*Cubic feet (m^3) measured at 70°F (21°C) and atmospheric pressure.

(g) The minimum separation between LP-Gas containers and liquefied hydrogen containers shall be in accordance with NFPA 50B, *Standard for Liquefied Hydrogen Systems at Consumer Sites.*

This requirement refers to the NFPA standard covering liquefied hydrogen storage, NFPA 50B. It is separate from the previous requirement as liquid hydrogen installations are less frequently encountered than gaseous hydrogen installations in conjunction with LP-Gas. Rather than incorporate requirements that would be infrequently used, the reference to NFPA 50B is made.

(h) Where necessary to prevent flotation due to possible high flood waters around aboveground or mounded containers, or high water table for those underground and partially underground, containers shall be securely anchored.

This requirement is applicable to installed containers, as anchorage would be impractical for noninstalled containers. In practice, anchorage has been applied only to ASME containers; the smaller DOT containers are not anchored because of their size. It is possible DOT cylinders being torn from their connections could be carried away in a flood. Even though their valves may not have been closed in anticipation of the flood, to the editor's knowledge, DOT containers have not contributed significantly to the overall flood damage.

Anchorage usually consists of concrete pads or foundations as the basic element and utilizes basic civil engineering criteria.

(i) Where LP-Gas containers are to be stored or used in the same area with other compressed gases, the containers shall be marked to identify their content in accordance with ANSI/CGA C-4, *Method of Marking Portable Compressed Gas Containers to Identify the Material Contained.*

This requirement is applicable to both installed and noninstalled containers. It is noted that the identification is to be made by marking with the name of the gas. Identification by color or other means is acceptable only if it is in addition to the name of the gas.

Steel LP-Gas cylinders and ASME containers are customarily painted for corrosion protection. While NFPA 58 does not stipulate paint colors, the color affects the rate of heat absorption from solar radiation and the consequent pressure in the container. This can result in operation of a pressure relief device in very warm climates.

Tests have shown a white paint with a titanium oxide pigment to reflect 90 to 95 percent of the solar light. Yellow (medium yellow chrome pigment) reflects about 80 percent. Aluminum reflects about 70 percent. Other colors reflect less than 15 percent and black reflects none.

(j) No part of an aboveground LP-Gas container shall be located in the area 6 ft (1.8 m) horizontally from a vertical plane beneath overhead electric power lines that are over 600 volts, nominal.

This reflects both the possibility of container failure through arc penetration and the hazard to firefighters applying water, and was added in the 1986 edition.

3-2.2.8 Impoundment in accordance with Section 9-2 shall be installed around refrigerated LP-Gas containers and those containing butane and its isomers.

NOTE: Because of the anticipated "flash" of nonrefrigerated LP-Gas when it is released to the atmosphere, dikes normally serve no useful purpose for nonrefrigerated installations.

This provision was revised in both the 1989 and 1992 editions with the addition of LP-Gas refrigerated storage to the standard, the relocation of refrigerated storage to Chapter 9, and the recognition that the previous text applied only to propane and not butane. The need to dike refrigerated LP-Gas storage containers has long been recognized because refrigerated liquid will not flash to vapor as will LP-Gas released from pressurized storage. A spill from a refrigerated container can puddle or run as a liquid on the ground. Butane, with a normal boiling point of about 31°F (–0.56°C), will not vaporize when spilled if the ambient temperature is below 31°F (–0.56°C). As this is a temperature that can be expected at some time of the year throughout the United States, butane is treated as a flammable liquid for diking purposes.

Liquid propane vaporizes rapidly when it escapes from a pressurized container and the ambient temperature is above its normal boiling point of about –44°F (–42). Much of the heat necessary for vaporization is contained in the liquid itself, and the remainder needed is readily available from the air, ground, or other material contacted by the liquid. Therefore, the establishment of a pool or flowing stream of LP-Gas over the ground or water is unlikely, and a dike has no function to perform.

3-2.2.9 Structures such as fire walls, fences, earth or concrete barriers, and other similar structures shall be avoided around or over installed nonrefrigerated containers.

Exception No. 1: Such structures partially enclosing containers shall be permitted if designed in accordance with a sound fire protection analysis.

Exception No. 2: Structures used to prevent flammable or combustible liquid accumulation or flow shall be permitted in accordance with 3-2.2.7(c).

Exception No. 3: Structures between LP-Gas containers and gaseous hydrogen containers shall be permitted in accordance with 3-2.2.7(f).

Exception No. 4: Fences shall be permitted in accordance with 3-3.6.

NOTE: The presence of such structures can create significant hazards, e.g., pocketing of escaping gas, interference with application of cooling water by fire departments, redirection of flames against containers, and impeding egress of personnel in an emergency.

This provision was inspired by a serious BLEVE of an aboveground propane container that had been enclosed in a roofed-over enclosure for aesthetic reasons. The enclosure not only contributed to ignition but made it difficult for the fire department to apply cooling water to the container. However, the Technical Committee was also aware of an increasing use of such methods to hide LP-Gas containers or to limit the travel of container pieces in the event of a BLEVE. Exception No. 1 recognizes that the problems associated with such structures can be prevented by design, which eliminates the problems cited in the note.

If a structure is desired to hide an LP-Gas tank, it is important to use materials that allow air to circulate freely. Examples of such materials are chain-link fence or materials that have significant openings on all sides. Wood can be used, but its flammability must be considered. Wood cannot be stacked around an LP-Gas container per 3-2.2.7(b), but its use in a structure is not prohibited. A light fence constructed of wood, if ignited, would probably be consumed before generating enough heat to affect an LP-Gas container.

3-2.2.10 ASME tanks filled on site, of 125 gal (0.5 m^3) water capacity or more, shall be located so that the filling connection and fixed liquid level gauge are at least 10 ft (3.0 m) from any external source of ignition (i.e., open flame, window A/C, compressor, etc.)

This provision was added to the 1995 edition to provide guidance on location of ASME containers over 125 gal with respect to sources of ignition. It was noted that spacing of ASME containers of less than 125 gal from sources of ignition are covered in Note (b)(2) to Table 3-2.2.2, but that no minimum was specifically provided in this section. While this requirement overlaps the provision of Table 3-7.2.2, that table may not be consulted for installations of containers of 125 to 2,000 gal, in all cases, as it is not to be applied to residential and commercial installations per 3-7.2.2. (It is impractical, if not impossible, to enforce ignition source control at a residential occupancy).

3-2.3 Location of Transfer Operations.

This section was added in the 1992 edition following a proposal submitted by the National Propane Gas Association that was developed by the Technology and Standards Committee of NPGA at the request of the NFPA LP-Gas Committee. The LP-Gas Committee had received several requests for clarification and interpretation of the application of NFPA 58 to propane vehicle filling stations, especially when combined with conventional gas stations. This section and several other sections were changed or relocated to recognize the differences between LP-Gas and gasoline in vehicle fueling and to provide a set of regulations that would be consistent with or could be adopted by NFPA 30A, *Automotive and Marine Service Station Code.*

3-2.3.1 Liquid shall be transferred into containers, including containers mounted on vehicles, only outdoors or in structures specially designed for the purpose.

(a) The transfer of liquid into containers mounted on vehicles shall not take place within a building but shall be permitted to take place under a weather shelter or canopy (*see 3-9.3.2*).

The reference to 3-9.3.2 guides the reader to more information on what the limits of a weather shelter are. A minimum of 50 percent of the perimeter must be open to the atmosphere.

(b) Structures housing transfer operations or converted for such use after December 31, 1972, shall comply with Chapter 7.

(c) The transfer of liquid into containers on the roofs of structures shall be prohibited.

This prohibition prevents the permanent installation of containers on the roofs of buildings. This option is not needed in the United States or Canada as densely populated cities do not use LP-Gas as a residential fuel. In other parts of the world it is a common practice to locate containers on the roofs of buildings. If an LP-Gas container must be located on a roof, it must be refilled by exchange with a full cylinder.

(d) The transfer hose shall not be routed in or through any building except those specified in 3-2.3.1(b).

Transfer hose contains liquid LP-Gas and leakage caused by failure of a hose inside a building could be catastrophic, hence this prohibition. In some areas where row houses are common and containers are located in back yards it results in a long hose for deliveries.

3-2.3.2 Containers located outdoors in stationary installations (*see definition*) in accordance with 3-2.2 and with the point of transfer located at the container shall be permitted to be filled at that location. If the point of transfer (*see definition*) is not located at the container, it shall be located in accordance with 3-2.3.3.

Containers fitted with filling connections that are actually container appurtenances—or that are very close to the container—are smaller containers, and lower transfer rates are involved. Under these circumstances, the siting criteria for the container in Table 3-2.2.2 are considered adequate for the transfer operation.

Once a container is installed the filler does not usually check the installation requirements each time the container is refilled. While this is understandable, certain requirements that often change must be checked. Specifically, Table 3-2.2.2, in Note (b)(3), requires that a container filling connection and vent be 10 ft (3 m) from any exterior source of ignition. A commonly found source of ignition is window-mounted air conditioners, which are frequently installed after a container is installed and should be checked for at each filling during warm seasons. If one is found within 10 ft (3 m) of a container, either the container or the air conditioner must be moved prior to filling and continued operation.

3-2.3.3 Containers not located in stationary installations (*see definition*) shall be filled at a location determined by the point of transfer (*see definition*) in accordance with Table 3-2.3.3.

The table applies to the location of the filling of portable containers (those not at a stationary installation). It also applies to containers located outdoors in stationary installation (3-2.3.2), where the filling connection is not located at the container, but is piped away from the container.

In the 1995 edition, Part 1 was revised to add mobile homes, recreational vehicles, and modular homes to buildings. This recognizes that the added items present a fire hazard similar to buildings. Part 10 identifies a safe spacing from points of LP-Gas transfer to flammable liquid dispensers and flammable liquids containers installed above- and belowground. This distance is consistent with the requirements of NFPA 30A, *Service Station Code*. It is considered by many in the LP-Gas distribution industry to be restrictive and the dispensing of LP-Gas as a motor fuel at gasoline service stations is discouraged. Representatives of the Technical Committee have met with the Service Station Code Committee and the requirements of NFPA 30A are under review.

(a) If the point of transfer is a component of a system covered by Section 3-8 or Chapter 8 or part of a system installed in accordance with standards referenced in 3-1.3, parts 1, 2, and 3 of Table 3-2.3.3 shall not apply to the structure containing the point of transfer.

(b) If LP-Gas is vented to the atmosphere under the conditions stipulated in 4-3.1, Exception No. 4, the distances in Table 3-2.3.3 shall be doubled.

(c) If the point of transfer is housed in a structure complying with Chapter 7, the distances in Table 3-2.3.3 shall be permitted to be reduced provided either the exposing wall(s) or the exposed wall(s) complies with 7-3.1(a).

3-2.4 Installation of Containers.

This section provides general provisions common to the installation of all containers. More specific details for the installation of horizontal aboveground, vertical, portable storage, vehicular tank, mounded, partially underground unmounded, and underground containers are given in subsequent sections.

Table 3-2.3.3 Distance Between Point of Transfer and Exposures

Part	Exposure	Min. Horizontal Distance, Feet (Meters)
1.	Buildings,[1] mobile homes, recreational vehicles, and modular homes with fire-resistive walls.[2]	10 (3.1)
2.	Buildings[1] with other than fire-resistive walls.[2]	25 (7.6)
3.	Building wall openings or pits at or below the level of the point of transfer.	25 (7.6)
4.	Line of adjoining property that can be built upon.	25 (7.6)
5.	Outdoor places of public assembly, including school yards, athletic fields, and playgrounds.	50 (15)
6.	Public ways, including public streets, highways, thoroughfares, and sidewalks.	
	(a) From points of transfer in LP-Gas dispensing stations and at vehicle fuel dispensers.	10 (3.1)
	(b) From other points of transfer.	25 (7.6)
7.	Driveways.[3]	5 (1.5)
8.	Mainline railroad track centerlines.	25 (7.6)
9.	Containers[4] other than those being filled.	10 (3.1)
10.	Flammable and Class II combustible liquid[5] dispensers and aboveground and underground containers.	20 (6.1)

[1] Buildings, for the purpose of the table, also include structures such as tents and box trailers at construction sites.

[2] Walls constructed of noncombustible materials having, as erected, a fire resistance rating of at least one hour as determined by NFPA 251, *Standard Methods of Fire Tests of Building Construction and Materials.*

[3] Not applicable to driveways and points of transfer at vehicle fuel dispensers.

[4] Not applicable to filling connections at the storage container or to dispensing vehicle fuel dispensers units 2,000 gal (7.6 m^3) water capacity or less when used for filling containers not mounted upon vehicles.

[5] NFPA 30 defines these as follows:

Flammable liquids include those having a flash point below 100°F (37.8°C) and having a vapor pressure not exceeding 40 lb psia (an absolute pressure of 2.068 mm Hg) at 100°F (37.8°C).

Class II combustible liquids include those having a flash point at or above 100°F (37.8°C) and below 140°F (60°C).

3-2.4.1 Containers shall be installed in accordance with the following:

(a) DOT cylinder specification containers shall be installed only aboveground, and shall be set upon a firm foundation, or otherwise firmly secured. Flexibility shall be provided in the connecting piping. (*See 3-2.8.6 and 3-2.8.9.*)

DOT specification containers are designed primarily for transportation purposes but with the understanding that they will be used as fuel supply storage. They are not designed for belowground installation. In an earlier edition of NFPA 58, an explanation was given that they could be installed in a niche in a slope or terrace wall as long as the container and regulator did not contact the ground and the compartment or recess was ventilated and drained. Flexibility in the connecting piping is generally obtained by the use of a pigtail (short length of copper tubing with POL[1] connectors).

(b) All containers shall be positioned so that the pressure relief valve is in direct communication with the vapor space of the container.

Positioning the container so that the relief valve is in communication with the vapor space ensures a minimal release of LP-Gas should abnormal conditions exist. Tilting or laying a cylinder on its side so that liquid LP-Gas can be withdrawn is not to be permitted. There are liquid withdrawal fittings available that operate without doing this and allow the proper position of the relief valve to be retained.

(c) Where physical damage to LP-Gas containers, or systems of which they are a part, from vehicles is a possibility, precautions shall be taken against such damage.

This is intentionally written as a performance provision, rather than providing specific guidelines for when protection is needed or the nature of the protection. The Technical Committee believes that it can neither anticipate all the ways a vehicle may potentially threaten a container, nor specify the types of protection required. These are left to the user of the standard or the authority having jurisdiction. Many users of the standard would prefer to have the requirement stated in a "how-to" format, providing specific structures required for different traffic situations. This is not possible, and the user must anticipate the possibility of reasonable physical damage.

Questions about this provision usually are the result of an incident and regard the intent of the Committee. For example, an incident occurred in which a tractor trailer, transporting 63 head of cattle, veered off a road and traveled almost 500 ft (152 m) before striking the piping associated with an 18,000-gal (68-m^3) and a 30,000-gal (114-m^3) LP-Gas container installed about 110 ft (34 m) from the edge of the road. Escaping LP-Gas ignited, and the torch flame impinged on the 18,000-gal (68-m^3) container, resulting in a BLEVE after about 1/2 hour.[2] In this case the installer could not have been

[1]The letters "POL" stand for "Prest-o-Lite," a trademark of a pioneer LP-Gas operator (Pyrofax-Union Carbide). Because these fittings have a left-hand thread, they have also come to stand for "Put-on-Left."

[2]Isner, MS. "BLEVE—LP-Gas Storage Tank." Woodruff, Utah, October 16, 1986. *Fire Investigation Report.* National Fire Protection Association, Quincy, MA.

reasonably expected to protect containers located a reasonable distance from a road from impact by a stray tractor trailer. Other incidents have occurred on construction sites where LP-Gas containers have been installed close to roadways or driveways used by delivery trucks and construction vehicles. In these cases the installer would be expected to provide protection for the anticipated vehicles that pass close to the container.

(d) The installation position of ASME containers shall make all container appurtenances accessible for their normally intended use.

Occasionally a storage container must be evacuated, before it is moved or for other reasons. There are fittings for this evacuation to eliminate the need to roll a container on its side to pump it out. (*See commentary on 2-2.3.3.*) Also, there are many other installation situations in which the container appurtenances may not be accessible unless attention is given to the container position before installation.

(e) Field welding on containers shall be limited to attachments to nonpressure parts, such as saddle plates, wear plates, or brackets applied by the container manufacturer. Welding to container proper shall comply with 2-2.1.7.

(f)* Aboveground containers shall be kept properly painted.

A-3-2.4.1(f) **Aboveground Storage Tank Paint Color.** Generally, a light reflecting color paint is preferred unless the system is installed in an extremely cold climate.

See commentary on 3-2.2.6(i) for further information on paint colors.

3-2.4.2 Horizontal ASME containers designed for permanent installation in stationary service aboveground shall be placed on substantial masonry or noncombustible structural supports on concrete or firm masonry foundations, and shall be supported as follows:

(a) Horizontal containers shall be mounted on saddles in such a manner as to allow expansion and contraction, and to prevent an excessive concentration of stresses. Structural steel supports shall be permitted to be used if in compliance with 3-2.4.2(b) or as follows:

Exception No. 1: Temporary use as provided in 3-2.4.2(a)2.b.

Exception No. 2: Isolated locations as provided in 3-2.4.2(b).

1. Containers of more than 2,000 gal (7.6 m^3) water capacity shall be provided with concrete or masonry foundations formed to fit the container contour, or if furnished with saddles in compliance with 2-2.5.1, shall be permitted to be placed on flat-topped foundations.

2. Containers of 2,000 gal (7.6 m^3) water capacity or less shall be permitted to be installed on concrete or masonry foundations formed to fit the container contour, or if equipped with attached supports complying with 2-2.5.2(a), shall be permitted to be installed as follows:

a. If the bottoms of the horizontal members of the container saddles, runners or skids are to be more than 12 in. (300 mm) above grade, fire-resistive foundations shall be provided. A container shall not be mounted with the outside bottom of the container shell more than 5 ft (1.5 m) above the surface of the ground.

b. For temporary use of no more than 6 months at a given location, fire-resistive foundations or saddles shall not be required provided that the outside bottom of the container shell is not more than 5 ft (1.5 m) above the ground and that flexibility in the connecting piping is provided. (*See 2-4.6.3.*)

3. Containers or container-pump assemblies mounted on a common base complying with 2-2.5.2(b) shall be permitted to be placed on paved surfaces or on concrete pads at ground level within 4 in. (102 mm) of ground level.

(b) Single containers complying with 2-2.5.1 or 2-2.5.2 shall be permitted to be installed in isolated locations, with nonfireproofed steel supports resting on concrete pads or footings, provided the outside bottom of the container shell is not more than 5 ft (1.5 m) above the ground level, with the approval of the authority having jurisdiction.

(c) Means of preventing corrosion shall be provided on that part of the container in contact with the saddles or foundations or on that part of the container in contact with masonry.

The primary considerations in the installation of horizontal ASME containers are to ensure that the integrity of the supports is retained and that damage under fire conditions is minimized. For large containers this is usually accomplished through masonry foundations and supports. This method may also be used for smaller containers [2,000 gal (7.6 m^3) water capacity or less] or, as an alternate, when nonfireproofed structural supports are provided, they may be placed on fire-resistive foundations as long as certain heights from the container bottom to the ground are maintained. These minimum heights, together with a provision that flexibility be provided in the piping, ensure that the container will not settle to an unsafe level.

Two exceptions are given for these smaller containers. In a temporary installation, as described in 3-2.4.2(a)(2)(b), a 5-ft (1.5-m) height is permitted and fire-resistive foundations and saddles are not required. The other exception is described in 3-2.4.2(b) for a permanent installation in an isolated location with the approval of the authority having jurisdiction. These higher elevation tanks are generally used to provide additional pump suction head when high flow rates are required, and are sometimes used for filling containers by gravity.

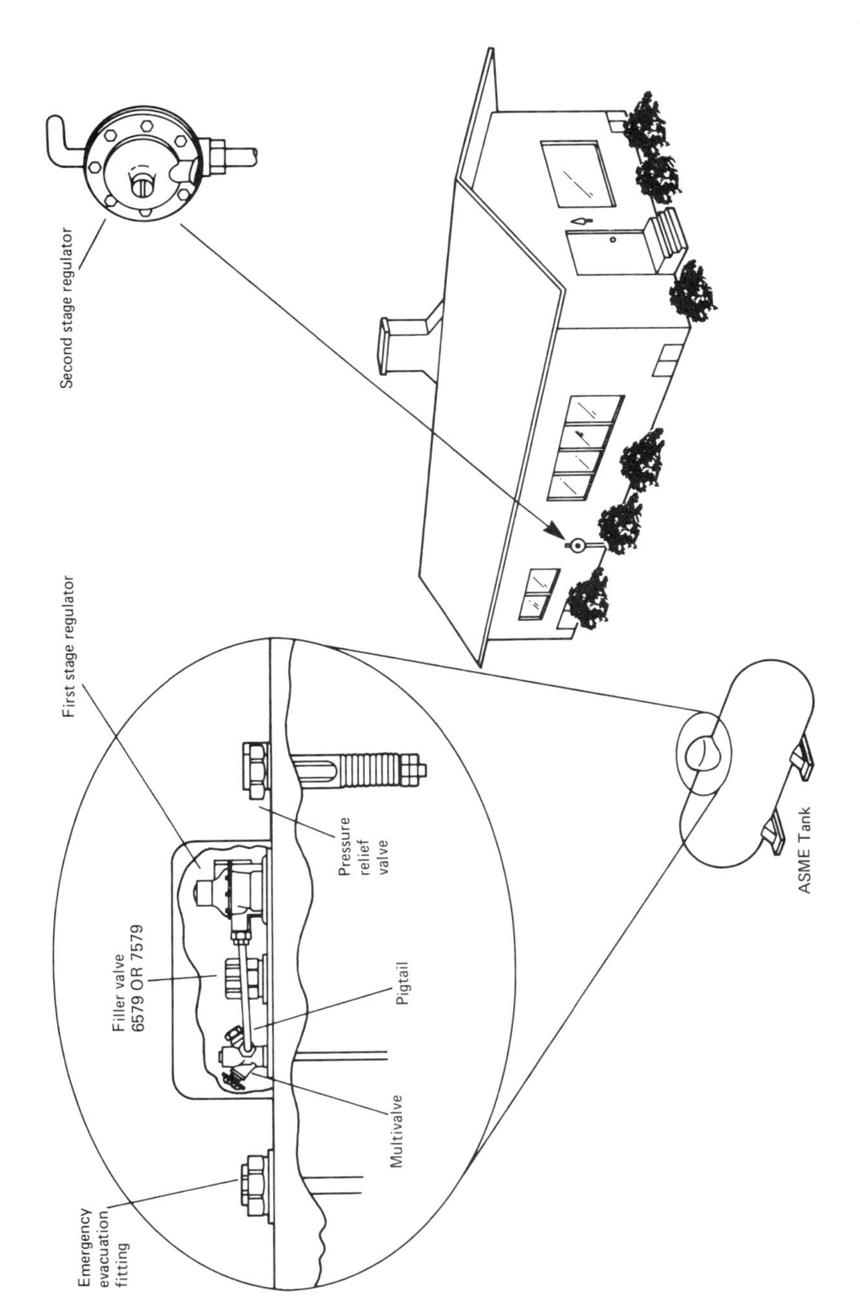

Figure 3.5 *Typical LP-Gas domestic container installation. (Drawing courtesy of Engineered Controls, Inc.)*

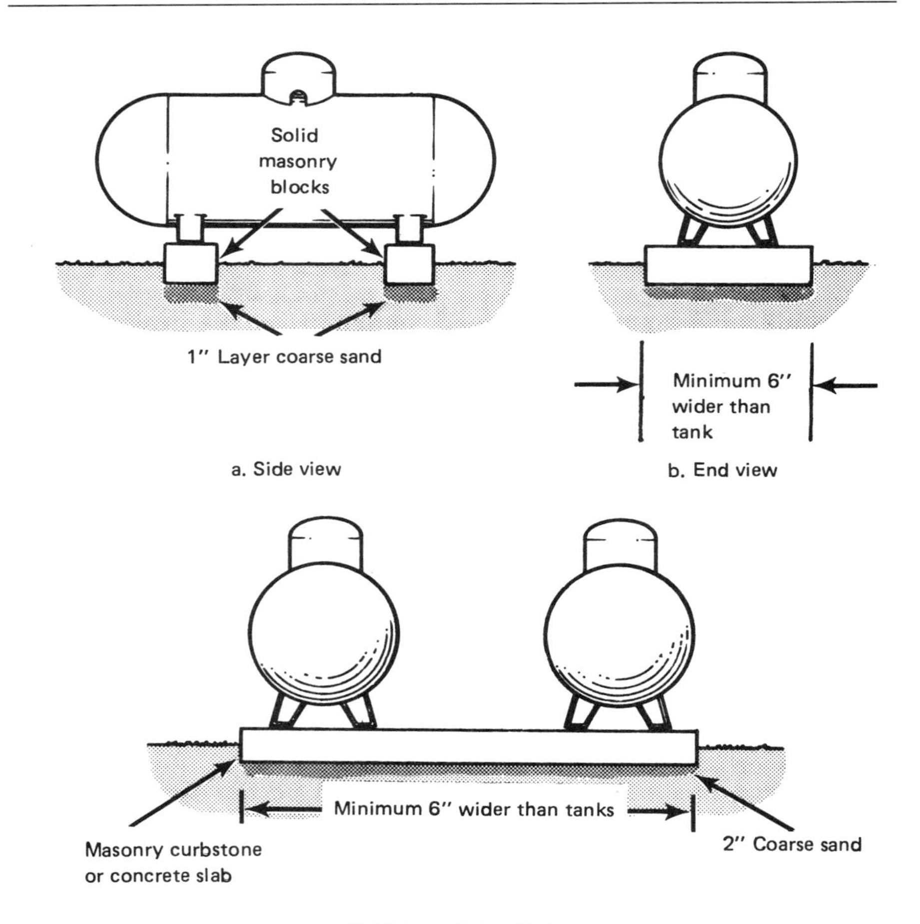

Figure 3.6 *Typical smaller ASME container aboveground installation. (Drawing courtesy of National Propane Gas Association.)*

(d) In locations where heavy snow can be expected to cover the container, the following additional requirements shall apply:

1. A stake or other marking shall be installed higher than the highest anticipated snow cover up to a height of 15 ft (4.6 m).
2. The installation shall prevent moving of the container by the forces anticipated as a result of snow accumulation.

This requirement was added in the 1995 edition. It accompanies revisions to 1-1.3.1(f) and 3-2.11. The new requirements are the result of a severe problem in the Sierra Mountains of Northern California during the winter of 1992–1993. Exceptionally heavy snowfall covered the area resulting in several accidents in which propane gas was released and ignited, causing multiple

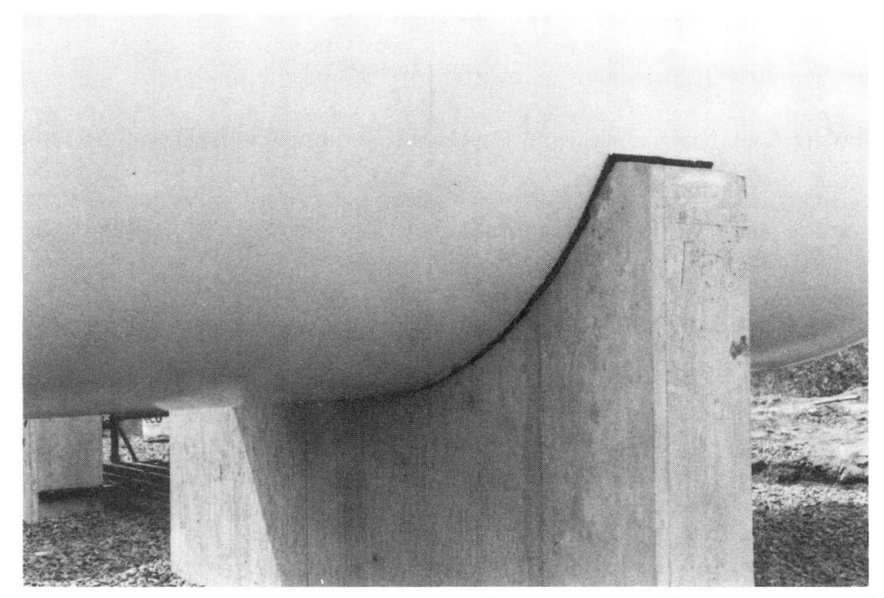

Figure 3.7 ASME storage container with concrete saddles. A 30,000-gal (114-m³) container supported 3 ft (1 m) above grade. This installation utilized a concrete saddle with a ½-in.-thick (13-mm) bitumastic felt pad installed between the steel container and concrete foundation. (Photo courtesy of H. Emmerson Thomas.)

deaths. The accidents resulted from the weight of the snow, which broke piping, allowing propane gas to migrate under the snow and enter buildings. The requirement applies to areas where very heavy snowfall can be expected, even if infrequently. In the Sierra Mountains, similar snowfalls have been recorded at approximately 10-year intervals. The new requirements serve two purposes, enabling the gas company to locate the container for refilling, and to ensure that snow removal operations will not further bury the container. If the highest expected snowfall would not cover LP-Gas containers for an extended period, the provision would clearly not apply.

3-2.4.3 Vertical ASME containers over 125 gal (0.5 m³) water capacity designed for permanent installation in stationary service aboveground shall be installed on reinforced concrete or steel structural supports on reinforced concrete foundations that are designed to meet the loading provisions established in 2-2.2.3.

(a) Steel supports shall be protected against fire exposure with a material having a fire resistance rating of at least 2 hr. Continuous steel skirts having only one opening 18 in. (457 mm) or less in diameter shall be permitted to have such fire protection applied only to the outside of the skirts.

Although they are not used as extensively as horizontal containers, vertical containers are useful for certain applications, e.g., as dispensers in service stations and where space is at a premium.

3-2.4.4 Single containers constructed as portable storage containers (*see definition*) for temporary stationary service in accordance with 2-2.5.4(a) shall be placed on concrete pads, paved surfaces, or firm earth for such temporary service (normally not more than 12 months at a given location) and shall meet the following requirements:

Temporary single portable storage container installations have applications such as fuel supplies for crop drying and road construction. Note that 1-4.2 requires that the authority having jurisdiction be notified of all temporary installations where the container exceeds 2,000 gal. This requirement would not apply to most portable container installations. Temporary installations of portable containers are considered to be those of 12 months or shorter duration, while a 6-month limit is applied to the larger quantities covered under 1-4.2

(a) The surface on which the containers are placed shall be substantially level and, if not paved, shall be cleared (and kept cleared) of dry grass and weeds, and other combustible material within 10 ft (3.0 m) of the container.

(b) Flexibility shall be provided in the connecting piping.

(c) If such containers are to be set with the bottoms of the skids or runners above the ground, nonfireproofed structural supports shall be permitted to be used for isolated locations with the approval of the authority having jurisdiction, and provided the height of the outside bottom of the container shell above the ground does not exceed 5 ft (1.5 m). Otherwise, fire-resistive supports shall be provided.

3-2.4.5 If the container is mounted on, or is part of, a vehicle as provided in 2-2.5.4(b), the unit shall be parked in compliance with the provisions of 3-2.2.2 for the location of a container of that capacity for normal stationary service, and shall be in accordance with the following:

(a) The surface shall be substantially level and if not paved shall be suitable for heavy vehicular use and shall be cleared (and kept cleared) of dry grass and weeds, and other combustible material within 10 ft (3.0 m) of the container.

(b) Flexibility shall be provided in the connecting piping.

If the container is mounted as part of a vehicle, it is actually a cargo container and is not considered a permanent storage container, except if connected for use and not transfer. In such a case, the provisions of 3-2.4.4 apply.

3-2.4.6 Portable containers of 2,000 gal (7.6 m^3) water capacity or less complying with 2-2.5.5 shall be permitted to be installed for stationary service as provided in 3-2.4.2(a)(2) for stationary containers.

This permits these particular DOT (CTC) portable containers to be considered permanent storage containers.

3-2.4.7 Mounded containers shall be installed as follows:

(a) Mounding material shall be earth, sand, or other noncombustible, noncorrosive materials such as vermiculite or perlite and shall provide minimum thickness of cover for the container of at least 1 ft (0.3 m).

(b) Unless inherently resistant to erosion, a suitable protective cover shall be provided.

A mounded container may be above or partially buried below the ground surface, but must be completely covered by at least one foot of cover. Prior to the 1992 edition, only earth or sand were permitted as mounding materials; the list of covering materials has since been expanded to include noncombustible, noncorrosive materials, such as vermiculite and perlite. These materials have significant advantages over sand and earth in certain installations. If it were desired to "mound" a container by erecting walls around it and filling the space with earth or sand, a very strong wall would have to be constructed to contain the heavy material. By permitting a lighter material, the option of mounding in this manner is encouraged. Of course, means have to be taken to ensure that these lighter materials remain in place, and that can be done by covering them with an upper layer of sand or by installing a fabric to stabilize the surface.

Many times it is not desirable to bury a tank completely because of the water table, rock formations, and other reasons. Sometimes mounding is for aesthetic purposes and is done at the request of the user. Mounding also provides insulation from low or high temperatures and is, therefore, a means of "special protection" (*see definition in Section 1-6*). Mounded containers should be protected from corrosion in the same manner required for underground containers.

(c) Tank valves and appurtenances shall be accessible for operation or repair, without disturbing mounding material.

This provision recognizes that valve accessibility is sufficiently important to be required in NFPA 58. The means of access are left to the designer or installer, and can be "tunnels" constructed of lightweight conduit, sleeves to permit access to top-mounted appurtenances, or other means.

(d)* Mounded containers shall be protected against corrosion in accordance with good engineering practice.

This provision was added in the 1992 edition to recognize that the probability of corrosion of mounded containers must be addressed prior to installation. Specific requirements were not added, as the methods of corrosion

Figure 3.8 Installation of three 80,000-gal (302-m³) storage containers mounded for special protection. These containers have an epoxy coal tar coating for corrosion protection in addition to cathodic protection. A 12-in. (300-mm) covering of sand was provided around each container and a 6-in. (150-mm) layer of gravel installed over the three containers. The containers were provided with extended manways and relief valve nozzles to permit installation of all tank trim above the top surface of the final layer of gravel. (Photo courtesy of H. Emmerson Thomas.)

protection vary with the corrosiveness of soils and the sophistication of the installer. References on corrosion are located in A-3-2.4.7(d), and an additional reference was added in the 1995 edition.

A-3-2.4.7(d) For information on corrosion protection of containers and piping systems see:

1. API Publication 1632-1983, *Cathodic Protection of Underground Petroleum Storage Tanks and Piping Systems.*

2. Underwriters Laboratories of Canada, ULC S603.1-M, 1982, *Standard for Galvanic Corrosion Protection Systems for Steel Underground Tanks for Flammable and Combustible Liquids.*

3. National Association of Corrosion Engineers Standard RP-01-69 (1983 Rev.), Recommended Practice, *Control of External Corrosion of Underground or Submerged Metallic Piping Systems.*

4. National Association of Corrosion Engineers Standard RP-02-85, Recommended Practice, *Control of External Corrosion on Metallic Buried, Partially Buried, or Submerged Liquid Storage Systems.*

5. Underwriters Laboratories Inc., UL 1746, *External Corrosion Protection Systems for Steel Underground Storage Tanks,* 1989.

3-2.4.8 ASME container assemblies listed for underground installation, including interchangeable aboveground-underground container assemblies, shall be permitted to be installed underground as follows:

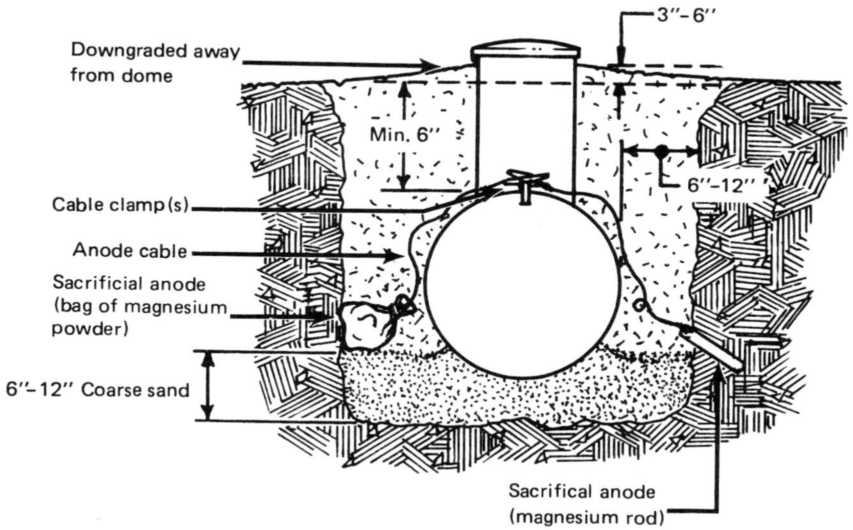

Figure 3.9 *Typical small ASME container underground installation. Cathodic protection is not always needed. (Drawing courtesy of National Propane Gas Association.)*

This section provides a complete set of rules for the installation of containers underground. Note that this covers containers listed for underground and interchangeable aboveground–underground containers only.

Care should be used to see that the ASME containers for underground installation are marked by the manufacturer as suitable for underground or for the interchangeable aboveground–underground type of service. A container built for aboveground service should never be installed underground. Likewise, a container built for underground installation must not be installed aboveground, as the relief valve may be undersized. Some people prefer underground or mounded tank installations for BLEVE protection. However, there are some procedures that have to be done differently with underground installations, than with aboveground installations, because the condition of the underground tank is not easily visible for checking.

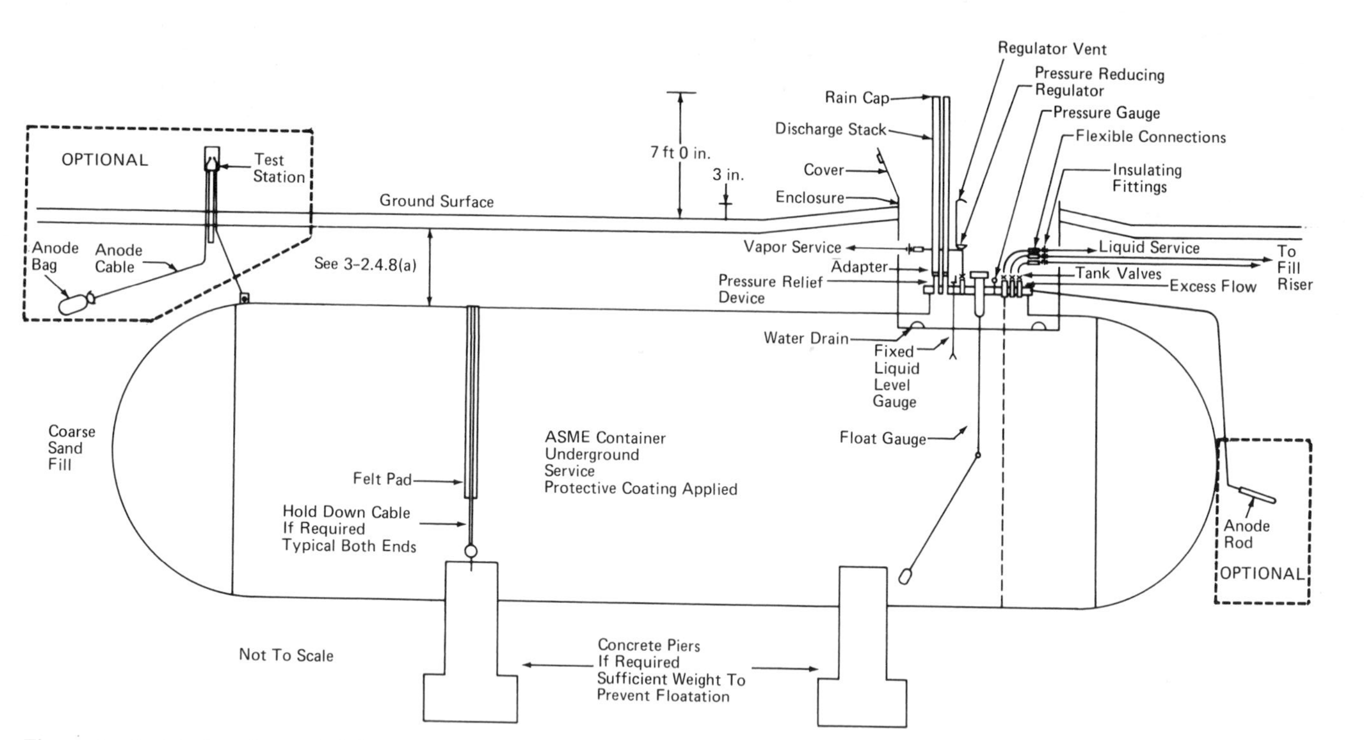

***Figure 3.10** Typical large ASME container underground installation. Cathodic protection is not always needed. (Drawing courtesy of Vision Energy Research.)*

It is very important that underground and mounded tanks be properly protected against corrosion by means of a suitable coating or cathodic protection. It is especially important not to allow any scarring of a protective coating while installing the tank, because just one small, unprotected spot can concentrate corrosion and lead to a leak. Tanks should never be moved by putting a chain or a cable around them without some cushioning material positioned between the chain or cable and the container coating. It is not necessary to have a concrete or metal saddle for an underground container installation, provided there is firm sand or earth for it to rest on. But there can be conditions where it is advisable to have a proper foundation or slab on which to place the underground container. When the backfill is placed, it should be free of rocks and similar abrasives, which can adversely affect the coating.

Backfill should be tamped so that there will not be settling. A major fire occurred when a vaporizer installed on untamped fill over a buried container settled and tipped, causing a piping break and liquid escape.

An underground tank should not be installed in a pit unless the pit is properly filled in around the tank with sand or suitable earth; otherwise, the unfilled space would provide a place for vapor to accumulate.

(a) The container shell shall be placed at least 6 in. (0.15 m) below grade unless the container might be subject to abrasive action or physical damage from vehicular traffic within a parking lot area, driveway, or similar area. In such a case, a noninterchangeable underground container shall be used and the container shell placed at least 18 in. (0.46 m) below grade [*see 3-2.4.8(c)*] or equivalent protection shall be otherwise provided, such as the use of a concrete slab, to prevent imposing the weight of a vehicle directly on the container shell. Protection of the fitting housing, housing cover, tank connections, and piping shall be provided to protect against vehicular damage.

A burial depth of 6 in. is sufficient when only pedestrian traffic is expected above or near the container. When heavier loads are expected from vehicles, the burial depth is increased to 18 in. to distribute the loads so that the container will not be harmed, or protection must be provided by other means. A civil engineer familiar with soils should be consulted when alternate means are used to ensure that the container is properly protected. When containers are installed in areas where gardening is expected, the deeper burial depth is a means to protect the container coating (if relied on for corrosion protection) from damage. Alternately, the container user should be notified, and periodically reminded, of the need to avoid digging in the area of the container.

(b) Where containers are installed underground within 10 ft (3.0 m) of where vehicular traffic can be reasonably expected, such as driveways and streets or within a utility easement subject to vehicular traffic, protection of the fitting housing, housing cover, tank connections, and piping shall be provided to protect against vehicular damage.

This requirement recognizes the importance of protecting the underground container appurtenances and their housing from vehicular traffic. This protects both the container and the initial length of pipe or tubing at the container.

(c) Approved interchangeable aboveground-underground container assemblies installed underground shall not be placed with the container shell more than 12 in. (0.30 m) below grade.

Interchangeable aboveground–underground containers are usually listed in accordance with UL Standard 644, *Container Assemblies for LP-Gas*.

(d) Any party involved in construction or excavation in the vicinity of a buried container shall be responsible for determining the location of and providing protection for the container and piping against their physical damage from vehicular traffic.

This indicates responsibility for protection of underground containers from damage during construction subsequent to the installation of the container. It is consistent with many local laws. As the container is not hidden and the dome is visible, it is not unreasonable to make anyone working in the area of a container responsible for prevention of damage to the container.

(e) The portion of the container to which the fitting cover or other connections are attached shall be permitted to be covered. The discharge of the regulator vent shall be above the highest probable water level.

The purpose of this provision is to prevent blockage of the container relief valve vent as a result of freezing. If this vent is blocked it can cause overpressure of downstream piping with serious consequences. Blocked regulator vents have caused fires in buildings.

(f)* Containers shall be protected against corrosion for the soil conditions at the container site by a method in accordance with good engineering practice. Precaution shall be taken to prevent damage to the coating during handling. Any damage to the coating shall be repaired before backfilling.

A-3-2.4.8(f) See A-3-2.4.7(d).

Underground containers must be protected from corrosion. This leaves the method of protection to the installer. Corrosion protection is a complex subject best left to a person familiar with the techniques and their limits. Usual means of protection of underground containers include thick coatings and passive cathodic protection systems where an anode (usually a bag containing a metallic powder) is electrically bonded to the container. Coatings (paints) appear to provide protection against corrosion by isolating the metal tank from the soil. Unfortunately it is not possible to insure that a coating is perfect and pinhole free. A small void in a coating will actually concentrate the electrochemical corrosion activity to the void and accelerate corrosion at that point. This can be prevented when a passive (or active) cathodic protection system is used as the metallic powder will corrode, rather than the tank. Active cathodic protection is the method usually used to protect underground pipelines. It utilizes an electric current and is usually beyond the means of the installer of LP-Gas containers. For reference materials on corrosion protection see A-3-2.4.7(d).

(g) Containers shall be set substantially level on a firm foundation and surrounded by earth or sand firmly tamped in place. Backfill shall be free of rocks or similar abrasives.

It is important that underground containers be installed such that they will not shift and place loads on the piping system.

NOTE: Firm earth can be used.

(h) Where a container is to be abandoned underground, the following procedure shall be followed:

1. As much liquid LP-Gas as possible shall be removed through the container liquid withdrawal connection.
2.* As much of the remaining LP-Gas vapor as possible shall be removed by venting it through a vapor connection; either by burning the vapor, or venting it to the open air at a safe location. The vapor shall not be vented at such a rapid rate as to exceed the vaporization rate of any residual liquid LP-Gas that remains after the liquid removal procedure of 3-2.4.8(h)(1).

A-3-2.4.8(h)2 If vapor is vented too rapidly, the pressure drop due to the refrigeration of the liquid can lead to the erroneous conclusion that no liquid remains in the container.

3. Where only vapor LP-Gas at atmospheric pressure remains in the container, the container shall be filled with water, sand, or foamed plastic, or shall be purged with an inert gas. The displaced vapor shall be permitted to be burned or vented to the open air at a safe location.

The safe abandonment of underground containers is of equal importance to proper installation. If a container were to be abandoned without the precautions required here, the container could corrode and result in gas leakage, which could cause a fire. Also, collapse of an underground container could cause ground subsidence.

Leakage of underground gasoline and heating oil tanks has caused much concern about underground tanks. Gasoline and heating oil are groundwater contaminants and have caused environmental damage. Propane is not a groundwater contaminant, however, and from an environmental viewpoint there is no need to remove underground propane containers.

3-2.4.9 Partially underground, unmounded ASME containers shall be installed as follows:

(a) The portion of the container below the surface, and for a vertical distance of at least 3 in. (75 mm) above the surface, shall be protected to resist corrosion as required for underground containers. [*See 3-2.4.8(f).*]

(b) Containers shall be set substantially level on a firm foundation, with backfilling to be as required for underground containers. [*See 3-2.4.8(g).*]

(c) Spacing provisions shall be as specified for aboveground containers in 3-2.2.2 and Table 3-2.2.2.

(d) The container shall be located so as not to be subject to vehicular damage or shall be adequately protected against such damage.

3-2.5 Installation of Container Appurtenances.

3-2.5.1 Pressure relief devices shall be installed on containers in accordance with this section and positioned so that the relief device is in direct communication with the vapor space of the container.

It is important that the pressure relief device be installed such that it is always in the vapor space in normal operation (*see commentary on 2-2.3.5*). In locating such devices, it is important to remember that what seems to be the normal filling point in a container is radically changed if the container is filled at a low temperature and then is subjected to a rising temperature, which expands the liquid. If the pressure relief device is not properly located, the liquid expansion will be enough to then place the pressure relief device in the liquid phase.

3-2.5.2 Pressure relief devices on portable DOT cylinder specification containers, or their equivalent in accordance with ASME construction, of 1,000 lb (454 kg) [120 gal (0.5 m^3)] water capacity or less shall be installed to minimize the possibility of relief device(s) discharge(s) impingement on the container.

Burning gas being discharged from a pressure relief device and impinging upon the container or an adjacent container is the most common reason for spread of fire from container to container and their subsequent BLEVEs. Because the relief valves on most DOT containers discharge horizontally when the container is in its normal position, this provision is difficult to implement where more than one or two containers are present. Where the container has a protective collar, the collar should have an opening opposite the relief valve discharge outlet for the gas to pass through.

3-2.5.3 Pressure relief devices on ASME containers of 125 gal (0.5 m^3) water capacity or more that are permanently installed in stationary service, portable storage containers (*see definition*), portable containers (tanks) of nominal 120 gal (0.5 m^3) water capacity or more, or cargo tanks shall be installed so that any gas released is vented away from the container upward and unobstructed to the open air. The following provisions shall also apply:

(a) Means such as rain caps shall be provided to minimize the possibility of the entrance of water or other extraneous matter (which might render the relief device inoperative or restrict its capacity) into the relief device or any discharge piping. If necessary, provision shall be made for drainage. The rain cap or other protector shall be designed to remain in place except when the relief device operates and shall permit the relief device to operate at sufficient relieving capacity.

Rain caps that direct gas discharge downward or to the side should not be used. Because any rain cap that remains in place during discharge will throttle the discharge to some extent, rain caps should be used only after determining their effect upon the relieving capacity of the relief device.

(b) On each aboveground container of more than 2,000 gal (7.6 m^3) water capacity, the relief device discharge shall be installed vertically upward and unobstructed to the open air at a point at least 7 ft (2 m) above the top of the container. The following also shall apply:

1. Relief device discharge piping shall comply with 3-2.5.3(f).
2. In providing for drainage in accordance with 3-2.5.3(a), the design of the relief device discharge(s) and attached piping shall:

   a. Protect the container against flame impingement that might result from ignited product escaping from the drain opening.
   b. Be directed so that a container(s), piping, or equipment that might be installed adjacent to the container on which the relief device is installed is not subjected to flame impingement.

These requirements call for the "discharge stacks" found on larger containers. The stacks must be vertical and straight (no bends). They are required so that in the event of an emergency the relief valve discharge, if ignited, will not impinge on the container and cause a BLEVE. Note that they are not required on containers of 2,000 gal (7.6 m^3) or less capacity. (*See Figure 3.28 where "discharge stacks" are shown.*)

(c) On underground containers of 2,000 gal (7.6 m^3) or less water capacity, the relief device shall be permitted to discharge into the manhole or housing, provided such manhole or housing is equipped with ventilated louvers, or their equivalent, of adequate area as specified in 2-3.7(d).

Exception: Pressure relief devices installed in dispensing stations covered in the exception to 3-2.5.3(d).

The exception refers to the exception to the next paragraph, which requires that underground containers in dispensing stations have their relief valve stacks terminate at least 10 ft (3 m) above ground level.

The discharge of a relief valve into the housing on underground containers recognizes that the relief devices seldom operate on underground containers and that the discharge rate on underground containers of 2,000 gal (7.6 m^3) and less water capacity is reduced.

(d) On underground containers of more than 2,000 gal (7.6 m^3) water capacity, the discharge from relief devices shall be piped vertically and directly upward to a point at least 7 ft (2 m) above the ground. Relief device discharge piping shall comply with 3-2.5.3(f).

Exception: On underground containers in dispensing stations the relief device discharge shall be piped vertically and directly upward to a point at least 10 ft (3.0 m) above the ground. Discharge piping shall comply with 3-2.5.3(f) and shall be adequately supported and protected against physical damage.

The relief stack requirements for underground containers larger than 2,000 gal (7.6 m^3) are identical to those for aboveground containers. The exception covers underground containers larger than 2,000 gal (7.6 m^3) located in dispensing stations where vehicle traffic is routine. In these stations, the relief valve stack discharge is required to be 10 ft (3 m) aboveground to ensure dispersal of LP-Gas in case of relief valve operation. Note that the exception requires that the stack be protected against physical damage. To accomplish this it may be necessary to install the stack with elbows to locate it away from traffic areas. If this is done, it is important to verify that the added resistance to flow will not reduce the relief valve capacity. The relief valve manufacturer may be able to provide assistance.

(e) The discharge terminals from relief devices shall be located to provide protection against physical damage. The discharge piping used shall be adequate in size to allow sufficient relief device relieving capacity. Such piping shall be metallic and have a melting point over 1500°F (816°C). Discharge piping shall be designed so that excessive force applied to the discharge piping will result in breakage on the discharge side of the valve rather than on the inlet side without impairing the function of the valve. Return bends and restrictive pipe or tubing fittings shall not be used.

The 7 ft (2 m) provision cited in 3-2.5.3(b) and (d) is intended to ensure that the discharge will be above the head of anyone near or on a container. This is increased to 10 ft (3 m) in service stations [3-2.5.3(e)] because individuals may be on a vehicle. These distances also facilitate vapor disposal at a height above most building openings and ignition sources.

It is required that discharge piping be constructed of metals with a melting point over 1500°F, the same specification for piping materials. This is to ensure integrity of the discharge piping in a fire condition. Discharge piping is not permitted to have restriction of any type, including pipe fittings. If it were necessary to extend the discharge pipe and fittings required by the installation, the provisions of 1-1.4, Alternate Materials, Equipment and Procedures, could apply if approved by the authority having jurisdiction. In such a case, the discharge piping could be designed in accordance with American Petroleum Institute Recommended Practice 520, *Sizing, Selection, and Installation of Pressure-Relieving Devices in Refineries*, to size the discharge piping system so that it would not restrict the proper operation of the relief valve.

The discharge of a relief valve into the housing on underground containers permitted by 3-2.5.3(c) recognizes that the relief devices seldom operate on underground containers and that the discharge rate on underground containers of 2,000 gal (7.6 m^3) and less water capacity is reduced. This variation, however, is not permitted in service stations because, even though operation may be unlikely, such locations can present considerable public exposure.

(f) Shutoff valves shall not be installed between relief devices and the container, or between the relief devices and the discharge piping.

Shutoff valves between containers and relief devices greatly facilitate maintenance and testing because it is not necessary to empty and purge the container. However, experience revealed that these valves were closed all too often. The arrangements cited will permit a shutoff valve to be removed and replaced while maintaining the desired protection at all times.

See Figure 2.12 for an illustration of equipment used for this purpose.

Exception: Specially designed relief device/shutoff valve combinations covered by 2-3.2.4(c), or where two or more separate relief devices are installed, each with an individual shutoff valve, and the shutoff valve stems are mechanically interconnected in a manner that will allow the rated relieving capacity required for the container from the relief device or devices that remain in communication with the container.

3-2.5.4 Pressure relief devices on portable storage containers (constructed and installed in accordance with 2-2.5.4 and 3-2.4.4 respectively) used temporarily in stationary-type service shall be installed in accordance with the applicable provisions of 3-2.5.3.

3-2.5.5 Additional provisions (over and above the applicable provision in 3-2.5.2 and 3-2.5.3) apply to the installation of pressure relief devices in containers used in connection with vehicles as follows:

(a) For containers installed on vehicles in accordance with Section 3-8 and Chapter 8.

(b) For cargo containers (tanks) installed on cargo vehicles in accordance with Section 6-3, see 6-3.2.1.

3-2.6 Regulator Installation.

The requirements for installation of regulators were extensively revised in the 1995 edition. Also see commentary following Section 2-5.8.

3-2.6.1 A two-stage regulator system or an integral two-stage regulator shall be required on all fixed piping systems that serve ½ psi (3.4 kPa) appliance systems [normally operated at 11 in. W.C. (2.7 kPa) pressure.] The regulators utilized in these systems shall meet the requirements of 2-5.8. This requirement includes fixed piping systems for appliances on RV (recreational vehicles), mobile home installations, manufactured home installations, catering vehicles, and food service vehicle installations. Single-stage regulators shall not be installed in fixed piping systems after June 30, 1997.

Two-stage pressure regulation will be required on LP-Gas fixed piping systems serving most buildings, mobile homes, recreational vehicles, and catering vehicles as of the effective date of the 1995 edition of NFPA 58. The Technical Committee believes that this will add a significant margin of safety to piping systems by preventing overpressure conditions that result in leakage of LP-Gas in buildings which could result in fires.

The last sentence of the paragraph is an unusual one for the standard, as it sets a time limit to the retroactivity requirement stated in 1-1.5. That statement permits the installation of equipment in manufacturers', distributors', and installers' inventory prior to a change in the standard. The last sentence specifically allows single-stage regulators to be installed only until June 30, 1997, just over two years following the effective date of the 1995 edition. After that date they cannot be installed in fixed piping systems serving most buildings, leaving only a few applications for them. The Technical Committee took this important step to prevent stockpiling of single-stage regulators as a way to postpone compliance with this important safety measure.

Exception No. 1: This requirement does not include small portable appliances and outdoor cooking appliances with input ratings of 100,000 Btu/hr (29 kw) or less.

Small portable appliances (under 100,000 Btu/hr), such as gas grills, portable heating appliances, portable cooking stoves (*see 3-4.8.4*) and similar appliances do not require two-stage regulation as they are attended appliances and the operator will be able to shut off the appliance in case of malfunction.

Exception No. 2: Gas distribution systems utilizing multiple second-stage regulators are permitted to use a high-pressure regulator installed at the container provided a first-stage regulator is installed downstream of the high-pressure regulator and ahead of the second-stage regulators.

This exception permits the use of high pressure where an integral two-stage or first- and second-stage regulator meeting these requirements is installed downstream. This ensures that the second-stage regulator does not receive an inlet pressure in excess of the 10 psi.

Exception No. 3: High-pressure regulators with an overpressure protection device and a rated capacity of more than 500,000 Btu/hr (147 kw) shall be permitted to be used in two-stage systems where the second-stage regulator incorporates an integral or separate overpressure protection device. This overpressure protection device shall limit the outlet pressure of the second stage regulator to 2.0 psi (14 kPa) when the regulator seat disc is removed and with an inlet pressure equivalent to the maximum outlet pressure setting of the high-pressure regulator.

Exception No. 4: Systems consisting of listed components that provide an equivalent level of overpressure protection.

This exception recognizes that the overpressure protection system utilizing a first-and second-stage regulator meeting the requirements of 2-5.8 and UL 144, is not the only way to prevent overpressure of piping systems above 2 psi. Any system that limits fixed piping system pressure to 2 psi or less, under failure conditions can be used provided that it is listed. Listing by an independent agency ensures an independent organization with an interest in safety has verified that the equipment will perform as intended.

3-2.6.2 First-stage or high-pressure regulators shall be directly attached or attached by flexible connectors to the vapor service valve of a container or to a vaporizer outlet. The regulators shall also be permitted to be installed with flexibility in the interconnecting piping of manifolded containers or vaporizers.

Exception: First-stage regulators installed downstream of high-pressure regulators.

In a single container or single vaporizer installation, the first-stage regulator can be connected directly to the container service valve or the vaporizer outlet (where used). As an alternate, a flexible connector such as a pigtail (short length of copper tubing with POL connectors) may be used between the container and the first-stage regulator or integral two-stage regulator.

In the case of manifolded containers or vaporizers, it is impractical to install a regulator at each container service valve or vaporizer. Pressure imbalances could occur, causing regulators to counteract each other and could result in an unstable pressure condition. Thus, a single regulator may be used with this type of system with interconnecting piping in the manifolded system.

Figures 3.11 and 3.12 Some regulator installation arrangements.

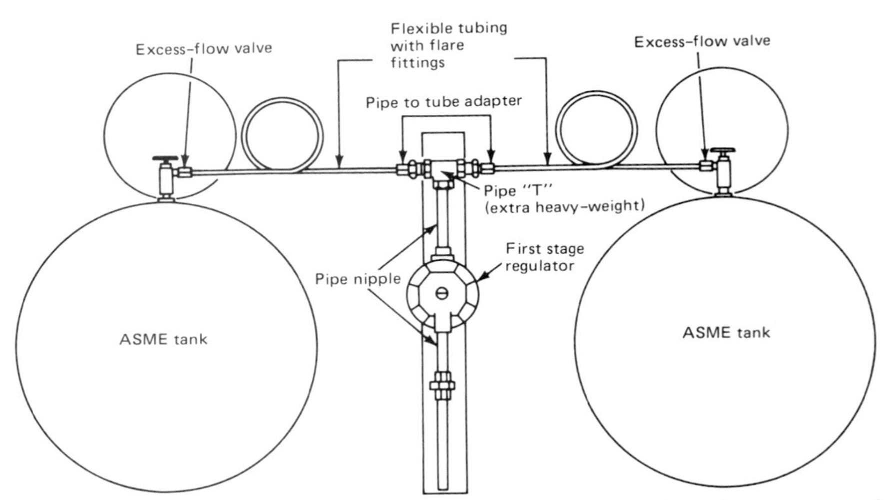

Figure 3.11 *Dual cylinder installation (stationary service). (Drawing courtesy of the National Propane Gas Association.)*

Figure 3.12 *Automatic change-over regulator. (Photo courtesy of Engineered Controls, Inc.)*

The requirement that the regulator be connected directly to the container shutoff valve can be met in a manifolded container installation by connecting the regulator to a tee that is directly connected to the container shutoff valves with only connectors as required for spacing or flexibility. [*See Figure 3.9.*] Flexibility in the piping system is necessary to provide for expansion or contraction.

A serious fire hazard could occur if LP-Gas vapor were allowed to recondense in long lengths of piping under cold weather conditions. Liquid LP-Gas would be present at the point of use where vapor usage was intended. To prevent this from happening, first-stage regulators must be located as close to the supply container or outlet of the vaporizer as practical.

3-2.6.3 First-stage and high-pressure regulators shall be installed outside of buildings.

The changes in UL 144, made concurrently with the changes to NFPA 58 require a high capacity internal relief valve in the second-stage regulator. In case of regulator failure, the relief valve will operate and vent LP-Gas to limit pressure in the building. The first-stage regulator is also provided with an internal relief valve so that any failure of this regulator can be identified. Failure of the first-stage regulator will not create an overpressure condition (pressure in excess of 2 psi) downstream of the second-stage regulator.

Exception No. 1: Regulators on portable containers installed indoors in accordance with Section 3-4.

Exception No. 2: Regulators on containers of less than 125 gal (0.5 m^3) for the purpose of being filled or in structures complying with Chapter 7.

Exception No. 3: Regulators on containers on LP-Gas vehicles complying with, and parked or garaged in accordance with, Chapter 8.

Exception No. 4: Regulators on containers used with LP-Gas stationary or portable engine fuel systems complying with Chapter 8.

Exception No. 5: Regulators on containers used with LP-Gas fueled industrial trucks complying with 8-3.6.

Exception No. 6: Regulators on containers on LP-Gas fueled vehicles garaged in accordance with Section 8-6.

Exception No. 7: Regulators on portable containers awaiting use, resale, or exchange when stored in accordance with Chapter 5.

Several exceptions are provided to the rule to prevent misapplication of the new requirements. These exceptions include:

- Special applications where first-stage pressure regulators must be located indoors (*see 3-4*)

- Engine fuel uses (*see Chapter 8*)
- Industrial applications
- Some gas distribution facilities

All the exceptions cover regulators used on other than fixed piping systems in residential and commercial buildings.

3-2.6.4 All regulators for outdoor installations shall be designed, installed, or protected so their operation will not be affected by the elements (freezing rain, sleet, snow, ice, mud, or debris). This protection shall be permitted to be integral with the regulator.

Exception: Regulators used for portable industrial applications.

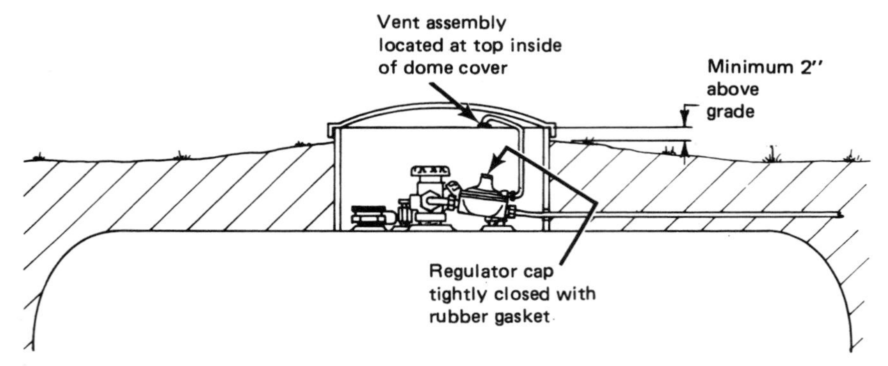

Figure 3.13 *Use of vent extension to protect regulator operation from elements on underground container. The termination of the vent assembly must be above grade level. (Drawing courtesy of the National Propane Gas Association.)*

To minimize the possibility of regulator vents being blocked by ice, it is imperative that the vent on regulators used outdoors (except for portable industrial uses) be protected against the elements. This can be done by either enclosing the regulator in housing, or mounting the regulator with the vent opening pointing vertically downward, if the vent has a drip lip as required by UL 144, with an inside diameter of not less than $^{11}/_{16}$ in. (17 mm) and an outside diameter at least $^{3}/_{4}$ in. (19 mm).

These dimensions evolved from a test project by Underwriters Laboratories Inc. where different designs were subjected to freezing rain and wind conditions. Current UL listed regulators are marked according to whether they must be installed under a protective cover, or in the downward position or equivalent. Also, it is important to ensure that the protective housing, if used, is installed correctly. Plastic closures have been used with the regulator upside down, and the housing has become filled with water and frozen, causing faulty operation.

3-2.6.5 The point of discharge from the required pressure relief device on regulating equipment installed outside of buildings in fixed piping systems shall be located not less than 3 ft (1 m) horizontally away from any building opening below the level of such discharge, and not beneath any building unless this space is well ventilated to the outside and is not enclosed for more than 50 percent of its perimeter. The point of discharge shall also be located not less than 5 ft (1.5 m) in any direction away from any source of ignition, openings into direct-vent (sealed combustion system) appliances, or mechanical ventilation air intakes.

Exception: This requirement shall not apply to vaporizers.

3-2.6.6 The discharge from the required pressure relief device on regulating equipment installed inside of buildings in fixed piping systems shall be vented with properly sized and supported piping to the outside air with the discharge outlet located not less than 3 ft (1 m) horizontally away from any building opening below the level of such discharge. The discharge outlet shall also be located not less than 5 ft (1.5 m) in any direction away from any source of ignition, openings into direct-vent (sealed combustion system) appliances, or mechanical ventilation air intakes.

It must be recognized that the two-stage regulation system required on fixed piping systems can protect against overpressure by venting LP-Gas outside of a building, in the event of component failure. This new provision addressed this infrequent occurrence, and further minimized the possibility of ignition and entry of flammable gas into the building.

The discharge from the relief valve vent of a second-stage regulator (not an appliance regulator) installed outdoors, or the discharge from the termination end of a vent pipeaway to the outdoors from an indoor second-stage regulator installation, cannot be less than 3 ft (1 m), horizontally, from a building opening (below the level of this discharge) and 5 ft (1.5 m) from a source of ignition or building ventilation inlet. In piping the vent of the relief valve discharge from second-stage regulators to the outside, a properly sized and supported piping system must be installed. For information on sizing vent lines from relief valve discharge of second-stage regulators, see API RP 520, *Sizing, Selection, and Installation of Pressure-Relieving Devices in Refineries.*

Exception No. 1: This provision shall not apply to appliance regulators otherwise protected, or to regulators used in connection with containers in buildings as provided for in 3-2.2.1, Exceptions Nos. 1, 2, 4, 5, and 6.

Exception No. 2: This requirement shall not apply to vaporizers.

3-2.6.7 Single-stage regulators shall be permitted to be used only on small portable appliances and outdoor cooking appliances with input ratings of 100,000 Btu/hr (29 kw) maximum.

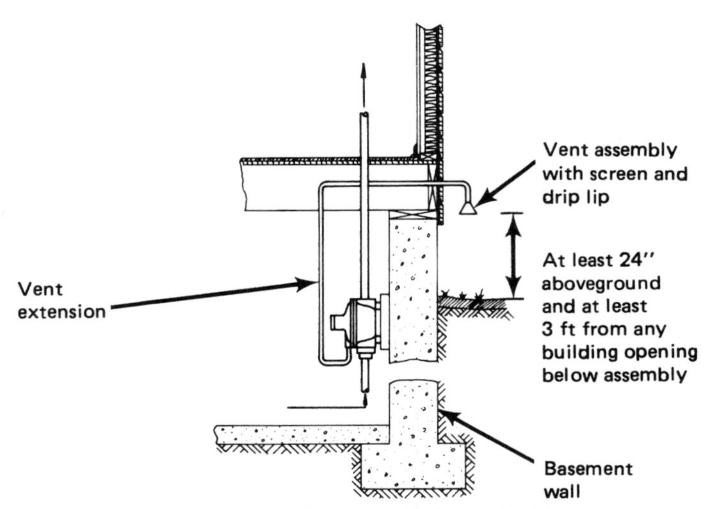

Figure 3.14 Venting regulator installed in a building. (Drawing courtesy of the National Propane Gas Association.)

Single-stage regulators are not completely prohibited. Their continued use on portable appliances is specifically permitted to prevent possible confusion.

3-2.7 Piping System Service Limitations. This subsection describes the physical state (vapor or liquid) and pressure at which LP-Gas shall be permitted to be transmitted through piping systems under various circumstances:

With the exception of 3-2.7(b) covering plastic piping, these provisions reflect the combustion explosion and fire hazard inside structures, as determined by the quantity of flammable mixtures likely to be formed as a result of escape of LP-Gas from a system. This quantity, in turn, is expressed in terms of the physical state (liquid or gas) and, for gas, the pressure which determines the amount of gas released through an opening in a piping system.

(a) LP-Gas liquid or vapor shall be permitted to be piped at all normal operating pressures outside of buildings.

(b) Polyethylene piping systems shall be limited to

1. Vapor service not exceeding 30 psi (208 kPa).
2. Installation outdoors and underground (*see 3-2.8.7*).

This restriction reflects the unsuitability of polyethylene piping where LP-Gas liquid may be present. The presence of liquid propane can lead to the development of excessive pressure and cause loss of strength. Polyethylene is restricted from inside buildings and aboveground installations; it has no resistance to fire and could even release LP-Gas and enhance a fire, also its physical properties could result in its being easily cut or severed. Note that polyethylene is often the preferred piping material in underground service due to its excellent corrosion resistance, compared to metallic piping materials. Polyethylene is completely prohibited in buildings, with no exceptions.

(c) LP-Gas vapor at pressures not exceeding 20 psi (138 kPa) shall be permitted to be piped into any building.

When piping vapor is at elevated pressures, condensation of LP-Gas vapor must be avoided, or liquid can be fed directly to the gas utilization equipment with potentially dangerous results. If the ambient temperature can fall below –5°F (–21°C) at any time, it may be necessary to heat-trace and insulate the outdoor portion of the piping. The condensation point of propane at different pressures is shown below. Piping at pressures exceeding 5 psi (34 kPa), and not exceeding 20 psi (138 kPa), in a building covered by NFPA 54, *National Fuel Gas Code*, paragraph 2-5.1, places additional constraints on the piping system, including approval by the authority having jurisdiction. (*See NFPA 54, 2-5.1.*)

Table 3.1 Vapor Pressure of Propane at Selected Temperatures

Pressure (psi)	Temperature (°F)	Pressure (kPa)	Temperature (°C)
20	−5	140	−21
40	20	276	−7
63	40	434	4

*For more information, see Figures B.1 and B.3.

(d) LP-Gas vapor at pressures exceeding 20 psi (138 kPa) or LP-Gas liquid shall not be piped into any building.

Exception No. 1: Buildings, or separate areas of buildings, constructed in accordance with Chapter 7, and used exclusively to:

(a) House equipment for vaporization, pressure reduction, gas mixing, gas manufacturing, or distribution.

(b) House internal combustion engines, industrial processes, research and experimental laboratories, or equipment or processing having a similar hazard.

NOTE: Complete compliance with Chapter 7 for buildings, or separate areas of buildings, housing industrial processes and other occupancies cited in 3-2.7(d), Exception No. 1(b), is not always necessary depending upon the prevailing conditions. Construction of buildings or separate areas of buildings housing certain internal combustion engines is covered in NFPA 37, *Standard for the Installation and Use of Stationary Combustion Engines and Gas Turbines.*

Exception No. 2: Buildings or structures under construction or undergoing major renovation, provided the temporary piping meets the provisions of 3-4.2 and 3-4.10.2.

Exception No. 3: In buildings or structures other than those covered by 3-2.7(d), Exceptions Nos. 1 and 2, in which liquid feed systems are used, liquid piping shall be permitted to enter the building or structure to connect to a vaporizer provided heavy walled seamless brass or copper tubing not exceeding 3/32-in. (2.4-mm) internal diameter and with a wall thickness not less than 3/64 in. (1.2 mm) is used.

The quantity of liquid that can escape from such piping is limited by the small pipe diameter. The likelihood of leakage is small because of the pipe wall thickness specified.

The occupancies listed in the three exceptions to 3-2.7(d) are those into which LP-Gas vapor over 20 psi (138 kPa) can be piped. In each case there is a need to provide LP-Gas vapor over 20 psi (138 kPa), which can create greater hazards than LP-Gas vapor at or under 20 psi (138 kPa) in the case of leakage. In each of the cases covered in the exceptions, LP-Gas is either the product being handled or a major raw material; installations are temporary and subject to increased supervision; or the operator will have extensive training in the safe handling of LP-Gas.

The note following Exception 1(b) refers to NFPA 37, *Standard for the Installation and Use of Stationary Combustion Engines and Gas Turbines*, which provides a complete source of information on installation of engines using all fuels.

3-2.8 Installation of Pipe, Tubing, Pipe and Tubing Fittings, Valves, and Hose.

Following are all the rules for installation of pipe, tubing, fittings (including valves), and hose. It is essential that these rules be followed because piping is exposed to a great number of hazards that could result in its failure, with accompanying risk of fire.

3-2.8.1 All metallic LP-Gas piping shall be designed and installed in accordance with ASME B31.3, *Chemical Plant and Petroleum Refinery Piping*. All welding and brazing of metallic piping shall be in accordance with ASME *Boiler and Pressure Vessel Code*, Section IX.

This new provision in the 1995 edition requires piping to be installed in accordance with the American Society of Mechanical Engineers Standard B31.3, *Chemical Plant and Petroleum Refinery Piping*, and for the piping to be installed in accordance with Section IX of the ASME Boiler and Pressure Vessel Code. Up to the 1995 edition, NFPA 58 has provided specific minimum pipe specifications and thicknesses to be used for different LP-Gas services. These can continue to be used as they meet or exceed the requirements of ASME B31.3. As B31.3 is a difficult standard to understand for those not involved in piping design and installation, most installers will probably continue to use the specific guidance of this section. The adoption of Section IX requires that all welders be qualified. This is done by completing a welding qualification test, which involves submitting of welding samples and their examination, in accordance with the procedures established in Section IX.

3-2.8.2 Metallic piping shall comply with the following:

(a) Piping used at pressures higher than container pressure, such as on the discharge side of liquid transfer pumps, shall be suitable for a working pressure of at least 350 psi (2.4 MPa).

The maximum design pressure for containers and piping is normally 250 psi. This is one of two exceptions to this pressure limit (the other is for vehicle fuel containers in certain locations); it recognizes that higher pressures can exist and specifically mentions pumps that generate higher pressures during normal operation.

(b) Vapor LP-Gas piping with operating pressures in excess of 125 psi (0.9 MPa), and liquid piping not covered by 3-2.8.2(a), shall be suitable for a working pressure of at least 250 psi (1.7 MPa).

At a temperature of 70°F (21°C), the pressure in a container of propane (the most volatile LP-Gas covered by NFPA 58) is in the range of 125 to 130 psi (about 0.9 MPa).

(c) Vapor LP-Gas piping subject to pressures of not more than 125 psi (0.9 MPa) shall be suitable for a working pressure of at least 125 psi (0.9 MPa).

Exception: Safety relief discharge piping (see 3-2.5.3).

3-2.8.3 Metallic pipe joints shall be permitted to be threaded, flanged, welded, or brazed using pipe and fittings complying with 2-4.2 and 2-4.4 as follows:

(a) When joints are threaded or threaded and back welded:

1. For LP-Gas vapor at pressures in excess of 125 psi (0.9 MPa), or for LP-Gas liquid, the pipe and nipples shall be Schedule 80 or
2. For LP-Gas vapor at pressures of 125 psi (0.9 MPa) or less, the pipe and nipples shall be Schedule 40 or heavier.

(b) Where joints are welded or brazed:

1. The pipe shall be Schedule 40 or heavier.
2. The fittings or flanges shall be suitable for the service for which they are to be used.
3. Brazed joints shall be made with a brazing material having a melting point exceeding 1,000°F (538°C).

(c) Gaskets used to retain LP-Gas in flanged connections in piping shall be resistant to the action of LP-Gas. They shall be made of metal or other suitable material confined in metal having a melting point over 1,500°F (816°C) or shall be protected against fire exposure. When a flange is opened, the gasket shall be replaced.

Exception No. 1: Aluminum O-rings and spiral wound metal gaskets shall be permitted.

Exception No. 2: Nonmetallic gaskets used in insulating fittings shall be permitted.

Gaskets are an essential component of flanged piping systems that maintain the integrity of piping systems by preventing leakage at flanges. This requirement ensures this integrity under normal and fire conditions. Replacement of gaskets after flange opening is important, as used (reinstalled) gaskets will often leak. The minimum melting point of 1500°F (816°C) for gaskets excludes the use of elastomeric (rubber) gaskets, unless they are protected from failure during a fire.

For additional information on pressure piping systems see ANSI B31, *Code for Pressure Piping*, which includes ANSI B313, *Chemical Plant and Refinery Piping*. This excellent code provides detailed requirements for all piping system components, but may be more than the typical LP-Gas installer needs or can easily comprehend.

Table 3.2 Use of Pipe Fittings in LP-Gas Service

Service	Schedule 40	Schedule 80
liquid	welded	threaded/welded
vapor over 125 psi (0.9 MPa)	welded	threaded/welded
vapor 125 psi (0.9 MPa) or less	threaded/welded	threaded/welded

NOTE: Brazing is permitted wherever welding is permitted.

3-2.8.4 Metallic tubing joints shall be permitted to be flared or brazed using tubing and fittings and brazing material complying with 2-4.3 and 2-4.4.

Refer to commentary following 2-4.4(a)(4).

3-2.8.5 Piping in systems shall be run as directly as is practical from one point to another, and with as few restrictions, such as ells and bends, as conditions will permit, giving consideration to provisions of 3-2.8.6.

Where condensation of vapor can occur, metallic and nonmetallic piping shall be pitched back to the container or suitable means provided for revaporizing the condensate.

The number of fittings (ells, tees, etc.) is usually minimized in a piping system when it is designed for cost and ease of installation. Frequently, however, piping is field run or modified, resulting in excessive fittings. The intent of the requirement is to minimize the number of fittings as each fitting is a potential source of leakage, and the more fittings the more resistance to flow and the slower the transfer. This can be especially critical in pump suction piping.

Although the number of fittings must be minimized, the need for flexibility in the piping system cannot be ignored, as the piping will shrink when liquid vaporizes, cooling to about –40°F (–40°C).

The temperature at which LP-Gas vapor will condense is dependent upon which LP-Gas is involved and its pressure. For example, at 10 psi (69 kPa) condensation of propane will occur at about –20°F (–29°C) and below; 20 psi (140 kPa) at about –5°F (–21°C) and below; and 60 psi (414 kPa) at about 30°F (–1°C) and below. (Additional data is available from LP-Gas regulator manufacturers.)

When installing piping between first- and second-stage regulators, precautions must be taken to ensure that any liquid that may condense does not reach the second-stage regulator. This could result in liquid passing through the regulator and vaporizing, in turn overfeeding the appliance and increasing the flame size, which can be a significant fire hazard. This can be prevented by sloping the piping away from the second-stage regulator or by other means.

3-2.8.6 Provision shall be made in piping including interconnecting of permanently installed containers, to compensate for expansion, contraction, jarring and vibration, and for settling. Where necessary, flexible connectors complying with 2-4.6 shall be permitted to be used (*see 3-2.8.9*). The use of nonmetallic pipe, tubing, or hose for permanently interconnecting such containers shall be prohibited.

3-2.8.7 Aboveground piping shall be supported and protected against physical damage. The portion of aboveground piping in contact with a support or a corrosion causing substance shall be protected against corrosion. Where underground piping is beneath driveways, roads, or streets, possible damage by vehicles shall be taken into account. Polyethylene pipe and tubing and thermoplastic compression-type mechanical fittings shall be installed outside underground with a minimum 18 in. (460 mm) of cover. The cover shall be permitted to be reduced to 12 in. (300 mm) if external damage to the pipe or tubing is not likely to result. If a minimum of 12 in. (300 mm) of cover cannot be maintained, the piping shall be installed in conduit or bridged (shielded). Underground polyethylene piping systems shall require assembled anodeless risers to terminate above ground. The horizontal portion of risers shall be buried at least 12 in. (300 mm) below grade and the casing material used for the risers shall be protected against corrosion in accordance with 3-2.12.

The polyethylene pipe or tubing shall be centered inside the aboveground portion of the riser casing to provide an annular air space around the pipe or tubing to prevent excessive temperature buildup in the pipe or tubing.

The factory-assembled riser shall be sealed and leak tested by the manufacturer and the field-assembled riser shall be sealed and leak tested by the installer.

The casing of the riser shall be constructed of ASTM A53 Schedule 40 steel pipe or ASTM A513 mechanical steel tubing with a minimum wall thickness of 0.073 in. (1.9 mm).

This section was revised in the 1992 and 1995 editions covering the installation of polyethylene piping and risers, which are used to make the transition between underground polyethylene piping and aboveground metallic piping. This additional guidance will assist installers in the uses and limitations of polyethylene piping, as its use is expected to continue to grow.

Aboveground piping must be well supported so that there will not be any sags in the piping. In addition, the pipe supports must be substantial so there will not be side pressure that can put stresses on the pipe. Encasing underground pipe with a larger pipe is a common method of providing protection for piping subject to vehicular loads.

Underground polyethylene piping systems must make a transition to metallic piping underground, as polyethylene is prohibited aboveground. This transition must be made by an anodeless riser, a field- or factory-assembled transition fitting of sufficient length, to go from the underground polyethylene pipe or tubing to the aboveground piping.

Exception: The belowgrade casing section of the riser shall be permitted to be flexible metal tubing with a minimum crush strength of 1000 lb (453.6 kg) and a tensile strength of 300 lb (136 kg) including the transition connection as tested by the manufacturer.

The exception was added in the 1992 edition to provide flexibility in the installation of polyethylene tubing. A riser will permit the tubing to be run without transition fittings that connect underground. The restrictions in the materials of construction are intended to ensure that the riser can protect the tubing and maintain its protection by resisting corrosion.

An electrically continuous corrosion-resistant tracer wire (min AWG 14) or tape shall be buried with the polyethylene pipe to facilitate locating. One end shall be brought above ground at a building wall or riser. The wire or tape shall not be in direct contact with the polyethylene pipe.

(a) Polyethylene piping that is installed in a vault or any other belowground enclosure shall be completely encased in gastight metal pipe and fittings that are protected from corrosion.

The requirement that polyethylene piping (which includes pipe, tubing, and fittings) be encased in gastight metal pipe and fittings when installed in a vault recognizes that a vault, while underground, is not encased in soil and must be protected.

(b) Polyethylene piping shall be installed so as to minimize thrust forces caused by contraction or expansion of the piping or by anticipated external or internal loading. The pipeline shall be designed and installed so that each joint will sustain these forces.

NOTE: Polyethylene will expand or contract 1 in. (25 mm) for every 10°F temperature change–for every 100 ft (30.5 m) of pipe.

(c) Polyethylene pipe shall be permitted to be inserted into an existing steel pipe only if done in a manner that will protect the polyethylene from being damaged during the insertion process. The leading end of the polyethylene being inserted shall also be closed prior to insertion.

(d) Polyethylene pipe that is not encased shall have a minimum wall thickness of 0.090 in. (2.3 mm).

Exception: Pipe with an outside diameter of 0.875 in. (22.2 mm) or less shall be permitted to have a minimum wall thickness of 0.062 in. (1.6 mm).

(e) Valve installation in plastic pipe shall be designed so as to protect the pipe against excessive torsional or shearing loads when the valve is being operated. Valve boxes shall be installed so as to avoid transmitting external loads to the valve or pipe. Valves shall be manufactured from thermoplastic materials fabricated from materials listed in ASTM D2513, *Specification for Thermoplastic Gas Pressure Pipe, Tubing and Fittings,* which have been shown to be resistant to the action of LP-Gas and comply with ASTM D2513, or from metals protected to minimize corrosion in accordance with 3-2.8.8. Valves shall be recommended for LP-Gas service by the manufacturer.

This section was added in the 1992 edition as part of the effort to bring the coverage of polyethylene pipe and tubing up to the state of the art.

(f) Each imperfection or damaged piece of polyethylene pipe shall be replaced by fusion or mechanical fittings. Repair clamps shall not be used to cover damaged or leaking sections.

3-2.8.8 Underground metallic piping shall be protected against corrosion as warranted by soil conditions (*see 3-2.12*).

LP-Gas piping shall not be used as a grounding electrode.

3-2.8.9 Flexible components used in piping systems shall comply with 2-4.6 for the service for which they are to be used, shall be installed in accordance with the manufacturer's instructions, and shall also comply with the following:

(a)* Flexible connectors in lengths up to 36 in. (1 m) (*see 2-4.6.3 and 2-4.6.4*) shall be permitted to be used for liquid or vapor piping, on portable or stationary tanks, to compensate for expansion, contraction, jarring, vibration, and settling.

(b) Hoses shall be permitted to be installed if flexibility is required for liquid or vapor transfer. The use of wet hose is recommended for liquid.

A-3-2.8.9(a) This is not to be construed to mean that flexible connectors must be used if provisions were incorporated in the design to compensate for these effects.

Flexibility is often necessary in LP-Gas piping to allow for thermal expansion and contraction of the piping and settling of containers and equipment. Flexible connections often fit in very well in an LP-Gas system but should not be used unless necessary because their expected life is generally not as long as that of permanent piping. Flexible connectors must be limited in length because the possibility of leakage increases with length. Flexibility can be designed into metallic piping systems, and an engineer experienced in piping design should be consulted if required.

3-2.8.10 On new installations, and by December 31, 1980, on existing installations,

(1) stationary single container systems of over 4,000 gal (15.1 m^3) water capacity, or

(2) stationary multiple container systems with an aggregate water capacity of more than 4,000 gal (15.1 m^3) utilizing a common or manifolded liquid transfer line, or

(3) railroad tank car transfer systems to fill trucks with no stationary storage involved

shall comply with the following:

(a) Where a hose or swivel-type piping 1½ in. (38 mm) or larger is used for liquid transfer or a 1¼-in. (32-mm) or larger vapor hose or swivel-type piping is used in this service, an emergency shutoff valve complying with 2-4.5.4 shall be installed in the fixed piping of the transfer system within 20 ft (6 m) of lineal pipe from the nearest end of the hose or swivel-type piping to which the hose or swivel-type piping is connected. Where either a liquid or vapor line has two or more hoses or swivel-type piping of the sizes designated, an emergency shutoff valve or a backflow check valve shall be installed in each leg of the piping.

Exception: Where the flow is only in one direction into the container, a backflow check valve shall be permitted to be used in lieu of an emergency shutoff valve if installed in the fixed piping downstream of the hose or swivel-type piping, provided the backflow check valve has a metal-to-metal seat or a primary resilient seat with a secondary metal seat not hinged with combustible material.

1. Emergency shutoff valves shall be installed so that the temperature sensitive element in the valve, or a supplemental temperature sensitive element [250°F (121°C) maximum] connected to actuate the valve, is not more than 5 ft (1.5 m) from the nearest end of the hose or swivel-type piping connected to the line in which the valve is installed.

2. Temperature-sensitive elements of emergency shutoff valve shall not be painted nor have any ornamental finishes applied after manufacture.

Formal Interpretation 79-1
Reference: 3-2.8.10(a)

Question: Is it the intent of 3-2.8.10(a) to require either an emergency shutoff valve or a backflow check valve in each leg of the piping when two or more hoses are used?
Answer: Yes.

Committee Comment: Unless these provisions are made, it would be possible for flow from one leg of the piping to escape through a leak in the other leg.

Issue Edition: 1979
Reference: 3168(a)
Date: November 1979 ■

Formal Interpretation 79-2
Reference: 3-2.8.10(a)

Question: In an LP-Gas installation subject to the provisions of 3-2.8.9 of NFPA 58 by virtue of the container capacity qualifications, the vapor piping used in liquid transfer operations is 1¼-in. nominal size. However, a vapor hose permanently affixed to the delivery end of this piping (by the use of a 1¼-in.- to 1-in. reducing elbow) is 1-in. nominal size. No backflow check valve is installed in this piping.
Is it the intent of 3-2.8.10(a) of NFPA 58 to require that an emergency shutoff valve be installed in the fixed vapor piping?
Answer: No.

Committee Comment: The Committee notes that, in the absence of either an emergency shutoff valve or a backflow check valve, 3-3.3.8(a) or (b) would require an excess-flow valve in the fixed vapor piping cited.

Issue Edition: 1979
Reference: 3168(a)
Date: December 1980 ■

(b) The emergency shutoff valve(s) or backflow check valve(s) specified in 3-2.8.10(a) shall be installed in the plant piping so that any break resulting from a pull will occur on the hose or swivel-type piping side of the connection while retaining intact the valves and piping on the plant side of the connection. Provisions for anchorage and breakaway shall be provided on the cargo tank side for transfer from a railroad tank car directly into a cargo tank.

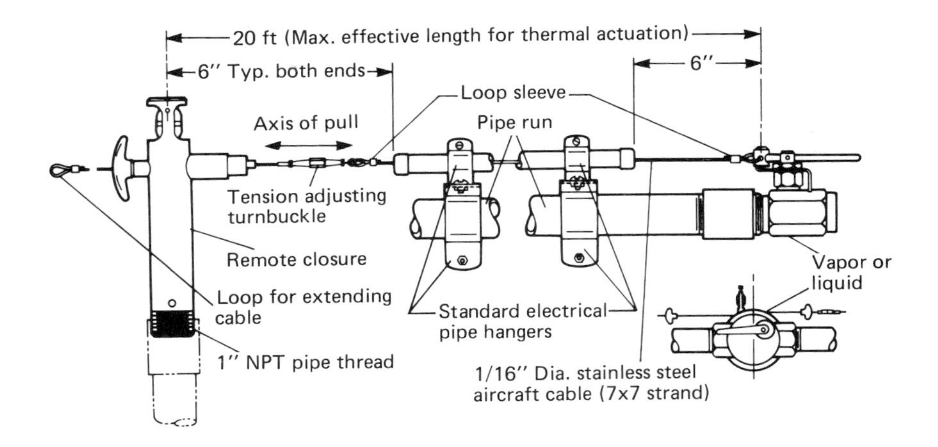

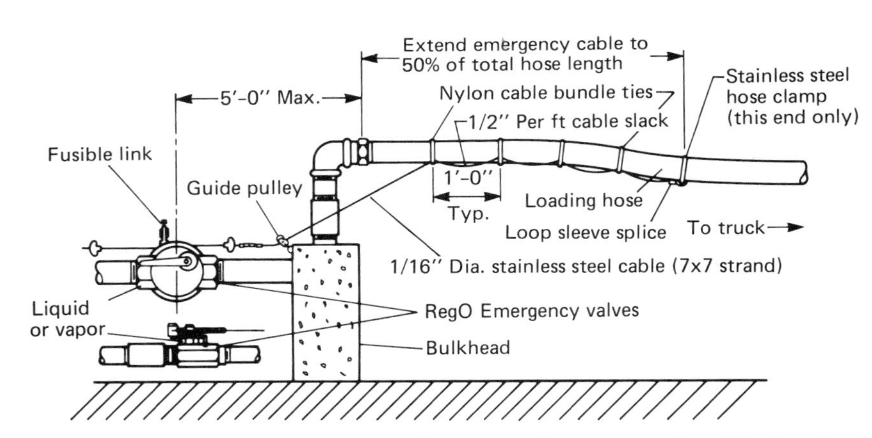

Figure 3.15 *Installation of mechanically operated emergency shutoff valve. (top) Conventional. (bottom) Incorporates hose pullaway protection. (Drawing courtesy of the National Propane Association.)*

This provision first appeared in the 1976 edition of the standard and was a rare incidence of retroactive application. It resulted from an extensive study of accidents—especially of those that resulted in a BLEVE of a stationary container.

The accident experience was dominated by release of liquid and vapor during liquid transfer operations as a result of hose failures, hose coupling failures, and piping and component failures. In many instances, these failures resulted from a cargo vehicle being driven away before the transfer hose was disconnected. In some cases, valves were torn out of the vehicle's female connections—specially brass valves inserted into steel fittings.

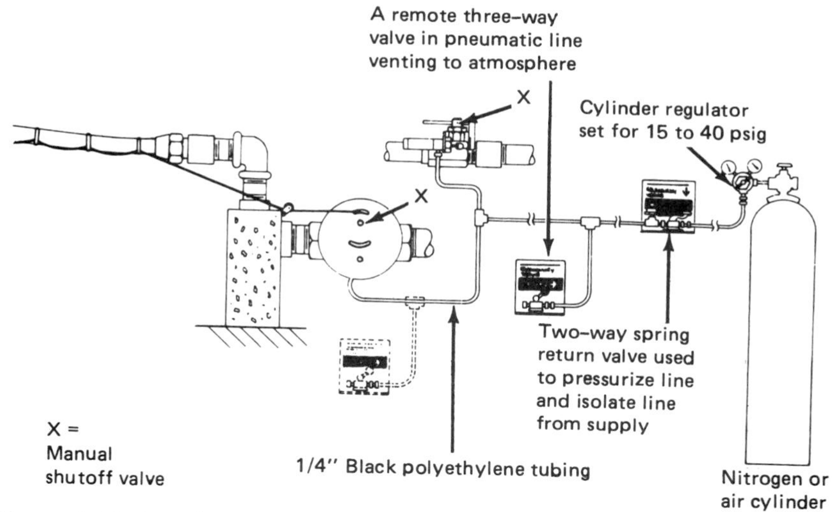

Figure 3.16 *Installation of pneumatically operated emergency shutoff valve. (Drawing courtesy of the National Propane Gas Association.)*

This provision essentially requires an automatic and manual means of stopping the escape of LP-Gas (other than by an excess-flow check valve) from either side of a leak on installations where both the capacity of the container (or containers) and the diameter of the hoses or swivel-type piping exceed those stipulated. There are installations where the container capacity qualifies, but not the size of the hoses or swivel-type piping. (*See Formal Interpretation 79-2 on page 213.*)

Where flow is in two directions, an emergency shutoff valve complying with 2-4.5.4 is used. In 1976, compact, simple, and economical valves having the desired characteristics were not available. As a result, the retroactive date had to be amended by Tentative Interim Amendments (TIAs) to provide time for these valves to be manufactured, tested, and listed (listing is not required but the manufacturers felt that it was highly desirable, and this facilitated approval by the authorities having jurisdiction). Where flow is only in one direction, a check valve can be used provided it has a degree of fire resistance.

Automatic actuation of an emergency shutoff valve occurs through sensing of heat from a fire [3-2.8.10(a)(1)] with the sensing element located near the source of leakage. In most instances, the sensing element is a point source fusible element or pressurized plastic tubing. The plastic tubing system is actually a line sensor capable of sensing fire at other locations (as could be a system using several fusible elements).

In addition to automatic actuation, an emergency shutoff valve must be installed so that it can be operated manually from a remote location and also at its installed location.

Subparagraph 3-2.8.10(b) reflects the cargo vehicle "pullaway" experience. The provision was modified in the 1992 edition to make it applicable to transfer from railroad tank cars directly into tank trucks with no intermediate storage. Changes were also made to 4-2.3.8 to provide for all aspects of this growing practice. The reported increase in use of trackside transfer results from abandonment of mainline track in some parts of the United States forcing LP-Gas distributors to send trucks to unload tank cars that formerly were able to be delivered to a bulk plant. The standard requires the same safety and pullaway protection for trackside transfer as for fixed installations.

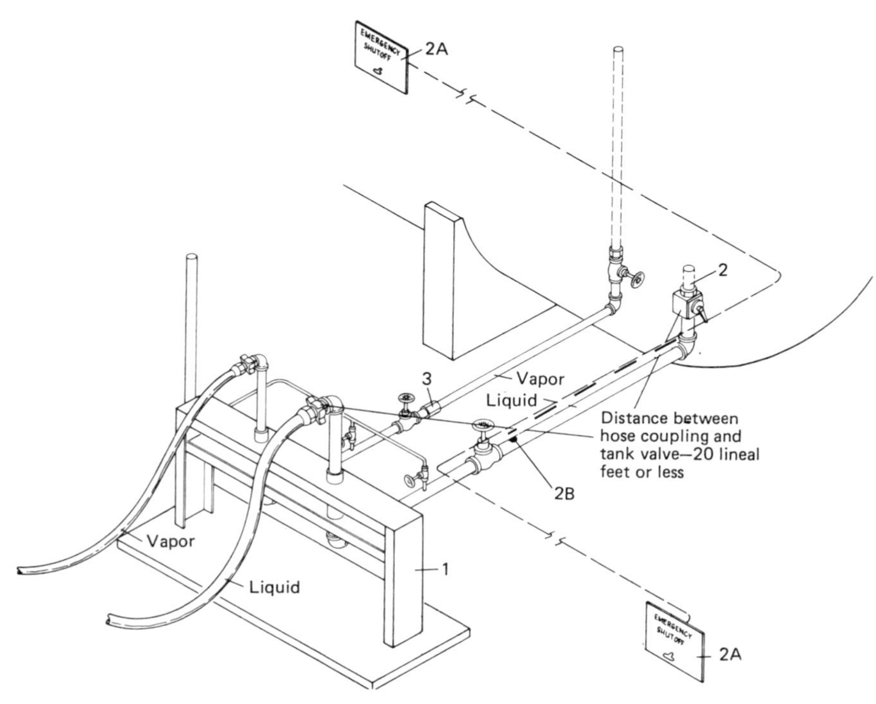

Figure 3.17 Installation of emergency shutoff valves on ASME storage containers. (1) Concrete bulkhead or equivalent anchorage, or use of a predictable breakaway point. (2) Emergency valve with means for manual shutoff at the valve; (2A) Remote shutoff control; (2B) Thermal release within 5 ft from the nearest end of the hose, or swivel-type piping. (3) Vapor line does not require an emergency valve if the hose diameter is less than $1^1/_4$ in. If pipe is larger, an emergency shutoff valve must also be installed in this line. (Drawing courtesy of National Propane Gas Association.)

Exception: Such anchorage shall not be required for tank car unloading.

NOTE: This can be accomplished by use of concrete bulkheads or equivalent anchorage or by the use of a weakness or shear fitting.

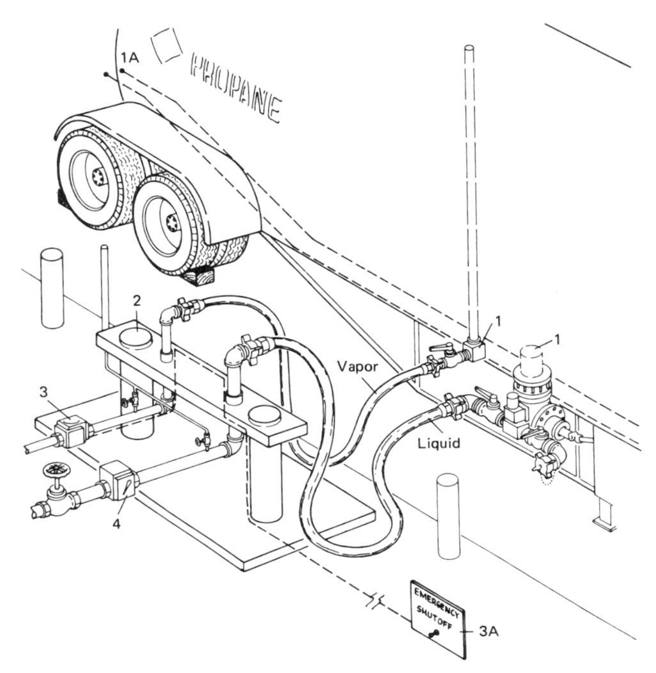

Figure 3.18 Installation of emergency shutoff valves at transport unloading station. Transports must have the following safety devices unless emergency valves are provided at transport end of hoses. (1) Internal valves, which must include a manual means for shutoff at internal valve, and thermal release (temperature sensitive element) at internal valve, within 5 ft of hose. (1A) Remote control to close internal valves. Note: Drawing does not include all DOT requirements.

Plant transport unloading area must have the following safety devices: (2) Concrete bulkhead or equivalent anchorage, with a predictable breakaway point to retain intact the valves and piping on the plant side of the connections; (3) Vapor line must have an emergency valve with means for manual shutoff, thermal release, and remote control provisions; (3A) Remote shutoff control; (4) Liquid line must have either a backflow check valve (may be part of sight flow unit) or an emergency valve with means for manual shutoff, thermal release, and remote shutoff control. (Drawing courtesy of National Propane Gas Association.)

3-2.8.11 Hose shall be permitted to be used on the low-pressure side of regulators to connect to other than domestic and commercial appliances as follows:

The prohibition of hoses in domestic and commercial applications essentially restricts them to agricultural and industrial applications. Because of their vulnerability to mechanical and thermal damage, and their limited service life, they should be used only where really necessary and where they can be readily inspected and maintained.

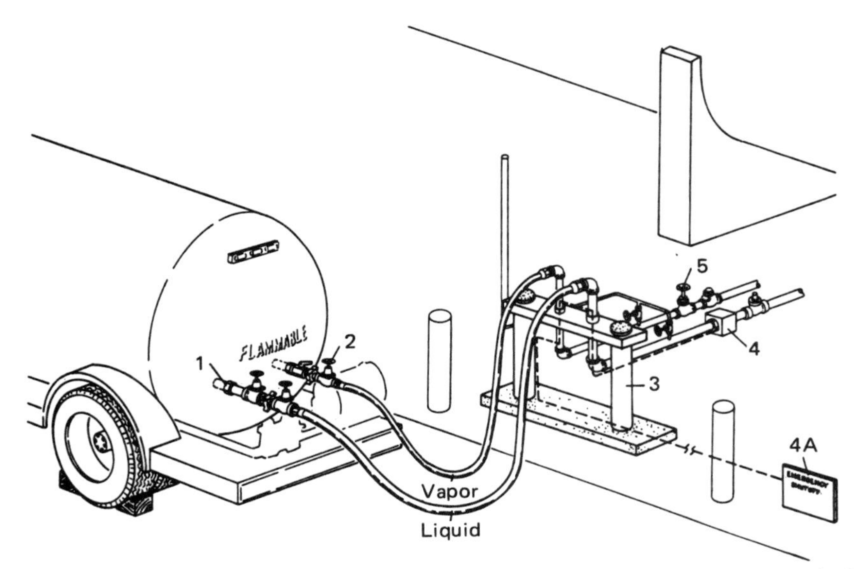

Figure 3.19 Installation of emergency shutoff valves at delivery cargo vehicle loading station. Cargo vehicle being filled directly into tank must have the following devices: (1) Liquid fill must have a backflow check valve mounted directly into tank, or an internal valve. (2) Vapor does not require emergency valve because hose diameter is less than 1¼ in.

Plant loading riser must have the following safety devices: (3) Concrete bulkhead, or equivalent anchorage, or use of a predictable breakaway point to retain intact the valve(s) on the plant side of the connection(s). (4) Liquid line must have an emergency valve with means for manual shutoff and thermal release. (4A) Remote shutoff control. (5) Vapor line does not need an emergency valve because the hose diameter is less than 1¼ in. (Drawing courtesy of National Propane Gas Association.)

(a) The appliance connected shall be of a portable type.

(b) For use inside buildings, the hose shall be of a minimum length, not exceeding 6 ft (1.8 m) [except as provided in 3-4.2.3(b)], and shall not extend from one room to another nor pass through any partitions, walls, ceilings, or floors (except as provided in 3-4.3.7). It shall not be concealed from view or used in concealed locations. For use outside buildings, hose length shall be permitted to exceed 6 ft (1.8 m) but shall be kept as short as practical.

(c) Hose shall be securely connected to the appliance. The use of rubber slip ends shall not be permitted.

(d) A shutoff valve shall be provided in the piping immediately upstream of the inlet connection of the hose. Where more than one such appliance shutoff is located near another, precautions shall be taken to prevent operation of the wrong valve.

(e) Hose used for connecting appliances to wall or other outlets shall be protected against physical damage.

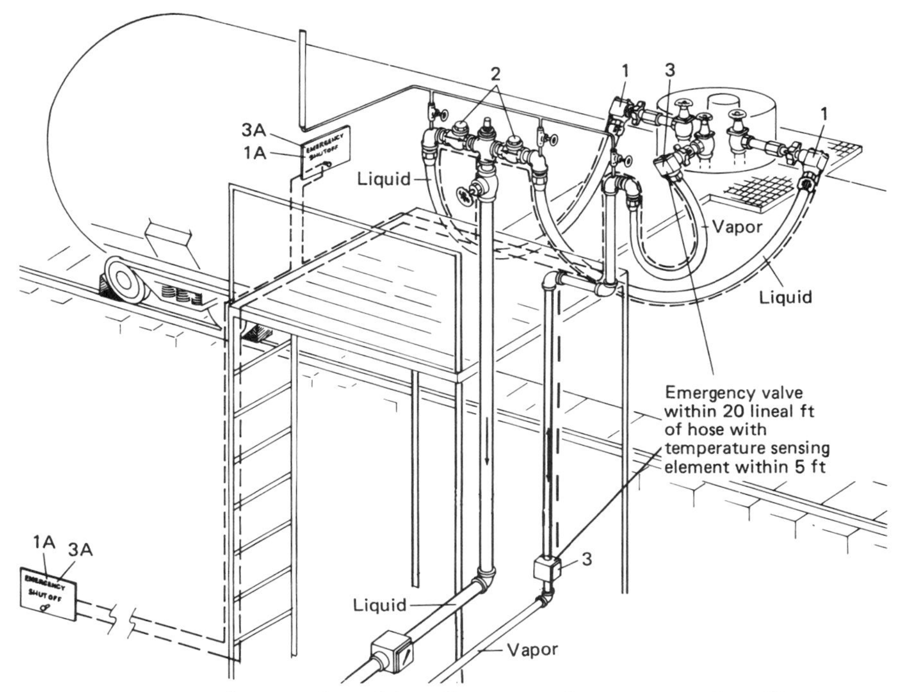

Figure 3.20 Installation of mechanically operated emergency shutoff valves at tank car loading or unloading station. As an alternate, emergency valves may be installed in the tank car unloading adapters [see Figure 3.21]. Tank car unloading risers must have the following safety devices: (1) Liquid hoses at tank car end must have an emergency valve with means for manual shutoff and thermal release. (1A) Remote shutoff control. (2) Riser ends of liquid hose connections must have backflow check valves. (3) Vapor hose must have an emergency valve at each end with means for manual shutoff and thermal release. (3A) Remote shutoff control. (4) When two hoses or swivel-type piping are used on tank car unloading riser, each leg of the piping should be protected by backflow check valve(s) or an emergency shutoff valve. (Drawing courtesy of National Propane Gas Association.)

3-2.9 Hydrostatic Relief Valve Installation. A hydrostatic relief valve complying with 2-4.7 or a device providing pressure-relieving protection shall be installed in each section of piping (including hose) in which liquid LP-Gas can be isolated between shutoff valves so as to relieve the pressure that could develop from the trapped liquid to a safe atmosphere or product-retaining section.

Over a temperature range from 30°F to 90°F (–1°C to 32°C), liquid propane will expand an average of about 1.6 percent for each 10 Fahrenheit degrees (5.5 Celsius degrees) it is heated. If the liquid is trapped in a length of pipe by, for example, being between two closed valves, and there is no

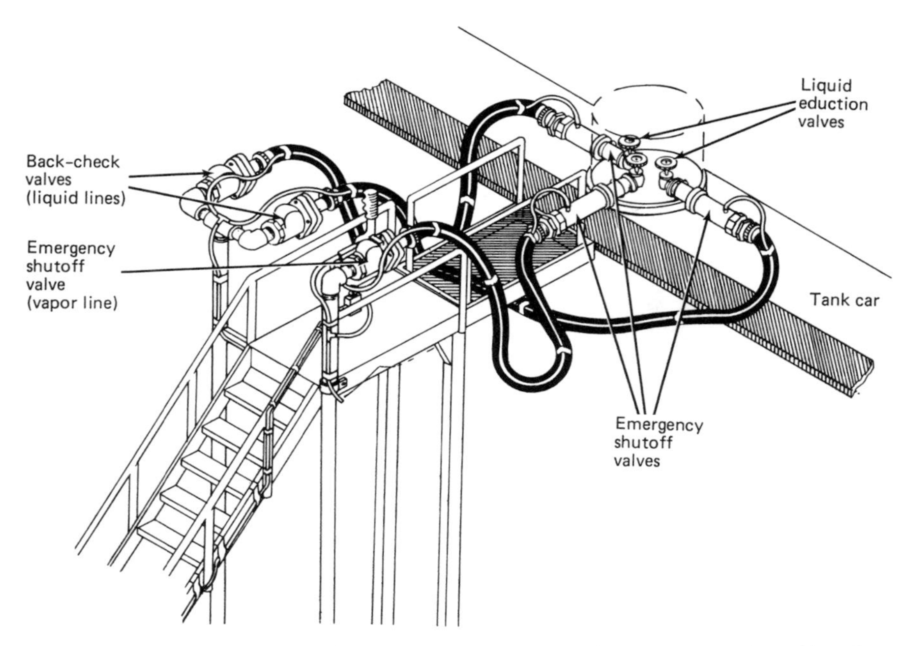

Figure 3.21 *Installation of pneumatically operated emergency shutoff valve/unloading adapter combination at tank car loading or unloading station. See Figure 2.36 for details of ESV/unloading adapter combination. (Drawing courtesy of National Propane Gas Association.)*

room for expansion, the pressure developed can be very high (thousands of psi) and pipe or valve failures can occur. Operation of a hydrostatic relief valve prevents this by discharging liquid. Unlike vapor pressure relief valves, the quantity that needs to be discharged is quite small. Being liquid, however, it represents a greater quantity of vapor and its low temperature presents a personnel hazard.

Some hydrostatic relief valves are constructed so that they use an elastomeric seal that can vulcanize over time and not reseal completely. If this occurs, and the valve operates, it may continuously leak a small amount of liquid after operation. This has not proven to be a hazard, but it may be a good practice to inspect hydrostatic relief valves on a regular basis.

3-2.10 **Testing Piping Systems.** After assembly, piping systems (including hose) shall be tested and proven free of leaks at not less than the normal operating pressure. Piping within the scope of NFPA 54 (ANSI Z223.1), *National Fuel Gas Code* [*see 1-1.3.1(f)*], shall be pressure tested in accordance with that code. Tests shall not be made with a flame.

Figure 3.22 *Rail car unloading using hoses and bobtail filling site with pullaway protection. Note the small tank in the foreground is not part of the liquid transfer operations, and the wheel chock sign at the tank car. (Photo courtesy of Ace Gas Company.)*

After assembly of piping (but before appliances are connected) the industry practice is to admit full container pressure to the system and check all connections for leaks with a soap or leak-testing solution. There are leak test solutions listed for this purpose. (Some soap solutions can be corrosive to piping.) It is important, particularly with copper and brass tubing or fittings, that the leak test solution contain no ammonia.

A widely used and very sensitive leakage test is that of using a tee block fitting with a pressure gauge between the container service shutoff valve and the first-stage regulator. This test is described in Appendix D of NFPA 54, National Fuel Gas Code. Essentially, it consists of admitting full container pressure to the system, closing the container shutoff valve, and lowering the pressure reading on the gauge 10 psi by bleeding off a small amount of gas in the system. A very small leak will be accentuated in the small volume involved at the pressure gauge location. If leakage occurs, the source is detected by checking all fittings and connections with a leak-testing solution.

Bearing in mind that a high proportion of LP-Gas systems from the outlet of the first-stage regulator are covered by NFPA 54, *National Fuel Gas Code*, reference should be made to Part 4 of that standard, "Gas Piping Inspection, Testing, and Purging."

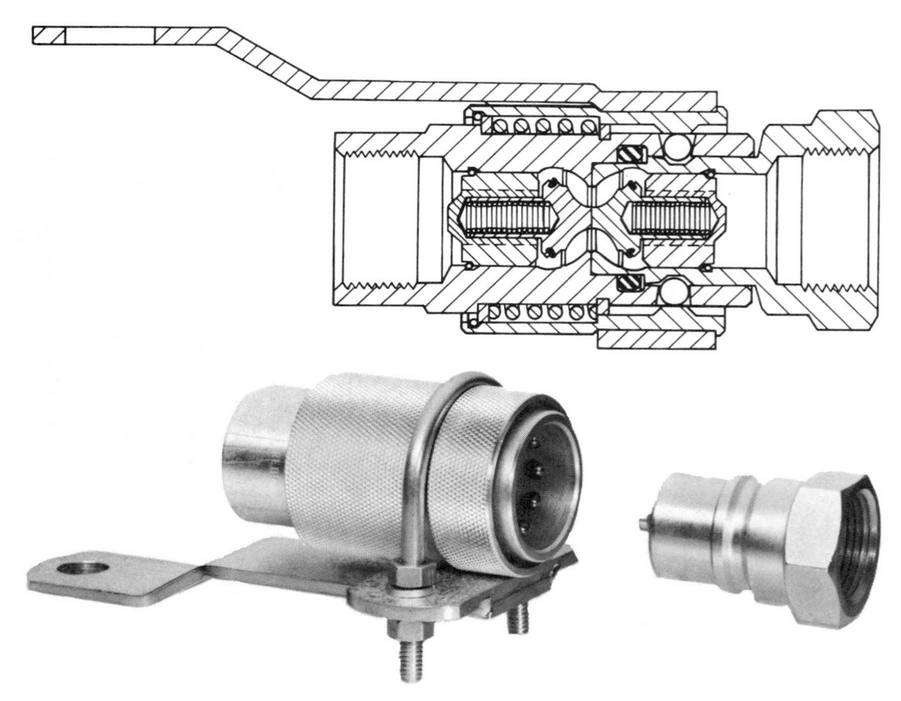

Figure 3.23 Pullaway valve. Valve closes automatically if vehicle is moved with hose still connected. This is a way to incorporate a "predictable breakaway point." (Photo and drawing courtesy of Engineered Controls, Inc.)

3-2.11 Areas of Heavy Snowfall. In areas where heavy snowfall can be expected, piping, regulators, meters, and other equipment installed in the piping system shall be protected from the forces anticipated as a result of accumulated snow.

This requirement was added in the 1995 edition in conjunction with revisions to 1-1.2(a), 1-1.3.1(f), and 3-2.4.2(d)2 as a result of failures of LP-Gas systems in the Sierra Mountains during a period of exceptionally heavy snowfall. This added requirement recognizes that piping, meters, and regulators outside of buildings can be damaged by the force of falling snow. Specific protection requirements have not been provided by the Technical Committee as they don't believe they are in a position to make one rule to cover all potential snowfalls. In the Sierra Mountain area, a model law to cover this subject has been proposed, and several municipalities have enacted specific rules in this area.

3-2.12 Corrosion Protection. All metallic equipment and components that are buried, mounded, or otherwise considered to be "in the ground" shall be protected and maintained to minimize corrosion.

Figure 3.24 Transfer from a tank car. (Photo courtesy of Plant Systems, Inc.)

This provision was added here in the 1995 edition to provide a corrosion protection requirement applicable to all LP-Gas installations covered in Chapter 3. Most soils can corrode metallic pipe. While protection has traditionally been provided by coatings, cathodic protection is increasingly being used.

If coating is used, fittings such as collars must be coated after installation. Coated piping should not be installed in soil where there are rocks that can scrape the coating and expose the pipe to corrosion. It is also important that, where dissimilar metals are connected together, an insulating fitting is installed so as to eliminate cathodic action on the piping. If this insulating fitting is not used, the piping system can be adversely affected rather rapidly.

NOTE: For information on protection of underground components see NACE RP-0169, *Control of External Corrosion on Underground or Submerged Metallic Piping Systems.*

Corrosion protection of all other materials shall be in accordance with accepted engineering practice.

3-2.13 Equipment Installation.

3-2.13.1 Pumps shall be installed as recommended by the manufacturer and in accordance with the following:

(a) Installation shall be made so that the pump casing shall not be subjected to excessive strains transmitted to it by the suction and discharge piping. This shall be accomplished by piping design, the use of flexible connectors or expansion loops, or by other effective methods, in accordance with good engineering practice.

(b) Positive displacement pumps shall be installed in accordance with 2-5.2.2.

1. The bypass valve or recirculating device to limit the normal operating discharge pressure to not more than 350 psi (2.4 MPa) shall discharge either into a storage container (preferably the supply container from which the product is being pumped) or into the pump suction.
2. If this primary device is equipped with a shutoff valve, an adequate secondary device designed to operate at not more than 400 psi (2.8 MPa) shall, if not integral with the pump, be incorporated in the pump piping. This secondary device shall be designed or installed so that it cannot be rendered inoperative and shall discharge either into the supply container or into the pump suction.

(c) A pump operating control or disconnect switch shall be located near the pump. Remote control points shall be provided as necessary for other plant operations such as container filling, loading or unloading of cargo vehicles and tank cars, or operation of motor fuel dispensers.

Because LP-Gas liquid under pressure in a container or pipeline will vaporize when the pressure is reduced, and the vapor thus produced can lead to cavitation in a pump, the types of pumps commonly used in installations covered by NFPA 58 are installed so that they have a positive suction head. In practice, this means that their suction inlets are well below the lowest liquid level in a container, and the suction piping is large enough to minimize friction loss.

The suction inlet pressures are substantial—e.g., up to 250 to 312.5 psi (1.7 to 2.2 MPa) for propane, therefore, discharge pressures can also be substantial. In order to permit the use of readily available and economical piping system components, the standard limits discharge pressures to a normal maximum of 350 psi (2.4 MPa) with transient maximums up to 400 psi (2.8 MPa).

These pressures, combined with the totally off or totally on operation in the typical installation, require provisions to prevent excessive vibration and strain from inlet and discharge connections.

Many pumps are started and stopped from a location remote from the pump—e.g., cylinder filling areas and dispensers. For the safety of anyone who might be working on a pump, 3-2.13.1(c) requires a means for preventing startup to be located near the pump.

3-2.13.2 Compressors shall be installed as recommended by the manufacturer and in accordance with the following:

(a) Installation shall be made so that the compressor housing shall not be subjected to excessive strains transmitted to it by the suction and discharge piping. Flexible connectors shall be permitted to be used where necessary to accomplish this.

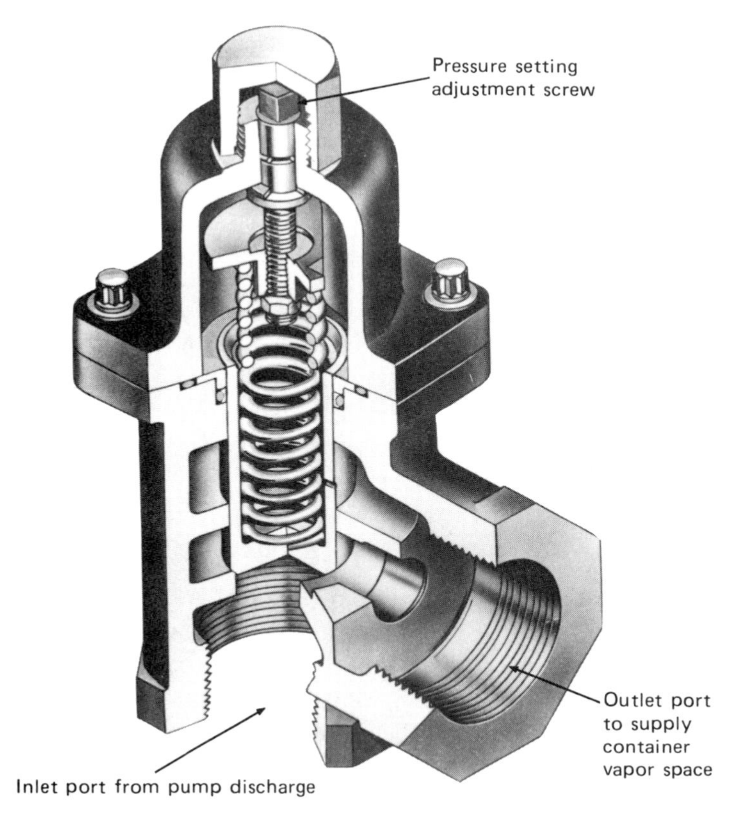

Figure 3.25 *Automatic pump bypass valve. (Photo courtesy of Fisher Controls Company.)*

(b) If the compressor is not equipped with an integral means to prevent the LP-Gas liquid entering the suction (*see 2-5.3.3*), a suitable liquid trap shall be installed in the suction piping as close to the compressor as practical.

Exception: Portable compressors used with temporary connections are excluded from this requirement unless used to unload railroad tank cars.

Note that the exception removes the requirement for a liquid trap to protect a compressor from drawing liquid that can damage the compressor from portable compressors with temporary connections. These are usually small, portable gasoline engine-driven units that are attended, and the operator provides the necessary safety. The exception does not apply to compressors used to unload tank cars, which are not small and are less likely to be as closely attended.

The use of a compressor piped with four-way valves for liquid transfer is shown. First, vapor is compressed from the receiver tank into the vapor

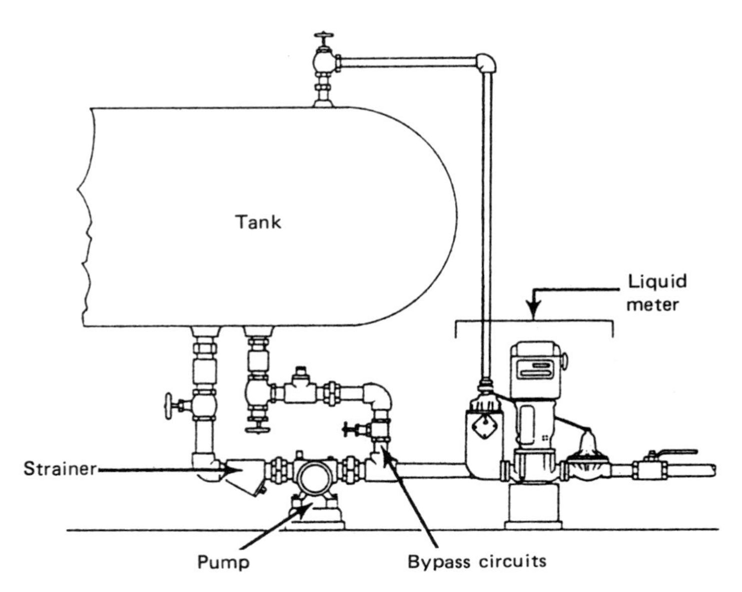

Figure 3.26 Typical pump and meter installation. (Drawing courtesy of the National Propane Gas Association.)

space of the supply tank until liquid flow ceases. Next, the valves are operated and vapor is taken from the supply tank, compressed, and discharged into the bottom of the receiver tank, where it bubbles up through the liquid, cools, and partially liquefies. This is continued until the supply tank pressure drops to 25–30 percent of the starting pressure.

(c) Engines used to drive portable compressors shall be equipped with exhaust system spark arresters and shielded ignition systems.

See commentary on 2-5.1.4.

3-2.13.3 Strainers shall be installed so that the strainer element can be serviced.

3-2.13.4 Liquid or vapor meters shall be installed as recommended by the manufacturer, and in compliance with the applicable provisions of the following:

(a) Liquid meters shall be securely mounted and shall be installed so that the meter housing is not subject to excessive strains from the connecting piping. If not provided in the piping design, flexible connectors shall be permitted to be used where necessary to accomplish this.

(b) Vapor meters shall be securely mounted and installed so as to minimize the possibility of physical damage.

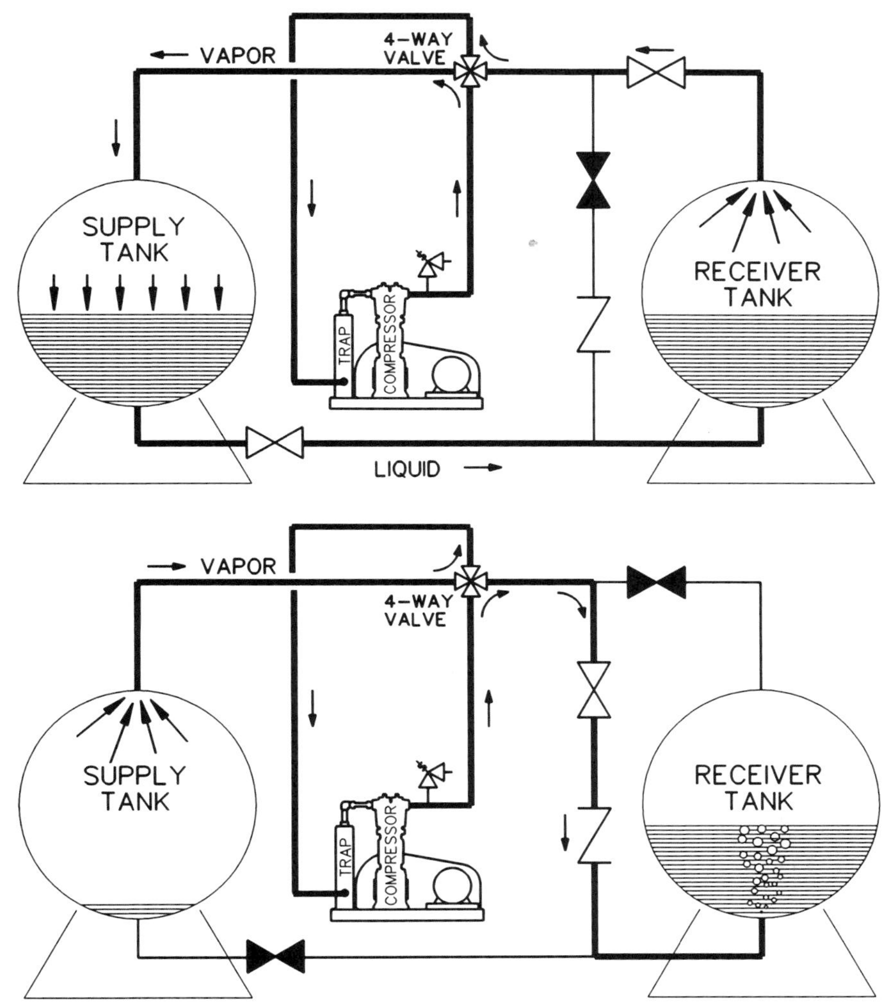

Figure 3.27 *Transferring with a compressor—liquid transfer (top) and vapor recovery (bottom). (Drawing courtesy of Blackmer.)*

3-3 Bulk Plant and Industrial LP-Gas Systems.

Section 3-3 is primarily a compilation of provisions scattered elsewhere in the standard that are applicable to facilities where larger quantities of LP-Gas are used and liquid transfer operations are frequent. To this extent, its purpose is one of convenience to those concerned with such facilities.

Much of Section 3-3 consists of references to other provisions that are commented upon therein.

3-3.1 Application. This section includes provisions for LP-Gas systems installed at bulk plants and industrial plants.

3-3.2 General. The location and installation of storage containers and the installation of container appurtenances, piping, and equipment shall comply with Section 3-2.

3-3.3 Installation of Liquid Transfer Facilities.

3-3.3.1 Points of transfer (*see definition*) or the nearest part of a structure housing transfer operations shall be located in accordance with 3-2.3.2 and 3-2.3.3.

3-3.3.2 Buildings used exclusively for housing pumps or vapor compressors shall be located in accordance with 3-2.3.3 considering the building as one that houses a point of transfer.

3-3.3.3 The track of the railroad siding or the roadway surface at the transfer points shall be relatively level. Adequate clearances from buildings, structures, or stationary containers shall be provided for the siding or roadway approaches to the unloading or loading points. Substantial bumpers shall be provided at the ends of sidings and as necessary to protect storage containers and points of transfer.

The primary reason for the track of railroad sidings and the roadway surface at tank truck transfer points to be level is so that the gauging device in the unit will be as accurate as possible and not be affected by the transport unit being on a slant. Also, it ensures that the entire contents of the rail car or tank truck can be unloaded, as the unloading connections are usually located in the center of the tank. It also helps prevent a "run-away" situation from occurring in the event wheel chocks become dislodged (or are not used).

3-3.3.4 Liquid transfer shall be permitted to be accomplished by pressure differential, by gravity, or by the use of pumps or compressors complying with Section 2-5.

3-3.3.5 Compressors used for liquid transfer normally shall take suction from the vapor space of the container being filled and discharge into the vapor space of the container from which the withdrawal is being made.

Compressors are used to transfer liquid from one container to another by creating a pressure difference, which causes liquid to flow (*see Figure 3.27*). It is important that a compressor draw from the vapor space of a container so that liquid is not drawn into the compressor, as liquid is not compressible and will damage the compressor. A compressor must also discharge into the vapor space of a container to obtain the maximum pressure for transfer.

3-3.3.6 Transfer systems using positive displacement pumps shall comply with 2-5.2.2.

3-3.3.7 Pumps and compressors shall be provided with an operating control or disconnect switch located nearby. Remote shutoff controls shall be provided as necessary in other liquid transfer systems.

This was added in the 1992 edition to provide safety requirements for liquid transfer facilities similar to the requirement for equipment installation in 3-2.13. While somewhat redundant, placing it here helps to ensure that this important safety requirement is not overlooked.

3-3.3.8 Safeguards shall be provided to prevent the uncontrolled discharge of LP-Gas in the event of failure in the hose or swivel-type piping. The provisions of 3-2.8.9 shall apply. For all other LP-Gas systems, the following shall apply:

(a) The connection, or connecting piping, larger than 1/2-in. (12-mm) internal diameter into which the liquid or vapor is being transferred shall be equipped with:

1. A backflow check valve, or
2. An emergency shutoff valve complying with 2-4.5.4, or
3. An excess-flow valve properly sized in accordance with 2-3.7(a)4.

(b) The connection, or connecting piping, larger than 1/2-in. (12-mm) internal diameter from which the liquid or vapor is being withdrawn shall be equipped with:

1. An emergency shutoff valve complying with 2-4.5.4, or
2. An excess-flow valve properly sized in accordance with 2-3.7(a)4.

3-3.3.9 Where a hose or swivel-type piping is used for loading or unloading railroad tank cars, an emergency shutoff valve complying with 2-4.5.4 shall be used at the tank car end of the hose or swivel-type piping.

3-3.3.10 Transfer hoses larger than 1/2-in. (12-mm) internal diameter shall not be used for making connections to individual containers being filled indoors.

3-3.3.11 If gas is to be discharged from containers inside a building, the provisions of 4-3.2.1 shall apply.

3-3.4 Installation of Gas Distribution Facilities.

3-3.4.1 This subsection applies to the installation of facilities used for gas manufacturing, gas storage, gas-air mixing and vaporization, and compressors not associated with liquid transfer.

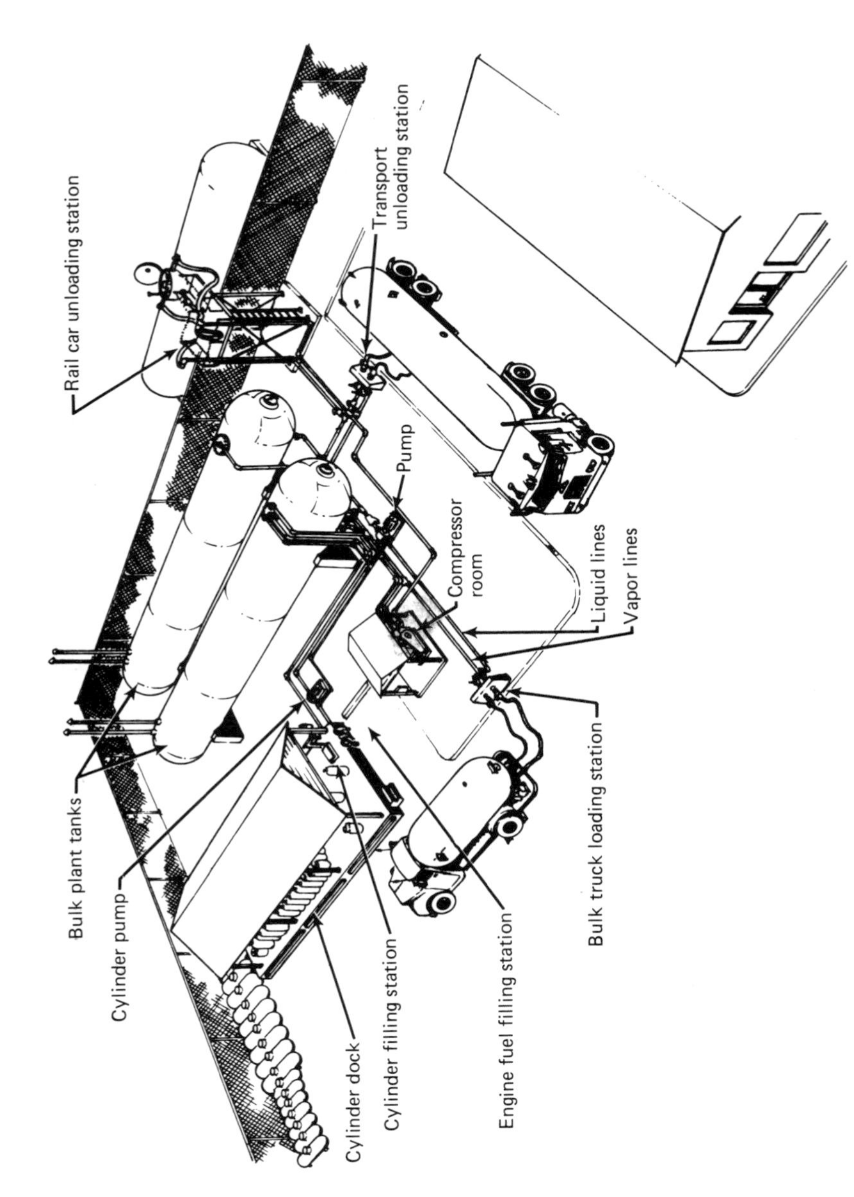

Figure 3.28 *LP-Gas bulk plant. (Drawing courtesy of National Propane Gas Association.)*

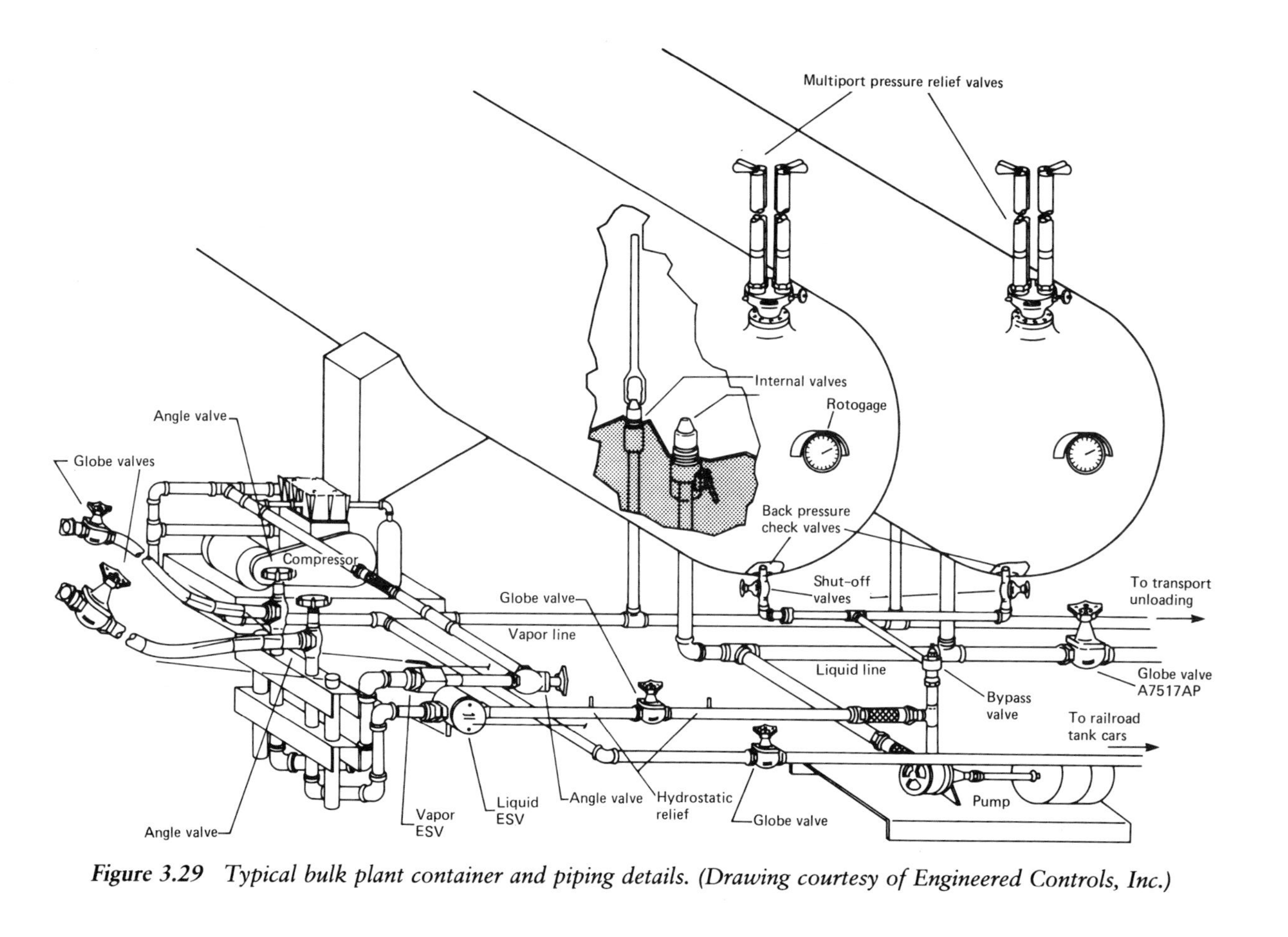

***Figure 3.29** Typical bulk plant container and piping details. (Drawing courtesy of Engineered Controls, Inc.)*

3-3.4.2 Separate buildings and attachments to or rooms within other buildings housing gas distribution facilities, constructed or converted to such use after December 31, 1972, shall comply with Chapter 7.

Exception No. 1: Facilities for vaporizing LP-Gas and gas-air mixing shall be designed, located, and installed in accordance with Section 3-6.

Exception No. 2: Facilities for storing LP-Gas in portable containers at industrial plants and distributing points shall comply with Chapter 5.

3-3.4.3 Separate buildings used for housing vapor compressors shall be located in accordance with 3-2.3.3 considering the building as one that houses a point of transfer.

3-3.4.4 The use of pits to house gas distribution facilities shall be permitted where automatic flammable vapor detecting systems are installed in the pit. Drains or blow-off lines shall not be directed into or in proximity of sewer systems.

3-3.4.5 If gas is to be discharged from containers inside a building, the installation provisions of 4-3.2 shall apply.

3-3.5 **Installation of Electrical Equipment.** Installation of electrical equipment shall comply with Section 3-7.

3-3.6 Protection against Tampering for Section 3-3 and 3-9 Systems. To minimize the possibilities for trespassing and tampering, the area that includes container appurtenances, pumping equipment, loading and unloading facilities, and container filling facilities shall be protected by one of the following methods:

(a) Enclosure with at least a 6-ft (1.8-m) high industrial-type fence, unless otherwise adequately protected. There shall be at least two means of emergency access from the fenced or other enclosure. Clearance shall be provided to allow maintenance to be performed, and a clearance of at least 3 ft (1.0 m) shall be provided to allow emergency access to the required means of egress. If guard service is provided, it shall be extended to the LP-Gas installation. Guard personnel shall be properly trained. (*See Section 1-5.*)

Exception: If a fenced or otherwise enclosed area is not over 100 sq ft (9 m^2) in area, the point of transfer is within 3 ft (1.0 m) of a gate and containers being filled are not located within the enclosure, a second gate shall not be required.

(b) As an alternate to fencing the operating area, suitable devices that can be locked in place shall be provided. Such devices, when in place, shall effectively prevent unauthorized operation of any of the container appurtenances, system valves, or equipment.

These security provisions are appropriate for most circumstances but are not intended to deter those intent upon doing damage. Tests and experience have shown that an LP-Gas container can withstand rifle and pistol projectiles.

Of the two methods described in 3-3.6, fencing is most common. Locking devices are used mostly in dispensing stations (*see definitions in Section 1-6*). While an industrial-type fence is not described in further detail, the chain-link type is the most common. A solid fence should not be used for the reasons cited in the commentary on 3-2.2.9.

Access to and from a fenced enclosure is for both operators/maintainers and emergency personnel, and two exits are required for safety in most cases. The exception in 3-3.6(a) recognizes that both the need to evacuate and the difficulty in doing so are lessened under the circumstances described. As an operational matter, all gates should be open when anyone is inside.

3-3.7 Lighting. If operations are normally conducted during other than daylight hours, adequate lighting shall be provided to illuminate storage containers, containers being loaded, control valves, and other equipment.

3-3.8 Ignition Source Control. Ignition source control shall comply with Section 3-7.

3-4 LP-Gas Systems in Buildings or on Building Roofs or Exterior Balconies.

3-4.1 Application.

3-4.1.1 This section includes installation and operating provisions for LP-Gas systems containing liquid LP-Gas located inside of, or on the roofs or exterior balconies of, buildings or structures. Systems covered include those utilizing portable containers inside of or on the roofs or exterior balconies of buildings, and those in which the liquid is piped from outside containers into buildings or onto the roof. These systems shall be permitted only under the conditions specified in this paragraph and in accordance with 3-4.1 and 3-4.2. Containers in use shall mean connected for use.

The exceptions to 3-2.2.1 list the only situations in which LP-Gas containers can be located inside of buildings. Section 3-4 covers one of these situations, the use of portable container systems inside of or on the roofs of buildings. In addition, this section also covers the piping of liquid LP-Gas into buildings or onto the roofs of buildings from containers located outside. This important requirement of the standard dates back to the first NFPA standard on LP-Gas in 1927 and has been part of the requirements ever since.

Prior to considering any installation of an LP-Gas container in a building, this section (3-4.1, Application) must be read in its entirety.

(a) The portable use of containers indoors shall be only for the purposes specified in 3-4.3 through 3-4.8. Such use shall be limited to those conditions where operational requirements make portable use of containers necessary and location outside is impractical.

The use of portable containers indoors (including balconies of buildings) is permitted only for the purposes specified in this section, and all other uses are prohibited. The use of portable cylinder systems having capacities larger than 1 lb (0.45 kg) of LP-Gas and the associated storage of such cylinders indoors are limited only to uses in construction and renovation of buildings, industrial applications, education, research, training, and temporary heating in cases of emergency. These limited applications acknowledge the good experience and presence of trained personnel in industrial uses and the temporary nature and lack of alternate fuel sources for certain appliances needed at certain times in buildings.

No other uses, including that for normal, routine comfort heating, are permitted. This philosophy was affirmed by the Technical Committee in preparing the 1986 edition of the standard by their rejection of a proposal to permit the use of portable nonvented indoor heating appliances fueled by integral 20-lb (9-kg) LP-Gas containers (cabinet heaters).

Where indoor use of LP-Gas cylinders is permitted in the section, the installer must first attempt to locate the container outdoors and may then install the container indoors, only if it is impractical to locate them outdoors. In determining whether an outdoor location is impractical, consultation with the authority having jurisdiction may be required.

(b) Installations using portable containers on roofs shall be as specified in 3-4.9.1. Such use shall be limited to those conditions where operational requirements make portable use of containers necessary and location not on roofs of buildings or structures is impractical.

Provisions for the installation of portable LP-Gas containers on roofs were incorporated into the standard to provide fuel for emergency generators and microwave relay stations. They have also been applied when providing LP-Gas containers on roofs serving penthouses. Again, such systems are to be used only when the use of portable containers is necessary and other outdoor locations are impractical.

(c) Installations using portable containers on exterior balconies shall be as specified in 3-4.9.2.

This requirement makes it clear that the use of portable containers on balconies and the transportation of containers within buildings are covered by the standard.

(d) Liquid LP-Gas shall be piped into buildings or structures only for the purposes specified in 3-2.7(d).

While LP-Gas vapor at pressures up to and including 20 psi (138 kPa) is permitted to be piped into buildings, there are certain limitations for piping liquid LP-Gas at pressures exceeding 20 psi (138 kPa) into buildings. These are covered in 3-2.7(d) and 3-4.10. Piping liquid LP-Gas into buildings is permitted only for the uses specified in the exceptions to 3-2.7(d).

3-4.1.2 Storage of containers awaiting use shall be in accordance with Chapter 5.

This reference clarifies that stored cylinders, including cylinders that are part of an appliance but are not connected to the appliance, are covered by the standard and that Chapter 5 applies.

3-4.1.3 Transportation of containers within a building shall be in accordance with 3-4.2.7.

3-4.1.4 These provisions shall be required in addition to those specified in Section 3-2.

3-4.1.5 Liquid transfer systems shall be in accordance with Chapter 4.

3-4.1.6 Engine fuel systems used inside buildings shall be in accordance with Chapter 8.

3-4.1.7 LP-Gas transport or cargo vehicles stored, serviced, or repaired in buildings shall be in accordance with Chapter 6.

3-4.2 General Provisions for Containers, Equipment, Piping, and Appliances.

Certain conditions are set out for specific applications of portable container systems in buildings in 3-4.3 through 3-4.8 and for permanent systems under 3-4.9. Subsection 3-4.2 specifies certain general provisions that pertain to all of these specific applications.

3-4.2.1 Containers shall comply with DOT cylinder specifications (*see 2-2.1.3 and 2-2.2.1*), shall not exceed 245 lb (111 kg) water capacity [nominal 100 lb (45 kg) LP-Gas capacity] each, shall comply with other applicable provisions of Section 2-2, and shall be equipped as provided in Section 2-3 [*see 2-3.3 and Table 2-3.3.2(a)*]. They shall also comply with the following:

Basic considerations are given for the type of container to be used. They must be DOT cylinders with a maximum capacity of 100 lb (45 kg) of

LP-Gas. This is the largest size that can be moved by personnel from a practical standpoint. In the past there have been some 500-gal (1.9-m^3) ASME tanks used at different floor levels on large construction sites, but this is at variance with NFPA 58 and must receive special approval. The NFPA 58 approach to the use of ASME bulk tanks in this instance would be to locate them at ground level on the outside and pipe the LP-Gas into the building in accordance with 3-4.10.

The importance of compliance with this requirement was highlighted by an incident where employees in a building were moving a 500-gal (1.9-m^3) ASME propane container with an industrial lift truck. The container was secured on the truck with a length of 2 × 4-in. (50 × 100-mm) wood. The container fell, shearing off a valve and leaving an opening of about $^{11}/_{16}$ in. (17 mm), through which liquid LP-Gas was released. After unsuccessfully attempting repairs, the workers attempted to turn off the electrical supply, left the building, and called the fire department. Just as the fire engine arrived, the propane ignited. The explosion killed five firefighters, two civilians, and injured 59–79 others.[3]

(a) Containers shall be marked as provided in 2-2.6.

(b) Containers with propane capacities greater than 2 lb (0.9 kg) shall be equipped as provided in Table 2-3.3.2(a). Excess-flow valve protection shall be provided for vapor service.

Table 2-3.3.2(a), Container Connections and Appurtenance Requirements for Containers Used on Domestic, Commercial, Industrial, Engine Fuel, and Over-the Road Mobile Applications, provides the requirements for all appurtenances on containers of 2,000 gal or less. Minimum requirements include a manual shutoff valve with an integral external pressure relief valve. Optional appurtenances include a fixed liquid level gauge and float gauge. A change in the 1995 edition added the requirement of excess-flow protection for vapor service for all cylinders used in buildings. This provides a higher level of safety by providing excess-flow protection in the event of a pipe or hose failure inside a building, which could lead to fire. This means that standard vapor service cylinders, including those used for gas grills, cannot be used in buildings unless they are fitted with an excess-flow valve.

(c) Valves on containers shall be protected in accordance with 2-2.4.1.

Paragraph 2-2.4.1 requires that all container valves be protected against physical damage by either recessing the valve into the container or protecting the valve with a ventilated cap or collar. Note that when a removable cap is used it must be in place when the container is not in use.

[3]"The Fatal Explosion in Buffalo." Vol. 51, No. 3. *Fire Command* March 1984; p. 28.

(d) Containers having water capacities greater than 2.7 lb (1.2 kg) filled with no more than 16.8 oz (0.522 kg) of LP-Gas, and connected for use shall stand on a firm and substantially level surface. If necessary, they shall be secured in an upright position.

(e) Containers and the valve protecting devices used with them shall be oriented to minimize the possibility of impingement of the pressure relief device discharge on the container and adjacent containers.

It is important that the container pressure relief discharge is directed through a hole in the cap or collar. In positioning a group of cylinders, attention should be given to ensure that pressure relief valves are not directed at adjacent containers.

3-4.2.2 Regulators, if used, shall be suitable for use with LP-Gas. Manifolds and fittings connecting containers to pressure regulator inlets shall be designed for at least 250-psi (1.7-MPa) service pressure.

3-4.2.3 Piping, including pipe, tubing, fittings, valves, and hose, shall comply with Section 2-4, except that a minimum working pressure of 250 psi (1.7 MPa) shall apply to all components. The following also shall apply:

(a) Piping shall be installed in accordance with the provisions of 3-2.8 for liquid piping or for vapor piping for pressures above 125 psi (0.9 MPa). [*See 3-2.8.2(b).*]

(b) Hose, hose connections, and flexible connectors used shall be designed for a working pressure of at least 350 psi (2.4 MPa), shall comply with 2-4.6, and shall be installed in accordance with 3-2.8.11. Hose length shall be permitted to exceed that specified by 3-2.8.11(b) but shall be as short as practical, although long enough to permit compliance with the spacing requirements (*see 3-4.3.3 and 3-4.3.4*) without kinking or straining hose or causing it to be close enough to a burner to be damaged by heat. See 3-4.9 for permanent roof installations.

General piping provisions are referred to in 3-2.8. However, special attention is given here to the hose used with portable container systems inside buildings. While basic provisions for hose are given through reference to 3-2.8.10 and 2-4.6, two exceptions are noteworthy. First, the hose must be designed for a 350-psi (2.4-MPa) working pressure. This is not so much for the pressures involved, but to ensure a stronger type of hose is used in this service, particularly at construction sites where rough usage may be encountered. (Note that this hose is required in 2-4.6.3 only for pressures over 5 psi for all other applications.) Second, a 6-ft (2-m) length is the general maximum size permitted, but longer lengths are allowed to permit some of the spacing requirements set out in this section. Too short a length could possibly decrease safety. With regard to permanent installations on roofs, the use of hose connections to containers is prohibited. See Formal Interpretation of 3-4.5.

Exception: Hoses at a pressure of 5 psi (34 kPa) or less used in agricultural buildings not normally occupied by the public.

This exception was added in the 1995 edition to provide an alternate to the hose designed for a pressure of 350 psi, to be used at low pressures (below 5 psi) in agricultural buildings, which are not normally occupied by the public. LP-Gas installations are frequently located in buildings used for poultry breeding, and the steel-braided 350 psi hose was a problem due to its stiffness.

3-4.2.4 Containers, regulating equipment, manifolds, pipe, tubing, and hose shall be located to minimize exposure to abnormally high temperatures (such as might result from exposure to convection and radiation from heating equipment or installation in confined spaces), physical damage, or tampering by unauthorized persons.

When containers are exposed to abnormally high temperatures, liquid LP-Gas expands and the container could become liquid-full, causing high hydrostatic pressures that result in the discharge of the container pressure relief valve. Heating equipment should be positioned so that infrared heaters do not focus on containers and convection heat is not directed at them. Also, particular attention should be given to installing cylinders in confined spaces where temperatures may build up, such as in pits or tunnels in construction areas.

Although indirectly related, appliance location in confined areas is equally important in order to provide sufficient air for combustion.

3-4.2.5 Heat-producing equipment shall be located and used to minimize the possibility of the ignition of combustibles.

Tarpaulins, plastic sheeting, wood scaffolding, etc., should not be positioned in such a way that they may be ignited by heaters.

3-4.2.6 Where containers are located on a floor, roof, or balcony, provisions shall be made to minimize the possibility of containers falling over the edge.

(a) Filling containers on roofs or balconies shall be prohibited. See 3-2.3.1(c).

3-4.2.7 Transportation (movement) of containers within a building shall comply with the following:

(a) Movement of containers having water capacities greater than 2.7 lb (1.2 kg) and filled with no more than 16.8 oz (0.522 kg) of LP-Gas within a building shall be restricted to movement directly associated with the uses covered by 3-4.3 through 3-4.9 and shall be conducted in accordance with these provisions and 3-4.2.7(b) through (d).

(b) Valve outlets on containers having water capacities greater than 2.7 lb (1.2 kg) and filled with no more than 16.8 oz (0.522 kg) of LP-Gas shall be tightly plugged or capped and shall comply with the provisions of 2-2.4.1.

(c) Only emergency stairways not generally used by the public shall be used, and precautions shall be taken to prevent the container from falling down the stairs.

(d) Freight or passenger elevators shall be permitted to be used when occupied only by those engaged in moving the container.

3-4.2.8 Portable heaters, including salamanders, shall be equipped with an approved automatic device to shut off the flow of gas to the main burner and to the pilot, if used, in the event of flame extinguishment or combustion failure. Such portable heaters shall be self-supporting unless designed for container mounting (*see 3-4.3.4*). Container valves, connectors, regulators, manifolds, piping, or tubing shall not be used as structural supports. The following shall also apply.

Portable heaters manufactured on or after May 17, 1967, having an input of more than 50,000 Btuh (53 MJ/h), and those manufactured prior to May 17, 1967, with inputs of more than 100,000 Btuh (105 MJ/h), shall be equipped with either:

Although standards exist for the listing of portable heaters by independent testing laboratories [e.g., ANSI Z83.7, *Gas-Fired Construction Heaters* (*see 2-6.2.1*)], not all heaters are tested and listed. This part of Section 3-4 is intended to provide basic requirements for the authorities to utilize in extending approvals.

(a) A pilot that must be lighted and proved before the main burner can be turned on, or

(b) An approved electric ignition system.

Exception: The provisions of 3-4.2.8 shall not be applicable to the following:

(a) Tar kettle burners, hand torches, or melting pots.

(b) Portable heaters with less than 7,500 Btuh (8 MJ/h) input if used with containers having a maximum water capacity of 2.7 lb (1.2 kg) and filled with no more than 16.8 oz (0.522 kg) of LP-Gas.

Except for tar kettle burners, hand torches, or melting pots, which are attended, and smaller heaters of less than 7,500 Btuh (8 MJ/h capacity connected to a 1-lb (0.45-kg) LP-Gas container, all portable heaters used indoors must have flame failure protection or, in the case of catalytic heaters, combustion failure protection. This protection is an approved automatic device to shut off the flow of gas to the main burner—and pilot, if used—in the event of such failures. Additionally, these portable heaters designed for container mounting must not use valves, piping, regulators, etc., as structural supports for the heater.

The possibility of operators being burned during ignition of heaters may exist with larger heaters designed to operate on higher inlet pressures if the provisions of this part are not followed. Unless proved pilot lights or electric ignition are used, a large flame rollout may occur. Pilots on these larger heaters may encounter problems of thermocouple premature failure or outages due to wind, etc. This can result in operators bypassing controls. Therefore, for older heaters (which, for the most part, have been replaced) and newer heaters of certain sizes, assurance must be made that delayed ignition of large volumes of gas will not occur.

3-4.3 Buildings Under Construction or Undergoing Major Renovation.

These are the basic provisions for transportation and use of portable container systems for construction or major renovation of buildings not occupied by the public. Prior approval is particularly noted when the building is partially occupied by the public. Sometimes occupancy is started before all construction is completed. Subsection 3-4.4 covers those instances where minor renovation is done while the building is frequented by the public.

Paragraph 3-4.3.1 specifically extends coverage to transportation and use of containers in buildings when used for construction and renovation.

3-4.3.1 Containers shall be permitted to be used and transported in buildings or structures under construction or undergoing major renovation where such buildings are not occupied by the public or, if partially occupied by the public, containers shall be permitted to be used and transported in the unoccupied portions with the prior approval of the authority having jurisdiction. Such use shall be in accordance with 3-4.3.2 through 3-4.3.8.

3-4.3.2 Containers, equipment, piping, and appliances shall comply with 3-4.2.

3-4.3.3 For temporary heating, such as curing concrete, drying plaster, and similar applications, heaters (other than integral heater-container units covered in 3-4.3.4) shall be located at least 6 ft (1.8 m) from any LP-Gas container.

3-4.3.4 Integral heater-container units specifically designed for the attachment of the heater to the container, or to a supporting standard attached to the container, shall be permitted to be used, provided they are designed and installed to prevent direct or radiant heat application to the container. Blower-type and radiant-type units shall not be directed toward any LP-Gas container within 20 ft (6.1 m).

3-4.3.5 If two or more heater-container units of either the integral or nonintegral type are located in an unpartitioned area on the same floor, the container(s) of each such unit shall be separated from the container(s) of any other such unit by at least 20 ft (6.1 m).

Paragraphs 3-4.3.3, 3-4.3.4, and 3-4.3.5 reflect the basic requirements set out in 3-4.2, but incorporate specific distances to accomplish them. A 6-ft (1.8-m) separation between nonintegral heaters and other LP-Gas containers is given, while a 20-ft (6-m) distance is required for integral heater container units. The latter are infrared or larger blower-type units, which have a more pronounced effect on heat transmission to nearby containers.

3-4.3.6 If heaters are connected to containers manifolded together for use in an unpartitioned area on the same floor, the total water capacity of containers manifolded together serving any one heater shall not be greater than 735 lb (333 kg) [nominal 300 lb (136 kg) LP-Gas capacity], and if there is more than one such manifold, it shall be separated from any other by at least 20 ft (6.1 m).

The 300-lb (136-kg) LP-Gas maximum for manifolded systems has a long history in NFPA 58. A 20-ft (6-m) distance is set out for the separation of separate manifolded systems in the same unpartitioned floor area.

3-4.3.7 On floors on which no heaters are connected for use, containers shall be permitted to be manifolded together for connection to a heater or heaters on another floor, provided:

(a) The total water capacity of the containers connected to any one manifold is not greater than 2,450 lb (1111 kg) [nominal 1,000 lb (454 kg) LP-Gas capacity], and

(b) Manifolds of more than 735 lb (333 kg) water capacity [nominal 300 lb (136 kg) LP-Gas capacity], if located in the same unpartitioned area, shall be separated from each other by at least 50 ft (15 m).

3-4.3.8 The provisions of 3-4.3.5, 3-4.3.6, and 3-4.3.7 shall be permitted to be altered by the authority having jurisdiction if compliance is impractical.

3-4.4 Buildings Undergoing Minor Renovation When Frequented by the Public. Containers shall be permitted to be used and transported for repair or minor renovation in buildings frequented by the public as follows:

(a) During the hours the public normally occupies the building, the following shall apply:

1. The maximum water capacity of individual containers shall be 50 lb (23 kg) [nominal 20 lb (9.1 kg) LP-Gas capacity], and the number of containers in the building shall not exceed the number of workers assigned to the use of the LP-Gas.
2. Containers having a water capacity greater than 2.7 lb (1.2 kg) and filled with no more than 16.8 oz (0.522 kg) LP-Gas shall not be left unattended.

(b) During the hours the building is not open to the public, containers shall be permitted to be used and transported within the building for repair or minor renovation in accordance with 3-4.2 and 3-4.3, provided that containers with a greater water capacity than 2.7 lb (1.2 kg) and filled with no more than 16.8 oz (0.522 kg) LP-Gas shall not be left unattended.

Renovation of buildings during the hours when the public is present requires special considerations. The maximum container size permitted is 20 lb (9 kg) LP-Gas capacity. The number of cylinders is not to exceed the number of workers assigned to them. The cylinders are not to be left unattended at any time. At other times, the provisions of 3-4.3 apply.

3-4.5 Buildings Housing Industrial Occupancies.

3-4.5.1 Containers shall be permitted to be used in buildings housing industrial occupancies for processing, research, or experimental purposes as follows:

(a) Containers, equipment, and piping used shall comply with 3-4.2.

(b) If containers are manifolded together, the total water capacity of the connected containers shall be not more than 735 lb (333 kg) [nominal 300 lb (136 kg) LP-Gas capacity]. If there is more than one such manifold in a room, it shall be separated from any other by at least 20 ft (6.1 m).

See commentary on 3-4.3.6.

(c) The amount of LP-Gas in containers for research and experimental use in the building shall be limited to the smallest practical quantity.

3-4.5.2 Containers shall be permitted to be used to supply fuel for temporary heating in buildings housing industrial occupancies with essentially noncombustible contents, if portable equipment for space heating is essential and a permanent heating installation is not practical, provided containers and heaters comply with and are used in accordance with 3-4.3.

3-4.6 Buildings Housing Educational and Institutional Occupancies.

Containers shall be permitted to be used in buildings housing educational and institutional laboratory occupancies for research and experimental purposes, but not in classrooms, as follows:

(a) The maximum water capacity of individual containers used shall be:

1. 50 lb (23 kg) [nominal 20 lb (9.1 kg) LP-Gas capacity] if used in educational occupancies.
2. 12 lb (5.4 kg) [nominal 5 lb (2 kg) LP-Gas capacity] if used in institutional occupancies.

Formal Interpretation 79-3
Reference: 3-4.5

Question: An LP-Gas–fired infrared space heater and an LP-Gas cylinder are located and used inside of a foundry. They are connected, through a regulator, by means of a hose. The pressure in the hose is less than 1 psi. What provisions in NFPA 58 characterize the hose that should be used?

Answer: This application is covered under 3-4.5, Buildings Housing Industrial Occupancies, of NFPA 58. It is, therefore, also subject to the provisions of 3-4.1 and 3-4.2.3 of Section 3-4, which, through references, characterize the type of hose to be used as follows:

Paragraph 3-4.2.3 provides: "Piping, including pipe, tubing, fittings, valves, and hose, shall comply with Section 2-4, except that a minimum working pressure of 250 psi (1.7 MPa) shall apply to all components. The following shall also apply:

(b) Hose, hose connections, and flexible connectors used shall be designed for a working pressure of at least 350 psi (2.4 MPa), shall comply with 2-4.6 . . . ."

Paragraph 2-4.6.1 requires that the hose be fabricated of materials resistant to the action of LP-Gas both as liquid and vapor and, if wire braid is used for reinforcement, it shall be corrosive resistant material such as stainless steel.

Paragraph 2-4.6.2 provides: "Hose and quick connectors shall be approved (*see Section 1-6, Approved*)."

Paragraph 2-4.6.3(a) reiterates the requirement in 3-4.2.3(b) for a 350 psi (2.4 MPa) working pressure regardless of the actual pressure, and stipulates hose marking and other pressure criteria applicable to the assembly of hose and hose connections.

Issue Edition: 1979
Reference: 334
Date: August 1982 ■

(b) If more than one such container is located in the same room, the containers shall be separated by at least 20 ft (6.1 m).

(c) Containers not connected for use shall be stored in accordance with Chapter 5.

In these occupancies, the maximum size container is 20 lb (9 kg) of LP-Gas for educational buildings and 12 lb (5.4 kg) LP-Gas for institutional occupancies, with a separation of 20 ft (6 m) if more than one container is located in the same room. Containers are not to be used in classrooms (there have been some portable demonstration cabinets proposed, but these are not recognized in NFPA 58). Storage must be in accordance with Chapter 5.

Formal Interpretation 89-2
Reference: 3-4.6

Question: Is it a violation of NFPA 58, 3-4.6, to install a 20-lb LP-Gas tank in a high school chemistry laboratory to supply Bunsen burners on the student lab tables, connected by permanently installed piping which complies with NFPA 58?

Answer: Yes.

Issue Edition: 1989
Reference: 3-4.6.1
Issue Date: May 22, 1990
Effective Date: June 10, 1990 ■

Note that the NFPA *Life Safety Code®* definition of an education occupancy covers grades through 12 (high school). Educational facilities beyond the 12th grade are considered to be the following occupancies:

Instructional building—Business occupancy

Classrooms under 50 persons—Business occupancy

Classrooms 50 persons and over—Assembly occupancy

Laboratories, instructional—Business occupancy

Laboratories, noninstructional—Industrial occupancy

For additional information see the *Life Safety Code* for proper occupancy classification of educational facilities beyond the 12th grade. Note that *Life Safety Code* classifications are for *Life Safety Code* purposes. The limits of propane cylinder size in NFPA 58 still apply.

The requirements of NFPA 45, *Standard on Fire Protection for Laboratories Using Chemicals*, apply to laboratories used for educational purposes above grade 12, and other laboratories.

Exception: Containers shall not be stored in a laboratory room.

3-4.7 Temporary Heating and Food Service Appliances in Buildings in Emergencies.

3-4.7.1 Containers shall be permitted to be used in buildings for temporary emergency heating purposes if necessary to prevent damage to the buildings or contents, and if the permanent heating system is temporarily out of service, provided the containers and heaters comply with and are used and transported in accordance with 3-4.2 and 3-4.3, and the temporary heating equipment is not left unattended.

This provision is strictly an emergency measure if the permanent heating system is temporarily out of service. It is not intended to apply to supplemental or zone heating. Also, the emergency heating equipment must be attended at all times. If someone must be hired to provide attendance, the problem with the permanent heating system is more likely to be solved promptly.

3-4.7.2 When a public emergency has been declared and gas, fuel, or electrical service has been interrupted, portable listed LP-Gas commercial food service appliances meeting the requirements of 3-4.8.4 shall be permitted to be temporarily used inside affected buildings. The portable appliances used shall be discontinued and removed from the building at the time the permanently installed appliances are placed back in operation.

This provision was added to the 1995 edition, following its addition as a Tentative Interim Amendment to the 1992 edition of the standard. It recognizes that LP-Gas is an excellent fuel for the portable cooking appliances covered in 3-4.8.4, and that these are required following natural disasters such as floods and hurricanes, where utilities supplying gas and electricity can be interrupted. Note that the portable appliances must be removed from the building, when permanent appliances are back in service. It has been reported to the editor that during the aftermath of Hurricane Hugo, many such appliances were supplied and used.

3-4.8 Use in Buildings for Demonstrations or Training, or Use in Small Containers.

3-4.8.1 Containers having a maximum water capacity of 12 lb (5.4 kg) [nominal 5 lb (2 kg) LP-Gas capacity] shall be permitted to be used temporarily inside buildings for public exhibitions or demonstrations, including use in classroom demonstrations. If more than one such container is located in a room, the containers shall be separated by at least 20 ft (6.1 m).

Formal Interpretation 89-3
Reference: 3-4.8.1

Question: Does the use of an approved portable cooking appliance utilizing a 2-lb LP-Gas container as its fuel supply for temporary table side cooking within a restaurant meet the intent of "public exhibition" as described in 3-4.8.1?
Answer: No.

Issue Edition: 1989
Reference: 3-4.8.1
Issue Date: March 19, 1991
Effective Date: April 8, 1991 ■

For temporary use in buildings for exhibitions or demonstrations, containers up to 5-lb (2-kg) LP-Gas capacity may be used. The use of a 20-lb (9-kg) container (e.g., for a barbecue grill) filled with 5 lb (2 kg) of LP-Gas is not permitted, as there is no easy way to verify that only 5 lb (2 kg) of LP-Gas is in the container. This provision (and 3-4.8.3) permits the demonstration of a portable cooking device with a 5-lb (2-kg) LP-Gas container at an indoor trade show to demonstrate the cooking equipment, but does not permit an identical device to be used at the show to prepare food for sale. For cooking in restaurants, 3-4.8.4 does permit the use of stoves fueled by a 10.0-oz butane container.

3-4.8.2 Containers shall be permitted to be used temporarily in buildings for training purposes related to the installation and use of LP-Gas systems, provided the following conditions are met:

(a) The maximum water capacity of individual containers shall be 245 lb (111 kg) [nominal 100 lb (45 kg) LP-Gas capacity], but not more than 20 lb (9.1 kg) of LP-Gas shall be placed in a single container.

(b) If more than one such container is located in the same room, the containers shall be separated by at least 20 ft (6.1 m).

(c) The training location shall be acceptable to the authority having jurisdiction.

(d) Containers shall be promptly removed from the building when the training class has terminated.

For training in buildings, 100-lb (45-kg) LP-Gas containers may be used, but they may only be filled with 20 lb (9 kg) of LP-Gas and approval of the authority having jurisdiction is required. Note that this differs from 3-4.8.1 (covering public exhibitions or demonstrations), where only small containers [12 lb (5 kg) maximum] are permitted.

3-4.8.3* Containers complying with UL 147A, *Standard for Nonrefillable (Disposable) Type Fuel Gas Cylinder Assemblies,* and having a maximum water capacity of 2.7 lb (1.2 kg) and filled with no more than 16.8 oz (0.522 kg) of LP-Gas shall be permitted to be used in buildings as part of approved self-contained torch assemblies or similar appliances.

This provision relates to and limits the use of LP-Gas containers of a capacity of up to 1 lb (0.45 kg). It permits the use of portable appliances fueled by butane and propane, such as curling irons and cigarette lighters, and the cylinders used to refill them. In the 1995 edition, the requirement that the 1-lb cylinders comply with a UL standard was added to establish minimum safety standards for these cylinders above those required by the Department of Transportation.

Note that the use of cylinders of this size with portable cooking appliances is not mentioned, and therefore prohibited in buildings, except as permitted in restaurants and in attended commercial food catering in the next paragraph, where a different UL standard is referenced.

Note that only certain small cylinders with strict limitations are allowed. Cylinders must be nonrefillable, be listed to UL 147B, and cannot contain more than 10.0 oz of butane.

3-4.9 Portable Containers on Roofs or Exterior Balconies.

Roof installations are generally used for microwave stations and emergency electric generating units. These systems can be installed only on roofs of buildings that are unlikely to sustain major structural failure from fire. Specific conditions for installation are outlined, but it is noteworthy that no container refilling can take place on roofs and that certain conditions are set forth for the movement of replacement cylinders to the roof location.

3-4.9.1 Containers shall be permitted to be permanently installed on roofs of buildings of fire-resistive construction, or noncombustible construction having essentially noncombustible contents, or of other construction or contents that are protected with automatic sprinklers (*see NFPA 220, Standard on Types of Building Construction*) in accordance with 3-4.2 and the following:

(a) The total water capacity of containers connected to any one manifold shall be not greater than 980 lb (445 kg) [nominal 400 lb (181 kg) LP-Gas capacity]. If more than one manifold is located on the roof, it shall be separated from any other by at least 50 ft (15 m).

(b) Containers shall be located in areas where there is free air circulation, at least 10 ft (3.0 m) from building openings (such as windows and doors), and at least 20 ft (6.1 m) from air intakes of air conditioning and ventilating systems.

(c) Containers shall not be located on roofs that are entirely enclosed by parapets more than 18 in. (457 mm) high unless:

1. the parapets are breached with low-level ventilation openings no more than 20 ft (6.1 m) apart, or
2. all openings communicating with the interior of the building are at or above the top of the parapets.

(d) Piping shall be in accordance with 3-4.2.3. Hose shall not be used for connection to containers.

(e) The fire department shall be advised of each such installation.

3-4.9.2 Containers having water capacities greater than 2½ lb (1 kg) [nominal 1 lb (0.5 kg)] LP-Gas capacity shall not be located on balconies above the first floor that are attached to a multiple family dwelling of three or more living units located one above the other.

Exception: Where such balconies are served by outside stairways and where only such stairways are used to transport the container.

This requirement was added in the 1989 edition of the standard to state clearly that the use of LP-Gas cylinders of a capacity greater than 1 lb (0.5 kg) is prohibited on most balconies. This was not a change in the standard, but rather a clarification to prevent misunderstanding on the part of users who did not thoroughly read the entire section.

3-4.10 Liquid Piped into Buildings or Structures.

Liquid LP-Gas may be piped into buildings at pressures higher than 20 psi (138 kPa) for certain applications listed in 3-2.7(d). The piping system must comply with 3-2.8 and other specific provisions of 3-4.10. These include maximum size of piping, protection against breakage and against exposure to high ambient temperatures, accessible shutoff valves, excess-flow valves, and use of hydrostatic relief valves. Protection against release of fuel when disconnecting is obtained either with an automatic quick-closing coupling or by shutting off the system and allowing the appliance to burn off the fuel.

3-4.10.1 Liquid LP-Gas piped into buildings in accordance with 3-2.7(d), Exception No. 1 shall comply with 3-2.8.

3-4.10.2 Liquid LP-Gas piped into buildings in accordance with 3-2.7(d), Exception No. 2 from containers located and installed outside the building or structure in accordance with 3-2.2 and 3-2.3 shall comply with the following:

(a) Liquid piping shall not exceed ¾ in. I.P.S. and shall comply with 3-2.7 and 3-2.8. If approved by the authority having jurisdiction, copper tubing complying with 2-4.3(c)(1) and with a maximum outside diameter of ¾ in. shall be permitted to be used. Liquid piping in buildings shall be kept to a minimum and shall be protected against construction hazards by:

1. Securely fastening it to walls or other surfaces to provide adequate protection against breakage.
2. Locating it so as to avoid exposure to high ambient temperatures.

(b) A readily accessible shutoff valve shall be located at each intermediate branch line where it leaves the main line. A second shutoff valve shall be located at the appliance end of the branch and upstream of any flexible appliance connector.

(c) Excess-flow valves complying with 2-3.3.3(b) shall be installed in the container outlet supply line, downstream of each shutoff valve, and at any point in the piping system where the pipe size is reduced. They shall be sized for the reduced size piping.

(d) Hose shall not be used to carry liquid between the container and building and shall not be used at any point in the liquid line except as the appliance connector. Such connectors shall be as short as practical and shall comply with 2-4.6, 3-2.8.9, and 3-2.8.11.

(e) Hydrostatic relief valves shall be installed in accordance with 3-2.9.

(f) Provision shall be made so that the release of fuel when any section of piping or appliances is disconnected shall be minimized by use of one of the following methods:

1. An approved automatic quick-closing coupling that shuts off the gas on both sides when uncoupled.
2. Closing the shutoff valve closest to the point to be disconnected and allowing the appliance or appliances on that line to operate until the fuel in the line is consumed.

For additional information involving the provisions of Section 3-4, refer to the following publications of the National Propane Gas Association:

NPGA 603, *How to Use LP-Gas Safely at Construction Sites*; NPGA 604, *Safe Use of LP-Gas for Heating Tar*; NPGA 605, *Safe Use of LP-Gas in Temporary Space Heating with Portable Containers*; NPGA 606, *Safe Use of LP-Gas with Portable Cylinders for Cutting, Brazing*; NPGA 800, *Recommended Procedures for the Temporary Use of LP-Gas in Places of Public Assembly Indoors*.

3-5 Installation of Appliances.

3-5.1 Application.

3-5.1.1 This section includes installation provisions for LP-Gas appliances fabricated in accordance with Section 2-6.

See commentary on Section 2-6.

3-5.1.2 Installation of appliances on commercial vehicles shall be in accordance with Section 3-8.

3-5.1.3 With the approval of the authority having jurisdiction, unattended heaters used for the purpose of animal or poultry production inside structures without enclosing walls shall not be required to be equipped with an automatic device designed to shut off the flow of gas to the main burners and to the pilot, if used, in the event of flame extinguishment or combustion failure.

This exception to 2-6.2.2 acknowledges the difficulty in getting an accumulation of LP-Gas in a structure having no walls. The provision is of considerable vintage, and it is doubtful that modern cost/benefit factors would result in the approval of the authority having jurisdiction.

3-5.2 Reference Standards. LP-Gas appliances shall be installed in accordance with this standard and other national standards that apply. These include:

(a) NFPA 37, *Standard for the Installation and Use of Stationary Combustion Engines and Gas Turbines.*

(b) NFPA 54 (ANSI Z223.1), *National Fuel Gas Code.*

(c) NFPA 61B, *Standard for the Prevention of Fires and Explosions in Grain Elevators and Facilities Handling Bulk Raw Agricultural Commodities.*

(d) NFPA 82, *Standard on Incinerators and Waste and Linen Handling Systems and Equipment.*

(e) NFPA 86, *Standard for Ovens and Furnaces.*

(f) NFPA 96, *Standard for Ventilation Control and Fire Protection of Commercial Cooking Operations.*

(g) NFPA 302, *Fire Protection Standard for Pleasure and Commercial Motor Craft.*

(h) NFPA 501A, *Standard for Fire Safety Criteria for Manufactured Home Installations, Sites, and Communities.*

(i) NFPA 501C, *Standard on Recreational Vehicles* (ANSI A119.2).

These references are included for the same reasons given in the commentary on 3-1.3.

3-6 Vaporizer Installation.

3-6.1 Application. This section shall apply to the installation of vaporizing devices covered in 2-5.4. It shall not apply to engine fuel vaporizers, or to integral vaporizing-burners such as those used for weed burners or tar kettles.

LP-Gas is stored as a liquid and most often used as a gas. LP-Gas will vaporize in a liquid storage container and reach an equilibrium pressure determined by the temperature of the container. As gas is withdrawn from the container, additional liquid will vaporize, but this process requires heat which is drawn from the media surrounding the tank, air or soil, depending on the installation. In installations where more vapor (and heat) is required than can be provided by the tank's surroundings, a vaporizer is used. There are several types of vaporizers:

- Direct-fired, which are similar to a boiler
- Indirect-fired, which use electricity, hot water, steam, or other nonflame heating media as a source of heat
- Vaporizing burners, where the heat of the flame vaporizes the liquid feeding the flame

Vaporizers used on engine fuel systems are not covered in this section as they are covered in Chapter 8. Integral vaporizing-burners are not covered in this section as they are portable units that do not present a hazard greater than an open flame. Vaporizing burners installed in a fixed location are covered in 3-6.5. Their construction is covered in 2-5.4.

From operational and maintenance standpoints, it should be recognized that some vaporizers—those used for standby systems and those used with high winter or low summer demand, such as shopping centers—are used only in the colder time of year. It is imperative that the vaporizers be given a thorough check and a test run before they are needed, so that they will be equipped to handle the vaporizing load. This is especially critical when the system is owned by the user, rather than the propane distributor, who may not be as familiar with the operation of the equipment. Periodic testing during idle periods is recommended as well as a program to alert the proper individuals to the need for this essential maintenance program. In the summer season, spiders, mud daubers, and so forth can get into the burner area, pilot area, and regulator vents and create operating problems. It is important to verify that the rain cap is always kept on the pressure relief valve outlet and that such outlet is piped to a proper safe point for discharge.

3-6.2 Installation of Indirect-Fired Vaporizers.

By definition, indirect vaporizers derive heat for operation from a remote source. The term "remote" is not defined. The strictest definition of "remote" would be any device not part of the unit itself. This could be a water or steam boiler mounted immediately adjacent to, or on the same skid or package with, the indirect vaporizer. In that event, 3-6.2.4 specifies that such a combination be sited in the same manner as direct-fired vaporizers since a source of ignition is present. This "source of ignition" can have positive or negative effects. For example, a source of ignition close to an indirect vaporizer installed outside has the effect of preventing a large buildup of gas from a leak because a leak would be ignited by the adjacent heat source before it became a more severe problem. With no source of ignition in the immediate area, on a calm day a large amount of gas could escape before finally reaching an ignition source. This could result in an unconfined vapor cloud explosion. On the other hand, if the indirect vaporizer were to be installed inside an enclosure, an adjacent source of ignition would cause a confined explosion, which is considerably more devastating. Accordingly, indirect vaporizers are generally used where the source of heat is a plant facility, such as steam or hot water from a plant heating or processing system.

3-6.2.1 Indirect-fired vaporizers shall comply with 2-5.4.2, and shall be installed as provided in this section.

3-6.2.2 Where an indirect-fired vaporizer is installed in a building or structure, the building or structure shall comply with the following:

***Figure** 3.32 Indirect-fired steam vaporizer. This vaporizer utilizes steam provided from a central steam supply, and the steam flow and pressure is regulated at the vaporizer. Dual condensate traps are provided on the unit to prevent condensate buildup. A pressure relief valve is installed in the vapor space of the unit, and the outlet of the relief valve is vented outside the building. A liquid carry-over control is provided to shut off the supply of liquid if the liquid level exceeds the height of the steam tubes. This unit is ASME Code stamped. (Photo courtesy of Allgas-Interex, division of Allgas Industries.)*

(a) Separate buildings or structures shall comply with Section 7-2.
(b) Attached structures or rooms shall comply with Section 7-3.

To prevent passage of vapor from a vaporizer room into an adjacent room, the connecting partition should be caulked to be gastight and contain no openings. When it is necessary to have a window in such a partition, it must be permanently closed and should be of shatterproof plastic or wired glass material. Service piping should leave the vaporizer room and go to the outside before reentering the adjacent room or structure. Electric conduit, which may be necessary for controls connecting between a vaporizer room and an adjacent control room, may be imbedded in a concrete floor common to the two rooms, provided that a conduit fitting for sealing is installed, and the installation is in accordance with Section 501-5 of NFPA 70, *National Electrical Code®*. Any conduit passing through the vaportight wall should be near the ceiling of the room, fastened to the wall in such a manner that expansion and contraction will not break any caulking or sealing material, or the conduit must be routed to the outside prior to going into the adjacent room.

(c) The building or structure shall not have any unprotected drains to sewers or sump pits. Pressure relief valves on vaporizers within buildings in industrial or gas manufacturing plants shall be piped to a point outside the building or structure and shall discharge vertically upward.

Because LP-Gases are heavier than air and will seek low places, it is important that there be no open drains, sewers, or sump pits in the building enclosure. Any drain that is piped away from a vaporizer room or location is suspect in that it might connect with a general drainage system and convey flammable gases to a source of ignition in another building or open area. Where it is necessary to provide a drain, it should be protected with a trap that will not permit vapors to pass through, or should be discrete to the vaporizer room and terminate outside, in open air, well away from other drains or sources of ignition.

The relief valve piping (which is required to discharge vertically upward) must include a rain cap, weep hole, or other protection to prevent the piping from filling with water and freezing. (Refer to Chapter 2 for further information on relief valves.)

3-6.2.3 If the heat source of an indirect vaporizer is gas fired and is located within 15 ft (4.6 m) of the vaporizer, the vaporizer and its heat source shall be considered to be a direct-fired vaporizer and subject to the requirements of 3-6.3.

This requirement prevents a user from defeating the intent of the standard by installing an indirect vaporizer with an adjacent heat source instead of a direct-fired vaporizer.

3-6.2.4 The installation of a heat source serving an indirect vaporizer that utilizes a flammable or combustible heat transfer fluid shall comply with one of the following:

(a) It shall be located outdoors, or

(b) If installed within a structure, the structure shall comply with Section 7-2.

(c) If installed within structures attached to, or rooms within buildings, the structures or rooms shall comply with Section 7-3.

These restrictions recognize the added hazard of flammable or combustible heat transfer fluid. Heat transfer fluids, such as Dowtherm and Therminol, are used in many process plants and may be available when steam is not. Not all heat transfer fluids are flammable, and if any question exists the manufacturer of the fluid should be contacted for information or the fluid should be tested. Heat transfer systems can have the fluid changed, and if this occurs, the flammability of the new fluid should be checked.

3-6.2.5 Gas-fired heating systems supplying heat for vaporization purposes shall be equipped with automatic safety devices to shut off gas to the main burners if ignition fails to occur.

This must be done to prevent raw, unburned gas from escaping into the building or the atmosphere. NFPA 8501, *Standard Single Burner Boiler Operation*, contains information on prevention of explosions in boilers and may provide useful information for gas-fired heating systems. Note that NFPA 8501 is not intended to cover small boilers and is applicable to boilers with a heat input of 12.5 MM Btu/hr and above. While this is much greater than most propane vaporizers, it can be used as a guide.

3-6.2.6 The installation of a heat source serving an indirect vaporizer that utilizes a noncombustible heat transfer fluid, such as steam, water, or a water-glycol mixture, shall be installed outdoors or shall comply with the following:

(a) A source of heat for an indirect vaporizer shall be permitted to be installed in an Industrial Occupancy (*see definition*) complying with Chapter 28 of NFPA *101,® Life Safety Code,®* and Section 6-3 of NFPA 54 (ANSI Z223.1), *National Fuel Gas Code*, where:

This was revised in the 1995 edition to simplify a previously complex requirement. It is an exception to the general rule that heat sources for vaporizers be installed outdoors or in a structure complying with Chapter 7 (with no sources of ignition) for industrial occupancies only. It permits existing boilers to be used as a source of heat for a vaporizer when condensate or circulating water is not returned to the boiler. If the condensate (or water) were returned and the vaporizer developed an internal leak, propane at high pressure could be released at the boiler relief valve, causing a fire.

1. The heat transfer fluid is steam or hot water and is not recirculated, and
2. A backflow preventer is installed between the vaporizer and the heat source.

(b) If the heat transfer fluid is recirculated after leaving the vaporizer, the heat source shall be installed in accordance with 3-6.2.4 and a phase separator shall be installed with the gas vented to a safe location.

This also permits the use of existing boilers, but requires a positive means of ensuring that any leaking propane vapors are prevented from entering the area where the boiler is installed.

3-6.2.7 Vaporizers employing heat from the atmosphere shall be installed outdoors and shall be located in accordance with Table 3-6.3.5.

This requirement was revised in the 1995 edition by deleting the requirement for an emergency shutoff valve that was added in the 1992 edition when the section was completely revised. Substantiation for the requirement of emergency shutoff valves was not clear, and the Committee deleted the requirement.

As pointed out in 2-5.4, some vaporizers use the atmospheric temperature for vaporizing, sometimes through tubing installed underground in order to maintain a relatively warm temperature to provide vaporizing heat in cold weather. In other cases, the tubing or piping is installed in basements or in crawl spaces to gain some heat for vaporization. It is not desirable to have such a vaporizer inside a building to prevent accumulation of gas, which could escape if the vaporizer were damaged. Because this type of vaporizer has a relatively large surface area, there is more chance for damage.

Exception: Atmospheric vaporizers of less than 1 qt (0.9 L) capacity shall be permitted to be installed inside an industrial building close to the point of entry of the supply line.

This exception was added in the 1995 edition to correct an error in the general revision of this section in the 1992 edition, where this exception was inadvertently deleted. A restriction to this exception was added to limit installation of these vaporizers to industrial buildings.

3-6.3 Installation of Direct-Fired Vaporizers.

3-6.3.1 Direct-fired vaporizers shall comply with 2-5.4.3 and shall be installed as provided in this section.

3-6.3.2 Direct-fired vaporizers shall be permitted to be installed outdoors or in separate structures constructed in accordance with Chapter 7.

Direct-fired vaporizers present a potential hazard that is not present in other LP-Gas equipment. Liquid is fed to the vaporizer and the surface is heated directly by a flame, which can reduce the normal service life of the metal. Therefore, installations must be outdoors or in a room or building without sources of ignition as provided in Chapter 7.

3-6.3.3 The housing for direct-fired vaporizers shall not have any drains to a sewer or a sump pit that is shared with any other structure or to a sewer. Pressure relief valve discharges on direct-fired vaporizers shall be piped to a point outside the structure or building.

3-6.3.4 Direct-fired vaporizers shall be permitted to be connected to the liquid space or to both the liquid or vapor space of the container, but in any case there shall be a manually operated shutoff valve in each connection at the container.

This was revised in the 1995 edition to revert to the text of the 1989 edition as the changes made in the 1992 edition general revision of this section proved impractical.

3-6.3.5 Direct-fired vaporizers of any capacity shall be located in accordance with Table 3-6.3.5.

Table 3-6.3.5

Exposure	Minimum Distance Required
Container	10 ft (3.0 m)
Container shutoff valves	15 ft (4.6 m)
Point of transfer	15 ft (4.6 m)
Nearest important buildings or group of buildings or line of adjoining property that may be built upon (except buildings in which vaporizer is installed; *see 3-6.3.2 and 3-6.3.3*).	25 ft (7.6 m)
Ch. 7 building or room housing gas-air mixer	10 ft (3.0 m)
Cabinet housing gas-air mixer outdoors	0 ft (0 m)

The distance requirements in Table 3-6.3.5 are based on experience. It should be noted that they vary depending on the type of exposure. For instance, a direct gas-fired vaporizer can be closer to a container than to the shutoff valves on the container because the possibility of leakage from the container shell is practically zero. Leakage potential is much greater from a connection to the container or at a valve. Also, with the making and breaking of connections at a transfer point, the hazard is greater than when the vaporizer is near the bare container without connections. The distance is greater from important buildings, a group of buildings, and the line of adjoining property that may be built upon to preclude mutual exposure in the event of a fire.

Figure 3.33 Installation of multiple direct-fired vaporizers. The vaporizers were installed 75 ft (23 m) from the storage tank. Each vaporizer includes inlet and outlet shutoff valves for servicing. A first-stage pressure regulator station is installed directly on the outlet of the vaporizer to reduce the gas pressure below 20 psi (138 kPa). (Photo courtesy of H. Emmerson Thomas.)

3-6.4 Installation of Tank Heaters.

3-6.4.1 Gas-fired tank heaters shall comply with 2-5.4.6 and shall be installed in accordance with this section.

The term "gas-fired tank heater" is descriptive. This simple form of vaporizer provides heat, via direct flame, to the bottom of the container.

3-6.4.2 Tank heaters shall be installed only on aboveground containers and shall be located in accordance with Table 3-6.4.2 with respect to the nearest important building, group of buildings, or line of adjoining property that can be built upon.

Note that the distance requirements in Table 3-6.4.2 are the same as the requirements for aboveground containers shown in Table 3-2.2.2, and that there are no references to the size of the tank heater but only to the size of the associated container. Table 3-6.4.2 ensures that spacing criteria are properly applied to direct-fired tank heaters. Of utmost importance is the point of transfer (filling connection, etc.) in relation to the pilot and burner on the

Table 3-6.4.2

Container Water Capacity	Minimum Distance Required
500 gal or less	10 ft (3.0 m)
501 to 2,000 gal	25 ft (7.6 m)
2,001 to 30,000 gal	50 ft (15 m)
30,001 to 70,000 gal	75 ft (23 m)
70,001 to 90,000 gal	100 ft (30.0 m)
90,001 to 120,000 gal	125 ft (38.1 m)

For SI Units: 1 gal = 3.785 L

tank heater. Most tank heaters on the market are used on containers of 1,000-gal (3.8-m^3) capacity and smaller, which have the point of transfer located at the top center of the container. It is essential that the gas supply to the pilot and burner be shut off while containers are being filled.

3-6.4.3 If the tank heater is gas fired, an automatic shutoff shall be provided on the fuel supply (including the pilot) that will operate if the container pressure exceeds 75 percent of the maximum design pressure specified in Table 2-2.2.2 or if the liquid level in the container falls below the top of the tank heater.

3-6.4.4 If the tank heater is of the electric immersion type, the heater shall be automatically de-energized when the pressure or level conditions specified in 3-6.4.3 are reached.

3-6.4.5 If the tank heater is similar in operation to an indirect vaporizer, the flow of the heat transfer fluid shall be automatically interrupted under the pressure or temperature conditions specified in 3-6.4.3 and the heat source shall comply with 3-6.2.6 and 3-6.2.7.

Note the safety devices required on direct-fired tank heaters. They reflect the special hazards of these devices, which are:

(1) Container overpressurization, which will result in relief valve operation and release of LP-Gas in proximity to an ignition source, and

(2) Damage to the container if the liquid level is allowed to fall below the level of the flame. This hazard is identical to that of fired boilers, where low water cutoff switches are mandated.

3-6.4.6 If a point of transfer is located within 15 ft (4.6 m) of a direct gas-fired tank heater, the heater burner and pilot shall be shut off during the product transfer and a caution notice shall be displayed immediately adjacent to the filling connections that reads:

"A gas-fired device that contains a source of ignition is connected to this container. Burner and pilot must be shut off before filling tank."

3-6.5 Installation of Vaporizing-Burners. Vaporizing-burners shall comply with 2-5.4.7 and shall be installed as follows:

(a) Vaporizing-burners shall be installed outside of buildings. The minimum distance between any container and a vaporizing-burner shall be in accordance with Table 3-6.5.

Table 3-6.5

Container Water Capacity	Minimum Distance Required
500 gal or less	10 ft (3.0 m)
501 to 2,000 gal	25 ft (7.6 m)
Over 2,000 gal	50 ft (15 m)

For SI Units: 1 gal = 3.785 L

(b) Manually operated positive shutoff valves shall be located at the containers to shut off all flow to the vaporizing-burners.

Vaporizing-burners are burners that are fed liquid LP-Gas and vaporize it internally prior to burning. It is important that vaporizing-burners not be used inside buildings because they are generally large-capacity units in which both liquid and vapor are present. The distance requirements in Table 3-6.5 reflect the potential for ignition of any gas leakage at the container and the hazard to the container if there is impinging flame from the vaporizing-burner.

3-6.6 Installation of Waterbath Vaporizers. Waterbath vaporizers shall comply with 2-5.4.4 and shall be installed as follows:

(a) If a waterbath vaporizer is electrically heated and all electrical equipment is suitable for Class 1, Group D, locations, the unit shall be treated as indirect-fired and shall be installed in accordance with 3-6.2.

(b) All others shall be treated as direct-fired vaporizers and shall be installed in accordance with 3-6.3.

3-6.7 Installation of Electric Vaporizers. Electric vaporizers, whether direct immersion or indirect immersion, shall be treated as indirect-fired and installed in accordance with 3-6.2.

3-6.8 Installation of Gas-Air Mixers.

Figure 3.34 Waterbath vaporizer with venturi gas-air mixer. Installation of a waterbath vaporizer with gas-fired burner. This vaporizer was installed in conjunction with a venturi gas-air mixer as a standby for natural gas. The vaporizer-burner control includes Factory Mutual and Industrial Risk Insurers approved flame safeguard controls. The vaporizer also includes a shutoff to prevent liquid carry-over in the event of burner failure or overcapacity. A water and antifreeze mixture with a rust inhibitor was installed in the waterbath of the vaporizer. The vaporizing coil within the vaporizer waterbath is ASME Code stamped. (Photo courtesy of H. Emmerson Thomas.)

Gas-air mixers are normally used to mix propane (or butane) and air to reduce the heating value of the gas to that of natural gas. This fuel is used as a substitute for natural gas when it is not available, which is common when natural gas is purchased on an interruptable basis, or when the price of propane or butane is favorable to natural gas. This is a common practice for large, industrial gas users. The use of gas-air mixers assists natural gas utilities by providing a quantity of natural gas that can be made quickly available for domestic use in times of severe domestic demand, such as very cold weather. The gas-air mixture is usually about 57 percent propane (or 45 percent butane), which is well above the upper limit of flammability of about 10 percent for propane (9 percent for butane).

3-6.8.1 Gas-air mixing equipment shall comply with 2-5.4.8 and shall be installed in accordance with this section. Piping and equipment installed with gas-air mixers shall comply with 3-2.6, 3-2.7, and 3-2.10.

(a) Where used without vaporizer(s), a mixer(s) shall be permitted to be installed outdoors or in buildings complying with Chapter 7.

(b) Where used with indirect-heated vaporizer(s), a mixer(s) shall be permitted to be installed outdoors, or in the same compartment or room with the vaporizer(s), in a building(s) complying with Chapter 7, or the mixer(s) shall be permitted to be installed remotely from the vaporizer(s) and shall be located in accordance with 3-6.2.

(c) Where used with direct-fired vaporizer(s), a mixer(s) shall be:

1. Listed or approved and installed in a common cabinet with the vaporizer(s) outdoors in accordance with 3-6.3.5.
2. Installed outdoors on a common skid with the vaporizer(s) in accordance with 3-6.3.
3. Installed adjacent to the vaporizer(s) to which it is connected in accordance with 3-6.3.
4. Installed in a building complying with Chapter 7 without any direct-fired vaporizer in the same room.

The ratio of gas to air in the mixture must be monitored, usually by measuring the specific gravity of the mixture. This can be accomplished by monitoring the relative pressures or, in larger installations, with an on-line meter that reads specific gravity or the heat content of the mixed gas, and shutdown of the gas and air are required if the upper flammable limit is approached. Note that the exception recognizes systems that use an aspirator to draw air into the gas with a venturi. These systems do not require shutoff of the air supply, as it is not drawn in when the flow of gas is stopped.

Figure 3.35 Vaporizer/gas-air mixers installed on common pad. (Photo courtesy of Alternate Energy Installations, Inc.)

Figure 3.36 *Test flare. Test flare normally installed with a gas-air system, which is used as standby for natural gas. This flare permits the system equipment to be checked during the nonoperational months and may also be used to train new people in the operation of the gas-air mixing system. (Photo courtesy of Allgas-Interex, division of Allgas Industries.)*

3-7 Ignition Source Control.

3-7.1 Application.

3-7.1.1 This section includes provisions for minimizing the possibility of ignition of flammable LP-Gas-air mixtures resulting from the normal or accidental release of nominal quantities of liquid or vapor from LP-Gas systems installed and operated in accordance with this standard.

3-7.1.2 The installation of lightning protection equipment shall not be required on LP-Gas storage containers.

NOTE: For information on lightning protection, see NFPA 780, *Lightning Protection Code.*

If a container or its piping is associated with a building or other structure equipped with lightning protection, the LP-Gas system may have to be integrated into the lightning protection system. Refer to NFPA 78, *Lightning Protection Code*—especially Chapter 3—in such instances.

3-7.1.3 Grounding and bonding shall not be required on LP-Gas systems.

NOTE: Since liquefied petroleum gas is contained in a closed system of piping and equipment, the system need not be electrically conductive or electrically bonded for protection against static electricity. For information on grounding and bonding for protection against static electricity, see NFPA 77, *Recommended Practice on Static Electricity.*

This nonmandatory note clearly states the intent of the Committee. It does not imply that a flammable LP-Gas–air mixture cannot be ignited by an electrostatic discharge. It does state that grounding and bonding are not required, because they are frequently requested by enforcement officials who are not familiar with LP-Gas but who are familiar with flammable liquids, such as gasoline, where grounding and bonding are required. This requirement does not prevent the installation of grounding and bonding, if desired by the installer or owner.

There have been a number of fires and explosions reported to the editor in which an electrostatic spark has been the source of ignition. In these cases, however, liquid LP-Gas was released at a high velocity, creating a mixture of liquid drops, vapor, air, and water drops (due to condensation of water vapor in the air from the refrigerating effect of vaporizing liquid). Such a mixed-phase discharge can generate an electrostatic change. This charge might be of sufficient energy to cause ignition of the mixture. See NFPA 77, *Recommended Practice on Static Electricity*, for a discussion of this phenomenon.

3-7.2 Electrical Equipment.

3-7.2.1 Electrical equipment and wiring shall be of a type specified by and shall be installed in accordance with NFPA 70, *National Electrical Code,*® for ordinary locations.

Exception: Fixed electrical equipment in classified areas shall comply with 3-7.2.2.

3-7.2.2 Fixed electrical equipment and wiring installed within classified areas specified in Table 3-7.2.2 shall comply with Table 3-7.2.2 and shall be installed in accordance with NFPA 70, *National Electrical Code.* This provision shall not apply to fixed electrical equipment at residential or commercial installations of LP-Gas systems or to systems covered by Section 3-8. The provision shall apply to vehicle fuel operations.

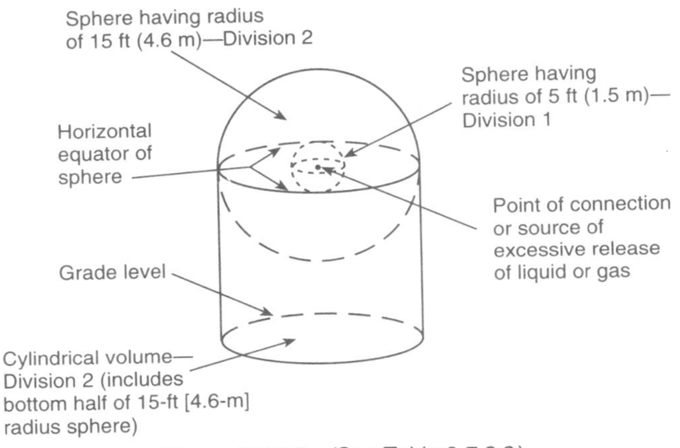

Figure 3-7.2.2 *(See Table 3-7.2.2.)*

The classified areas in Table 3-7.2.2 are based on experience with the types of LP-Gas system installations covered by NFPA 58. Reference is made to NFPA 70, *National Electrical Code*, for the definition of Division 1 and Division 2 areas. These are defined in Article 500 of the *National Electrical Code*, which is reprinted here (1993 edition) for the convenience of the reader (with reference to gases only):

> **500-5. Class I Locations.** Class I locations are those in which flammable gases are or may be present in the air in quantities sufficient to produce explosive or ignitible mixtures. Class I locations include those specified in (a) and (b) below.
>
> (a) **Class I, Division 1.** A Class I, Division 1 location is a location: (1) in which ignitible concentrations of flammable gases can exist under normal operating conditions; or (2) in which ignitible concentrations of such gases may exist frequently because of repair or maintenance operations or because of leakage; or (3) in which breakdown or faulty operation of equipment or processes might release ignitible concentration of flammable gases, and might also cause simultaneous failure of electric equipment.

Table 3-7.2.2

Part	Location	Extent of Classified Area[1]	Equipment Shall Be Suitable for National Electrical Code, Class 1, Group D[2]
A	Unrefrigerated Containers Other than DOT Cylinders and ASME Vertical Containers of Less than 1,000 lb Water Capacity.	Within 15 ft in all directions from connections, except connections otherwise covered in Table 3-7.2.2.	Division 2
B	Refrigerated Storage Containers.	Within 15 ft in all directions from connections otherwise covered in Table 3-7.2.2.	Division 2
		Area inside dike to the level of the top of the dike.	Division 2
C	Tank Vehicle and Tank Car Loading and Unloading.[3]	Within 5 ft in all directions from connections regularly made or disconnected for product transfer.	Division 1
		Beyond 5 ft but within 15 ft in all directions from a point where connections are regularly made or disconnected and within the cylindrical volume between the horizontal equator of the sphere and grade. (See Figure 3-7.2.2.)	Division 2
D	Gauge Vent Openings Other than Those on DOT Cylinders and ASME Vertical Containers of Less than 1,000 lb Water Capacity.	Within 5 ft in all directions from point of discharge.	Division 1
		Beyond 5 ft but within 15 ft in all directions from point of discharge.	Division 2

(continued)

Table 3-7.2.2 (continued)

Part	Location	Extent of Classified Area[1]	Equipment Shall Be Suitable for National Electrical Code, Class 1, Group D[2]
E	Relief Device Discharge Other than Those on DOT Cylinders and ASME Vertical Containers of Less than 1,000 lb Water Capacity, and Vaporizers.	Within direct path of discharge.	Division 1 Note: Fixed electrical equipment should preferably not be installed.
F	Pumps, Vapor Compressors, Gas-Air Mixers and Vaporizers (other than direct-fired or indirect-fired with an attached or adjacent gas-fired heat source).		
	Indoors without ventilation.	Entire room and any adjacent room not separated by a gastight partition.	Division 1
		Within 15 ft of the exterior side of any exterior wall or roof that is not vaportight or within 15 ft of any exterior opening.	Division 2
	Indoors with adequate ventilation.[4]	Entire room and any adjacent room not separated by a gastight partition.	Division 2
	Outdoors in open air at or above grade.	Within 15 ft in all directions from this equipment and within the cylindrical volume between the horizontal equator of the sphere and grade. (See Figure 3-7.2.2.)	Division 2

(continued)

Table 3-7.2.2 (continued)

Part	Location	Extent of Classified Area[1]	Equipment Shall Be Suitable for National Electrical Code, Class 1, Group D[2]
G	Vehicle Fuel Dispenser.	Entire space within dispenser enclosure, and 18 in. horizontally from enclosure exterior up to an elevation 4 ft above dispenser base. Entire pit or open space beneath dispenser.	Division 1
		Up to 18 in. aboveground within 20 ft horizontally from any edge of enclosure. *Note: For pits within this area, see Part H of this table.*	Division 2
H	Pits or Trenches Containing or Located Beneath LP-Gas Valves, Pumps, Vapor Compressors, Regulators, and Similar Equipment.		
	Without mechanical ventilation.	Entire pit or trench.	Division 1
		Entire room and any adjacent room not separated by a gastight partition.	Division 2
		Within 15 ft in all directions from pit or trench when located outdoors.	Division 2
	With adequate mechanical ventilation.	Entire pit or trench.	Division 2
		Entire room and any adjacent room not separated by a gastight partition.	Division 2
		Within 15 ft in all directions from pit or trench when located outdoors.	Division 2
I	Special Buildings or Rooms for Storage of Portable Containers.	Entire room.	Division 2

(continued)

Table 3-7.2.2 (continued)

Part	Location	Extent of Classified Area[1]	Equipment Shall Be Suitable for National Electrical Code, Class 1, Group D[2]
J	Pipelines and Connections Containing Operational Bleeds, Drips, Vents, or Drains.	Within 5 ft in all directions from point of discharge.	Division 1
		Beyond 5 ft from point of discharge, same as Part F of this table.	
K	Container Filling.		
	Indoors with adequate ventilation.[4]	Within 5 ft in all directions from a point of transfer.	Division 1
		Beyond 5 ft and entire room.	Division 2
	Outdoors in open air.	Within 5 ft in all directions from a point of transfer.	Division 1
		Beyond 5 ft but within 15 ft in all directions from point of transfer and within the cylindrical volume between the horizontal equator of the sphere and grade. (See Figure 3-7.2.2.)	Division 2

SI Conversions for Table 3-7.2.2
18 in. = 256 mm
4 ft = 1.2 m
5 ft = 1.5 m
15 ft = 4.6 m
20 ft = 6.1 m
1000 lb = 454 kg

[1]The classified area shall not extend beyond an unpierced wall, roof, or solid vaportight partition.

[2]See Article 500 "Hazardous (Classified) Locations" in NFPA 70 (ANSI) for definitions of Classes, Groups, and Divisions.

[3]When classifying extent of hazardous area, consideration shall be given to posssible variations in the spotting of tank cars and tank vehicles at the unloading points and the effect these variations of actual spotting point may have on the point of connection.

[4]Where specified for the prevention of fire or explosion during normal operation, ventilation is considered adequate where provided in accordance with the provisions of this standard.

[5]Fired vaporizers, calorimeters with open flames, and other areas where open flames are present either intermittently or constantly shall not be considered electrically classified areas.

Note: This classification usually includes locations where liquefied flammable gases are transferred from one container to another; gas generator rooms and other portions of gas manufacturing plants where flammable gas may escape; and all other locations where ignitible concentrations of flammable gases are likely to occur in the course of normal operations.

(b) **Class I, Division 2.** A Class I, Division 2 location is a location: (1) in which flammable gases are handled, processed, or used, but in which the gases will normally be confined within closed containers or closed systems from which they can escape, only in case of accidental rupture or breakdown of such containers or systems, or in case of abnormal operation of equipment; or (2) in which ignitible concentrations of gases are normally prevented by positive mechanical ventilation, which might become hazardous through failure or abnormal operation of the ventilating equipment; or (3) that is adjacent to a Class I, Division 1 location, and to which ignitible concentrations of gases might occasionally be communicated unless such communication is prevented by adequate positive-pressure ventilation from a source of clean air, and effective safeguards against ventilation failure are provided.

Note 1: This classification usually includes locations where flammable gases are used, but which, in the judgment of the authority having jurisdiction, would become hazardous only in case of an accident or some unusual operating condition. The quantity of flammable material that might escape in case of accident, the adequacy of ventilating equipment, the total area involved, and the record of the industry or business with respect to explosions or fires are all factors that merit consideration in determining the classification and extent of each location.

Note 2: Piping without valves, checks, meters, and similar devices would not ordinarily introduce a hazardous condition even though used for flammable gases. Locations used for the storage of liquefied or compressed gases in sealed containers would not normally be considered hazardous unless subject to other hazardous conditions also.

Electrical conduits and their associated enclosures separated from process fluids by a single seal or barrier shall be classified as a Division 2 location if the outside of the conduit and enclosures is an unclassified location.

The *National Electrical Code* does not classify areas, and this is left to the standard covering the installation. Therefore, the area is electrically classified in accordance with Table 3-7.2.2, not the *National Electrical Code.*

When using Table 3-7.2.2 it is important to understand that footnote 2 to the table refers to NFPA 70, *National Electrical Code*, for the definitions of Division 1 and 2 areas, reprinted above. In simple terms, the *National Electrical Code* defines a Division 1 area as one where combustible gases are normally present during operation, and a Division 2 area as one where combustible gases are present only under abnormal conditions. Therefore, a point of transfer where a hose is connected for filling is a Division 1 area because escape of some liquid is normal when the hose is disconnected.

The exception for residential and commercial installations incorporated into 3-7.2.2 recognizes that it is not possible to control sources of ignition in these occupancies, while ignition sources in classified areas in industrial occupancies are more easily controlled. In addition, these structures contain only ordinary electrical equipment (switches, motors) and use ordinary wiring methods. Consideration of both the fire risk and the expense of the specialized electrical equipment specified for installations in classified areas justify this exception.

There has been some confusion about the requirements of Parts E and F in Table 3-7.2.2 with regard to relief devices on vaporizers. This has been clarified in the 1995 edition by making Part E specifically applicable to relief valves on vaporizers for the first time. Part E covers all relief valves on containers except DOT and vertical containers under 1,000 lb water capacity (which are used as DOT 420s in some areas), including vaporizers. Part F covers area classification around vaporizers. It includes indirect or electric vaporizers. Note that direct-fired vaporizers and indirect-fired vaporizers, with an attached or adjacent gas-fired heat source, are not covered in Part F of the table, and therefore, no area classification is required. This is because the constant presence of an ignition source (the flame) makes prohibition of other ignition sources unnecessary. If a relief valve is provided on a direct-fired or indirect-fired vaporizer with an attached or adjacent gas-fired heat source, it is covered under Part E. In such a case, no electrical area classification would be required, if the relief valve is directed away from sources of ignition and it is locked or piped to a point at least 5 ft from a source of ignition.

To clear up misconceptions regarding Part F, the terms "pumps," "vapor compressors," "gas-air mixers," and "vaporizers" refer only to those items handling LP-Gas. A water pump or air compressor in a room or building separate from the gas equipment is not covered by this requirement. Also, gas-air mixers mixing noncombustible gases, or noncombustible gas with air, are not covered by this provision, nor are vaporizers handling noncombustible refrigerants and medical and industrial gases.

The exception for Section 3-8 concerns electrical installations on vehicles. In addition to similarities to residential occupancies noted earlier, a vehicle powered by an internal combustion engine has inherent nonelectrical ignition sources (e.g., an exhaust system), and obviating its electrical system as an ignition source would be of limited value.

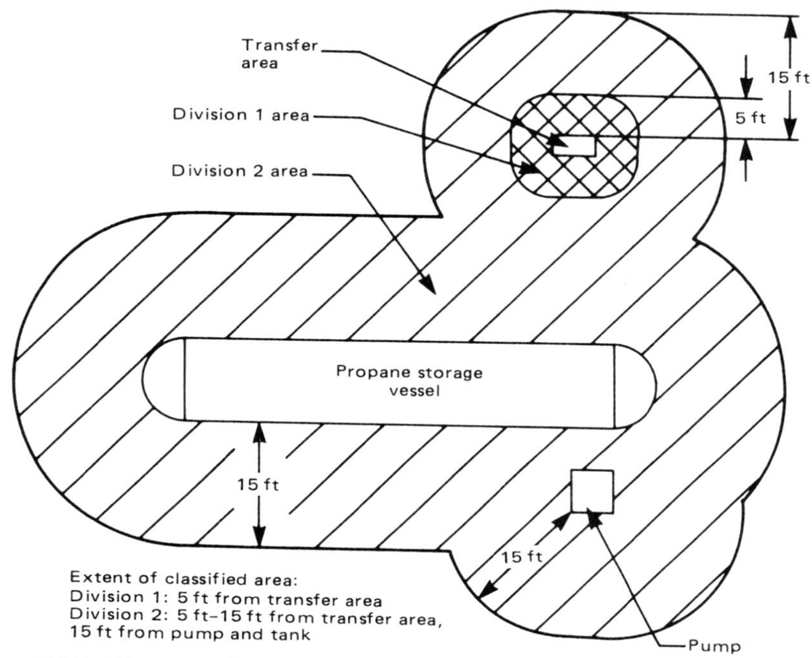

Figure 3.37 Plan view showing extent of classified area. (Drawing courtesy of Plant Systems, Inc.)

3-7.2.3 Electrical equipment installed on LP-Gas cargo vehicles shall comply with 6-1.1.4.

3-7.3 Other Sources of Ignition.

3-7.3.1 Open flames or other sources of ignition shall not be permitted in pump houses, container filling rooms, or other similar locations. Direct-fired vaporizers or indirect-fired vaporizers attached or installed adjacent to gas-fired heat sources shall not be permitted in pump houses or container filling rooms.

3-7.3.2 Open flames (except as provided in Section 3-6), cutting or welding, portable electric tools, and extension lights capable of igniting LP-Gas shall not be permitted within classified areas specified in Table 3-7.2.2 unless the LP-Gas facilities have been freed of all liquid and vapor, or special precautions have been taken under carefully controlled conditions.

NFPA 51B, *Standard for Fire Prevention in Use of Cutting and Welding Processes*, provides a procedure for permitting cutting, welding, and similar operations in areas where they would normally not be permitted because of the threat of fire. The American Petroleum Institute, the American Gas

Association, and the American Welding Society also have publications pertinent to this provision.

In the 1995 edition, Section 3-7.4, Control of Ignition Sources during Transfer, was deleted as it duplicated Section 4-3.2.2 due to an error in printing the 1992 edition.

3-8 LP-Gas Systems on Vehicles (Other than Engine Fuel Systems).

3-8.1 Application.

3-8.1.1 This section applies to non-engine fuel systems on commercial, industrial, construction, and public service vehicles such as trucks, semitrailers, trailers, portable tar kettles, road surface heating equipment, mobile laboratories, clinics, and mobile cooking units (such as catering and canteen vehicles). LP-Gas systems on such vehicles shall be permitted to be either vapor-withdrawal or liquid-withdrawal type. Included are provisions for installations served by exchangeable (removable) container systems and by permanently mounted containers.

3-8.1.2 This section shall not apply to:

(a) Systems installed on mobile homes.

(b) Systems installed on recreational vehicles [*see 3-1.3(d)*].

(c) Tank trucks, truck transports (trailers and semitrailers), and similar units used to transport LP-Gas as cargo, which are covered by Chapter 6.

(d) LP-Gas engine fuel systems on the vehicles covered by Section 3-8 and those cited in 3-8.1.2, which are covered by Chapter 8.

This section is the counterpart of Chapter 8 for vehicle propulsion engine systems: it covers all other applications of LP-Gas using systems mounted on vehicles. There are four exceptions listed in 3-8.1.2, namely, mobile homes, recreational vehicles, cargo tank vehicles, and certain engine fuel systems. Standards for installations on mobile vehicles for cooking and heating first appeared in the 1950 edition of NFPA 58. These were later separated into two different chapters: (1) systems for mobile homes and travel trailers and (2) systems for commercial vehicle uses. These provisions formed the basis for NFPA 501B, *Standard for Mobile Homes*. NFPA 501B was discontinued when the federal agency of Housing and Urban Development (HUD) used NFPA 501B and issued their *Mobile Home Construction and Safety Standards*, Part 280 CFR 24, in 1976. NFPA 501C, *Standard on Fire Safety Criteria for Recreational Vehicles*, utilized the former provisions of NFPA 58 extensively for standards on LP-Gas systems. Cargo tank systems are covered in Chapter 6 of NFPA 58. Engine fuel systems are covered in Chapter 8. Thus, vehicular systems covered by the section are now limited to the types of applications listed in 3-8.1.1.

3-8.2 Construction, Location, Mounting, and Protection of Containers and Systems.

3-8.2.1 Containers shall comply with Section 2-2, and appurtenances used to equip them for service shall comply with Section 2-3. In addition, the following shall apply:

Provisions for the construction of containers parallel to those for engine fuel containers in Chapter 8, where ASME container design pressure is 312.5 psi (2.2 MPa) for those installed in enclosed spaces [otherwise 250 psi (1.7 MPa) design pressure may be used], maximum sizes of containers are specified, etc. Paragraph (g) is particularly significant with respect to the need for protection of the container appurtenances and their connections. Refer to Table 2-3.3.2 (a), Columns 6 or 7, for container appurtenances for DOT or ASME containers of 2,000 gal water capacity or less, respectively. The columns cover Mobile Containers, which Note 2 to the table defines as "containers that are permanently mounted on a vehicle and are connected for uses other than engine fuel." A standard stationary container without the protection specified in 2-2.4 should not be used, as these appurtenances would be vulnerable in the event of a vehicular accident and would not have the appurtenances required in Table 2-3.3.2 (a). Refer to paragraph 2-3.3.2 (b) for appurtenances for containers over 2,000 gal.

(a) ASME containers shall be constructed for a minimum 250 psi (1.7 MPa) design pressure.

(b) Containers installed in enclosed spaces on vehicles (including recesses or cabinets covered in 3-8.2.2) shall be constructed as follows:

1. DOT cylinder specification containers shall be designed and constructed for at least a 240-psi (1.6-MPa) service pressure.
2. ASME containers shall be constructed for at least a 312.5-psi (2.2-MPa) design pressure.

(c) Portable (removable) containers shall comply with 2-2.4.

(d) Permanently mounted containers shall comply with 3-8.2.3(c).

(e) LP-Gas fuel containers used on passenger-carrying vehicles shall not exceed 200 gal (0.8 m^3) aggregate water capacity.

(f) Individual LP-Gas containers used on other than passenger-carrying vehicles normally operating on the highway shall not exceed 300 gal (1.1 m^3) water capacity. This shall not beconstrued as applying to the use of LP-Gas from the cargo tanks of vehicles covered by Chapter 6.

Exception: Containers on road surface heating equipment shall not exceed 1,000 gal (3.8 m^3) water capacity.

This exception was added in the 1989 edition when it became evident that the previous limit of 300 gal was overly restrictive. When the requirement was written in an early edition of the standard, a 300-gal (1.1-m^3) container was the largest the Technical Committee could envision being mounted on a vehicle. In recent years equipment has been developed that melts road surfacing after removal from roadways to recover the asphalt content for immediate reuse. Such equipment has a very high propane consumption rate and the 300-gal (1.1-m^3) limit on container size, along with the constraints of truck-mounted equipment limiting multiple containers, forced frequent shutdown for refilling (and the required refilling vehicle to stand by during operation).

The 1,000-gal (3.8-m^3) limit on container size is based on the Committee's knowledge of units in safe operation.

(g) Containers designed for stationary service only, and not in compliance with 2-2.4, shall not be used.

3-8.2.2 Containers utilized for the purposes covered by this section shall not be installed, transported, or stored (even temporarily) inside any vehicle covered by Section 3-8, except as provided in 3-8.2.3(d), Chapter 6, or as provided by applicable DOT regulations. The LP-Gas supply system, including the containers, shall be permitted to be installed on the outside of the vehicle, or in a recess or cabinet vaportight to the inside of the vehicle but accessible from and vented to the outside, with the vents located near the top and bottom of the enclosure, and 3 ft (1 m) horizontally away from any opening into the vehicle below the level of the vents.

Containers are to be located on the outside of the vehicle or in a recess or compartment that is vaportight to the inside of the vehicle. The exceptions are those systems installed in the interior of a vehicle similar to engine fuel containers, which are covered in 8-2.7, and the transportation of spare portable containers as outlined in Chapter 6 or by applicable DOT regulations. For example, portable cylinders used in connection with food warmers on delivery vehicles should be located outside or in a compartment and piped to the appliance inside. Requirements for supply systems on vehicles subject to DOT regulations are contained in Section 393.77 of the *Bureau of Motor Carrier Safety Regulations*, Part 393, CFR 49, and Section 177.834(d) of the *Hazardous Materials Regulations*, Part 177, CFR 49.

3-8.2.3 Containers shall be mounted securely on the vehicle, or within the enclosing recess or cabinet, and located and installed so as to minimize the possibility of damage to containers, their appurtenances, or contents as follows:

(a) Containers shall be installed with road clearance in accordance with 8-2.6(e).

(b) Fuel containers shall be mounted securely to prevent jarring loose and slipping or rotating, and the fastenings shall be designed and constructed to withstand, without permanent visible deformation, static loading in any direction equal to four times the weight of the container filled with fuel. Where containers are mounted within a vehicle housing, the securing of the housing to the vehicle shall comply with this provision. Any hoods, domes, or removable portions of the housing or cabinet shall be provided with means to keep them firmly in place in transit. Field welding shall comply with 3-2.4.1(e).

(c) All container valves, appurtenances, and connections shall be adequately protected to prevent damage from accidental contacts with stationary objects, from loose objects, stones, mud, or ice thrown up from the ground or floor, and from damage due to overturn or similar vehicular accident. In the case of permanently mounted containers, this provision shall be met by the location on the vehicle, with parts of the vehicle furnishing the protection. On portable (removable) containers, the protection for container valves and connections shall be attached permanently to the container. (*See 2-2.4.1 and 2-2.4.2.*) Such weather protection necessary to ensure safe operation shall be provided for containers and systems mounted on the outside of vehicles.

(d) Containers mounted on the interior of passenger-carrying vehicles shall be installed in compliance with 8.2.7. Pressure relief valve installations for such containers shall comply with 8-2.6(i).

These provisions also serve to ensure the container mounting arrangement is strong enough to stay intact, and to protect the container and its appurtenances and connections from damage caused by collisions, road debris, and weather. They are identical to those for engine fuel supply systems.

As with vehicle propulsion engine fuel containers, protection of exterior containers and appurtenances against material thrown up from the road is set forth. However, with nonengine fuel systems a pressure regulator may be mounted at the container for vapor withdrawal systems. The vent on this regulator must not be blocked with slush, etc., thrown up from the road. This may be prevented with a splash guard or by locating the regulator in a compartment. A blocked vent may lead to abnormally high pressures in the utilization system. Appurtenances on a container, particularly those mounted below the vehicle, should be installed so that they are readily accessible.

3-8.2.4 Containers installed on portable tar kettles alongside the kettle, on the vehicle frame, or on road surface heating equipment shall be protected from radiant or convected heat from open flame or other burners by the use of a heat shield or by the location of the container(s) on the vehicle to prevent the temperature of the fuel in the container from becoming abnormally high. In addition, the following shall apply:

(a) Container location, mounting, and protection shall comply with 3-8.2.3(a), (b), and (c) except that the protection for DOT container valves shall not be required to be attached to the container permanently; however, the protection shall comply with 2-2.4.1(a) and (b);

(b) Piping shall comply with 3-8.2.7(a), (b), (d), (e), (g), (h), and (i);

(c) Flexible connections shall comply with 2-4.6.1, 2-4.6.2, and 2-4.6.3;

(d) Container valves shall be closed when burner is not in use;

(e) Containers shall not be refilled while burners are in use as provided in 4-2.3.2(b).

Although somewhat in conflict with NFPA 58's basic rule of not setting out provisions for specific applications, tar kettles are one of the more extensive vehicular applications of LP-Gas. This paragraph provides direction as to where to find the pertinent information in the standard.

Tar kettles are generally standardized equipment today; however, attention should be given to these requirements to make certain they are applied. Cylinders should be mounted such that no wear takes place (chains can cut grooves in cylinders due to vehicular motion). Most tar kettles use liquid burners so the right type of cylinder must be used. The word "liquid" is stamped on the valve handle or on the cylinder next to the valve. Valve protection for portable cylinders covered by Section 3-8 must be permanently attached, such as by the use of a collar welded to the container. For portable containers used with tar kettles, however, these need not be permanently attached. The type of collar used is threaded to the cylinder valve boss.

3-8.2.5 Container appurtenances shall be installed in accordance with the following:

(a) Container pressure relief devices shall be located and installed as follows:

1. On containers mounted in the interior of vehicles complying with 8-2.7, the pressure relief valve installation shall comply with 8-2.6(i).
2. Pressure relief valve installations on containers installed on the outside of vehicles shall comply with 8-2.6(i) and 3-8.2.2.

(b) Connections and appurtenances on containers shall be in compliance with 2-3.3.1 through 2-3.3.3 and Table 2-3.3.2(a).

(c) Main shutoff valves on container for liquid and vapor shall be readily accessible.

(d) Containers to be filled volumetrically shall be equipped with liquid level gauging devices as provided in 2-3.4. Portable containers shall be permitted to be designed, constructed, and fitted for filling in either the vertical or horizontal position or, if of the portable universal type [*see 2-3.4.2(c)(2)*], in either position. The container shall be in the appropriate position when filled or, if of the portable universal type, shall be permitted to be loaded in either position, provided:

1. The fixed level gauge indicates correctly the maximum permitted filling level in either position.
2. The pressure relief devices are located in, or connected to, the vapor space in either position.

(e) All container inlets and outlets, except pressure relief devices and gauging devices, shall be labeled to designate whether they communicate with the vapor or liquid space. Labels shall be permitted to be affixed to valves.

(f) Containers from which only vapor is to be withdrawn shall be installed and equipped with connections to minimize the possibility of the accidental withdrawal of liquid.

The three types of fuel container installations affect how the pressure relief valve is installed and located. These are: (1) portable containers in cabinet or recess, (2) containers mounted on the exterior, and (3) containers installed in the interior of passenger-carrying vehicles. Paragraph 3-8.2.2 specifies what to do with cabinets and recesses (excluding cabinets in the interior of passenger vehicles), 3-8.2.5(a)(2) describes what has to be done with exterior-mounted containers, and 8-2.6(i) indicates what must be done with interior-mounted containers on passenger vehicles. Subparagraph (4) of 8-2.6(i) draws attention to protecting the pressure relief valve outlet (valve or discharge piping) from being plugged with water (which can freeze), dirt, asphalt, etc.

3-8.2.6 Regulators shall comply with 2-5.1.3(d) and (e) and 2-5.8 and shall be installed in accordance with 3-2.6. If the regulators are installed in an enclosed space, the discharge from the required pressure relief device shall be vented to the outside air in accordance with 3-8.2.2.

Provisions for venting regulators installed in cabinets to the outside are available through the louvers in the cabinet door, whereas an enclosed space does not have access to the outside. Therefore, provisions for venting this space must be provided.

(a) A two-stage regulator system or an integral two-stage regulator is required for vapor withdrawal systems in accordance with 3-2.6.1 and protected from the elements in accordance with 3-2.6.4. The regulator shall be installed with the pressure relief vent opening pointing vertically downward to allow for drainage of moisture collected on the diaphragm of the regulator. Regulators not installed in compartments shall be equipped with a durable cover designed to protect the regulator vent opening from sleet, snow, freezing rain, ice, mud, and wheel spray.

(b) If a vehicle-mounted regulator(s) is installed at or below the floor level, it shall be installed in a compartment that provides protection against the weather and wheel spray. The compartment shall be of sufficient size to allow tool operation for connection to and replacement of the regulator(s), shall be vaportight to the interior of the vehicle, shall have a 1-sq in. (6.5-cm^2) minimum vent opening to the exterior located within 1 in. (25 mm) of the bottom of the compartment, and shall not contain flame- or spark-producing equipment. A regulator vent outlet shall be at least 2 in. (51 mm) above the compartment vent opening.

Two-stage regulation is required and reference is made to Section 3-2.6, which was revised in the 1995 edition to mandate two-stage regulation in fixed piping systems. Two-stage pressure regulation had been required in this section previously. The same requirements are used in NFPA 501C, *Standard on Fire Safety Criteria for Recreational Vehicles*, where propane is also used as a fuel in vapor form.

3-8.2.7 Piping shall comply with Section 2-4, and the following requirements with respect to material and design and shall be installed in accordance with 3-2.8. The following shall also apply to piping systems on vehicles covered by Section 3-8.

Exception: Steel tubing shall have a minimum wall thickness of 0.049 in. (1.2 mm).

(a) A flexible connector or a tubing loop shall be installed between the regulator outlet and the piping system to protect against expansion, contraction, jarring, and vibration strains.

(b) In the case of removable containers, flexibility shall be provided in the piping between the container and the gas piping system or regulator.

(c) Flexible connectors shall comply with 2-4.6 and shall be installed in accordance with 3-2.8.9(a). Flexible connectors of more than 36 in. (91 cm) overall length, or fuel lines that essentially are made completely of hose, shall be used only with the approval of the authority having jurisdiction.

(d) The piping system shall be designed, installed, supported, and secured in such a manner as to minimize the possibility of damage due to vibration, strains, or wear, and to preclude any loosening while in transit.

(e) Piping (including hose) shall be installed in a protected location. If outside, piping shall be under the vehicle and below any insulation or false bottom. Fastening or other protection shall be installed to prevent damage due to vibration or abrasion. At each point where piping passes through sheet metal or a structural member, a rubber grommet or equivalent protection shall be installed to prevent chafing.

(f) Gas piping shall be installed to enter the vehicle through the floor directly beneath, or adjacent to, the appliance served. If a branch line is required, the tee connection shall be located in the main gas line under the floor and outside the vehicle.

(g) Exposed parts of the piping system shall be either of corrosion-resistant material or adequately protected against exterior corrosion.

(h) Hydrostatic relief valves, complying with 2-4.7, shall be installed in isolated sections of liquid piping as provided in 3-2.9.

(i) Piping systems, including hose, shall be tested and proven free of leaks in accordance with 3-2.10.

(j) There shall be no fuel connection between a tractor and trailer or other vehicle units.

Steel tubing in vehicular installations has been specified as having a minimum 0.049-in. (1.2-mm) wall thickness to provide strength against vibration and also an additional tolerance for corrosion. This came about as a result of experience with steel tubing during World War II, when copper tubing was not available. Subparagraphs (e) and (f) are the basic approach to piping of all vehicular-type installations (i.e., keeping the main lines and branch connections outside). Even though flow control protection is required as per Table 2-3.3.2(a), no fuel lines are to be connected between two vehicular units in order to avoid compounding problems involved in collisions, overturns, disconnection of vehicles, etc. These types of risks are regarded as greater than the corresponding risks in a system completely on one vehicle.

3-8.3 Equipment Installation. Equipment for installation on vehicles shall comply with Section 2-5 with respect to design and construction and shall be installed in accordance with 3-2.13 and the following:

(a) Installation shall be made in accordance with the manufacturer's recommendations and, in the case of listed or approved equipment, as provided in the listing or approval.

(b) Equipment installed on vehicles shall be considered as part of the LP-Gas system on the vehicle and shall be protected against vehicular damage as provided for container appurtenances and connections in 3-8.2.3(c).

3-8.4 Appliance Installation.

In addition to provisions for cargo heaters here, reference should be made to the U.S. Department of Transportation (DOT) requirements for those vehicles subject to their jurisdiction. Safety shutoffs are required on all heating appliances except tar kettle burners, hand torches, melting pots, and small heaters using a torch-type cylinder, as these are attended appliances. The basic concept of isolating the combustion system of appliances—except ranges in vehicle interiors where passengers might be—is consistent with all other vehicle standards on this subject (i.e., those for recreational vehicles). No portable or conventional room heaters should be used inside. Isolation may be accomplished through the use of direct-vent-type heaters and water heaters, or through separation by installation in a compartment with provisions for outside air. The range is an attended appliance and need not be isolated, but it should never be used for comfort heating.

Note that although extensive requirements for installation of appliances fueled by LP-Gas are included in NFPA 54, *National Fuel Gas Code*, that standard is limited to those appliances normally connected to a fixed piping system. The appliances covered under NFPA 58 are those not normally connected to a fixed piping system.

3-8.4.1 The term "appliances" as used in this subsection shall include any commercial or industrial gas-consuming device except engines.

3-8.4.2 All gas-consuming devices (appliances), other than engines, installed on vehicles shall be approved as provided in 2-6.1, shall comply with 2-6.2, and shall be installed as follows:

(a) Wherever the device or appliance is of a type designed to be in operation while the vehicle is in transit, such as a cargo heater or cooler, suitable means to stop the flow of gas in the event of a line break, such as an excess-flow valve, shall be installed. Excess-flow valves shall comply with 2-4.5.3 and 2-3.3.3(b).

(b) All gas-fired heating appliances shall be equipped with safety shutoffs in accordance with 2-6.2.5(a) except those covered in 3-4.2.8(b).

(c) For installations on vehicles intended for human occupancy, all gas-fired heating appliances, except ranges and illuminating appliances, shall be designed or installed to provide for a complete separation of the combustion system from the atmosphere inside the vehicle. Combustion air inlets and flue gas outlets shall be listed as components of the appliance.

(d) For installations on vehicles not intended for human occupancy, unvented-type gas-fired heating appliances shall be permitted to be used to protect the cargo. Provision shall be made to provide air for combustion [*see 3-8.4.2(f)*] and to dispose of the products of combustion to the outside.

(e) Appliances installed within vehicles shall comply with the following:

1. If in the cargo space, they shall be readily accessible whether the vehicle is loaded or empty.
2. Appliances shall be constructed or otherwise protected to minimize possible damage or impaired operation due to cargo shifting or handling.
3. Appliances shall be located so that a fire at any appliance will not block egress of persons from the vehicle.

(f) Provision shall be made in all appliance installations to ensure an adequate supply of outside air for combustion.

(g) A permanent caution plate shall be provided, affixed to either the appliance or the vehicle outside of any enclosure and adjacent to the container(s), and shall include the following items:

CAUTION

(1) Be sure all appliance valves are closed before opening container valve.

(2) Connections at the appliances, regulators, and containers shall be checked periodically for leaks with soapy water or its equivalent.

(3) Never use a match or flame to check for leaks.

(4) Container valves shall be closed when equipment is not in use.

3-8.5 General Precautions.

3-8.5.1 Containers on vehicles shall be filled or refilled as provided by 4-2.2. Requalification requirements for continued use and reinstallation of containers shall be in accordance with 2-2.1.4.

3-8.5.2 Mobile units containing hotplates and other cooking equipment, including mobile kitchens and catering vehicles, shall be provided with at least one approved portable fire extinguisher rated in accordance with NFPA 10, *Standard for Portable Fire Extinguishers*, at not less than 10-B:C.

3-8.6 Parking, Servicing, and Repair. Vehicles with LP-Gas fuel systems mounted on them for purposes other than propulsion shall be permitted to be parked, serviced, or repaired inside buildings, in accordance with the following:

This section provides specific coverage for the parking, servicing, and repair of vehicles with nonengine fuel LP-Gas systems on them. Note that this section does not cover recreational vehicles (which are covered by NFPA 501C), mobile homes (which are covered under federal regulations), or vehicles used to transport LP-Gas as cargo (which are covered under Chapter 6).

(a) The fuel system shall be leak-free, and the container(s) shall not be filled beyond the limits specified in Chapter 4.

(b) The container shutoff valve shall be closed except when fuel is required for test or repair.

(c) The vehicle shall not be parked near sources of heat, open flames, or similar sources of ignition, or near unventilated pits.

(d) Vehicles having containers with water capacities larger than 300 gal (1.1 m^3) shall comply with the requirements of Section 6-6.

Note that these requirements are similar to those for vehicles using LP-Gas as an engine fuel (8-6), and that vehicles having a total LP-Gas storage of more than 300 lb (136 kg) are required to follow the more detailed requirements of Chapter 6.

3-9 Vehicle Fuel Dispenser and Dispensing Stations.

This section was added in the 1992 edition to provide specific information in one place on vehicle fuel dispensers and dispensing stations/propane vehicle fueling stations. Further revisions were made in the 1995 edition to update the section and enhance its useability. The use of propane as an engine fuel is reported to the editor to be increasing to meet environmental requirements; the number of propane fueling stations is also projected to increase significantly.

This section was created following recognition that the requirements for dispensing were difficult to follow because they were dispersed throughout Chapter 3, and were therefore not being followed completely. The Committee has held discussions with the NFPA Automotive and Marine Service Station Committee, which is responsible for NFPA 30A, *Automotive and Marine Service Station Code*, with the intent of coordinating the requirements of both documents. This section may be extracted (copied) into NFPA 30A or may move to NFPA 30A in a future edition.

Figure 3.38 Dispensing station for propane vehicles. (Photo courtesy of Ace Gas Company.)

3-9.1 Application. This section includes location, installation, and operating provisions for vehicle fuel dispensers and dispensing stations. The general provisions of Section 3-2 shall apply unless specifically modified in this section.

3-9.2 Location.

3-9.2.1 Location shall be in accordance with Table 3-2.3.3.

Table 3-2.3.3 specifies distances from points of transfer and exposures. This table applies to the location of the filling of portable containers, i.e., those not at a stationary installation, which includes all containers brought to a dispensing station for filling. It provides separation distances from points of transfer—the hose end—and several important exposures including public ways, buildings with and without fire-resistive walls, mainline railroad track centerlines, outdoor places of public assembly, and flammable liquids dispensers.

3-9.2.2 Vehicle fuel dispensers and dispensing stations shall be located away from pits in accordance with Table 3-2.3.3 with no drains or blow-offs from the unit directed toward, or within 15 ft (4.6 m) of, a sewer systems opening.

3-9.3 General Installation Provisions.

General rules for installation of both portable cylinder filling and vehicle fuel dispensers are located here.

3-9.3.1 Vehicle fuel dispensers and dispensing stations shall be installed as recommended by the manufacturer.

3-9.3.2 Installation shall not be within a building but shall be permitted to be under weather shelter or canopy, provided this area is adequately ventilated and is not enclosed for more than 50 percent of its perimeter.

This section provides guidance on the extent of weather shelter that can be installed without being considered an indoor location. If an installation is considered indoor, compliance with Chapter 7, *Buildings or Structures Housing LP-Gas Distribution Facilities*, is mandated.

3-9.3.3 Control for the pump used to transfer LP-Gas through the unit into containers shall be provided at the device in order to minimize the possibility of leakage or accidental discharge.

3-9.3.4 An excess-flow check valve complying with 2-3.3.3(b) or an emergency shutoff valve complying with 2-4.5.4 shall be installed in or on the dispenser at the point at which the dispenser hose is connected to the liquid piping. A differential back pressure valve shall be considered as meeting this provision.

This requirement was revised in the 1995 edition by deleting the requirement for construction of dispensing hose, which does not belong in Chapter 3 as the chapter covers installation only. The requirements for excess-flow check valves and emergency shutoff valves are provided in 2-3.3.3 (b) and 2-4.5.4, respectively. A differential back pressure valve is a special excess-flow valve for use with a meter that is normally supplied with an LP-Gas meter. The valve senses the difference in pressure before and after the meter, and closes if they are significantly different. This ensures that only liquid (not vapor) passes through the meter for accurate reading. If the dispensing hose should break or become disconnected, the pressure downstream of the meter will drop, and the valve will close. See Figure 2.27 for a graphical explanation of the operation of a differential back pressure valve. This provision is essentially an exception to the requirement for an excess-flow valve or an emergency shutoff valve in liquid transfer lines in 3-3.3.

3-9.3.5 Piping and the dispensing hose shall be provided with hydrostatic relief valves as specified in 3-2.9.

3-9.3.6 Protection against trespassing and tampering shall be in accordance with 3-3.6.

3-9.3.7 A manual shutoff valve and an excess-flow check valve of suitable capacity shall be located in the liquid line between the pump and dispenser inlet where the dispensing device is installed at a remote location and is not part of a complete storage and dispensing unit mounted on a common base.

This requirement was revised in the 1995 edition to make it applicable only when the pump and dispenser inlet are not installed together, where there is a greater opportunity for pipe failure.

3-9.3.8 All dispensers shall either be installed on a concrete foundation or be part of a complete storage and dispensing unit mounted on a common base and installed in accordance with 3-2.4.2(a)3 and 2-2.5.2(a) and (b). Protection shall be provided against physical damage.

This new requirement provides for a solid anchorage of a dispenser. It helps to ensure that the breakaway device mandated in 3-9.4.2 will operate properly. Physical protection is also required and is often needed in areas where vehicle traffic is normal.

3-9.3.9 A listed quick-acting shutoff valve shall be installed at the discharge end of the transfer hose.

3-9.3.10 A clearly identified and easily accessible switch(es) or circuit breaker(s) shall be provided at a location not less than 20 ft (6.1 m) nor more than 100 ft (30.5 m) from dispensing device(s) to shut off the power in the event of a fire, accident, or other emergency. The marking for the switch(es) or breaker(s) shall be visible at the point of liquid transfer.

This requirement was revised in the 1995 edition by providing a specific minimum and maximum distance for the shutoff switch of the dispenser, as well as the addition of marking requirements. This was recommended by an enforcement official who pointed out that the previous requirement, "at some point from the dispensing station," was vague and unenforceable.

3-9.4 Installation of Vehicle Fuel Dispensers.

Additional requirements for vehicle fuel dispensers, which supplement those provided in Section 3-9.3, are located in this section.

3-9.4.1 Hose length shall not exceed 18 ft (5.5 m). All hose shall be listed. When not in use, hose shall be secured to protect it from damage.

Exception: Hoses longer than 18 ft (5.5 m) shall be permitted where approved by the authority having jurisdiction.

This is modeled after NFPA 30A, 4-2.6. The 18-ft limit on hose length comes from the U.S. Bureau of Weights and Measures. The exception permits exceeding the normal maximum hose length, when needed, with approval.

3-9.4.2 A listed emergency breakaway device complying with UL 567, *Standard Pipe Connectors for Flammable and Combustible Liquids and LP-Gas*, and designed to retain liquid on both sides of the breakaway point, or other devices affording equivalent protection approved by the authority having jurisdiction, shall be installed.

This is based on NFPA 30A, 4-2.7, and provides flexibility to the authority having jurisdiction. Pullaway incidents have occurred with both liquid and propane engine fuel dispensers. Reference to the UL standard was added in the 1995 edition to provided guidance on the device.

3-9.4.3 Dispensing devices for liquefied petroleum gas shall be located at least 10 ft (3.0 m) from any dispensing device for Class I liquids.

This requirement was revised in the 1995 edition to remain consistent with NFPA 30A. The previous requirement in NFPA 30A, and 3-9.4.7 in the 1992 edition of NFPA 58, stated that "Vehicle fuel dispensers shall not be located on the same island as a gasoline dispenser."

This has been revised here and in NFPA 30A, however NFPA 30A requires a 20-ft separation, while the reduced distance of 10 ft is required here. The LP-Gas Committee has based the 10-ft distance on the Canadian requirement. The editor has been advised that the use of LP-Gas in vehicle service stations has been much greater in Canada than in the United States. Discussions between the committees responsible for the two standards continue, and it is hoped that future editions will resolve this difference.

3-10 Fire Protection.

3-10.1 **Application.** This section contains provisions for fire protection to augment the leak control and ignition source control provisions in this standard.

With the exception of Section 3-10, all other provisions in NFPA 58 are leak control and ignition source control provisions reflecting equipment design, fabrication, installation, maintenance, and facility personnel performance. If these provisions were complied with initially and maintained subsequently, there would be little need for Section 3-10. Equipment doesn't always function as intended and people make mistakes. For these reasons, fire protection is required to augment the other provisions.

As noted in the commentary on 3-2.2.2, consideration of fire protection can often be a significant factor in determining the actual location of a facility.

3-10.2 General.

3-10.2.1 The wide range in size, arrangement, and location of LP-Gas installations covered by this standard precludes the inclusion of detailed fire protection provisions completely applicable to all installations. Provisions in this section are subject to verification or modification through analysis of local conditions.

When it is considered that an installation covered by NFPA 58 can range from one or two 100-lb (45-kg) LP-Gas capacity cylinders at a single-family residence in a rural area that are filled once a month (or maybe only exchanged and not filled at the installation), to several 30,000-gal (114-m^3) LP-Gas containers at a bulk plant or industrial plant (and larger storage facilities at marine and pipeline terminals) in a heavily populated or congested area with liquid transfer and associated transportation operations being conducted more or less continually, it is apparent that the need for and character of fire protection provisions can vary widely. In fact, these can be determined only by studying each installation and the hazards associated with it.

3-10.2.2* The planning for effective measures for control of inadvertent LP-Gas release or fire shall be coordinated with local emergency handling agencies, such as fire and police departments. Such measures require specialized knowledge and training not commonly present in the training programs of emergency handling agencies. Planning shall consider the safety of emergency personnel.

The supplement "Guidelines for Conducting a Firesafety Analysis" contains a thorough discussion of the factors needed to plan for an LP-Gas emergency.

Training facilities are expensive to install and operate, therefore there are only a few in the United States. Three of the largest, conducted on an ongoing basis each year, are by the Texas A & M Fire College, the Massachusetts Firefighting Academy, and Nassau County, on Long Island, New York. The need for this hands-on training is being recognized increasingly, however, and new facilities can be anticipated.

Two NFPA films on propane are available: "LP-Gas: Emergency Planning and Response," which is a useful tactical training instrument, and "BLEVE Update," which provides current information on BLEVEs. The National Propane Gas Association has produced an audio-visual slide production, "Handling LP-Gas Leaks and Fires."

A-3-10.2.2 The National Fire Protection Association, American Petroleum Institute, and National Propane Gas Association publish material, including visual aids, useful in such planning.

3-10.2.3* Fire protection shall be provided for installations having storage containers with an aggregate water capacity of more than 4,000 gal (15.1 m^3) subject to exposure from a single fire. The mode of such protection shall be determined through a competent fire safety analysis (*see 3-2.2.9*).

NOTE: Experience has indicated that hose stream application of water in adequate quantities as soon as possible after the initiation of flame contact is an effective way to prevent container failure from fire exposure. The majority of large containers exposed to sufficient fire to result in container failure have failed in from 10 to 30 min after the start of the fire where water was not applied. Water in the form of a spray can also be used to control unignited gas leakage.

A-3-10.2.3 A fire safety analysis should include the following:

(a) An analysis of local conditions of hazard within the container site.

(b) Exposure to or from other properties, population density, and congestion within the site.

(c) The probable effectiveness of plant fire brigades or local fire departments based on adequate water supply, response time, and training.

(d) Consideration for the adequate application of water by hose stream or other method for effective control of leakage, fire, or other exposures.

The first consideration in any such analysis shall be an evaluation of the total product control system including emergency internal and shutoff valves having remote and thermal shutoff capability and pullaway protection.

Unless the circumstances described in Exception No. 1 (*see commentary*) exist, it must be recognized that a degree of exposure exists to the facility, its employees, and the persons and property of neighbors, and that fire protection is needed beyond that specifically required by NFPA 58. This fire protection could be provided by a plant fire brigade or a municipal fire department.

If a facility were located within an area served by one or more emergency services such as public fire departments, police departments, and ambulance services, facility management has a right to expect assistance from these sources in the event of an emergency. Management has a duty to obtain this assistance without requiring the emergency personnel to accept undue risks. The resolution of this risk factor is a key element in the fire safety analysis process. In a few locations, especially where many chemical and petrochemical facilities operate, agreements state that public emergency response services do not enter the facilities, rather the facilities maintain trained personnel at all times and provide their own mutual aid. In these cases the facilities do not require or expect assistance from publicly funded emergency services.

The citation of a 4,000 gal (15 m^3) total water capacity reflects the Committee's concern of the greater severity of an accident with larger storage quantities involved. The exception for smaller facilities reduces the burden on regulatory officials, installers, and the facility operators. It is not intended that smaller facilities be totally excluded, however, and there may well be circumstances where smaller facilities should be considered.

The second paragraph of 3-10.2.3 was revised in the 1995 edition to change the first consideration from the application of water to an evaluation of the total product control system. Emergency shutoff valves and remotely controlled internal shutoff valves are required in the standard, and this change mandates their review as the first step in the analysis which must consider all factors. Since the requirement of product control equipment was added in the 1976 edition of the standard, the number of incidents involving bulk plants is believed to have decreased. Data is available begining in 1980. Rolling three-year averages from 1980–1992, show a marked decline in the number of reported fires in bulk plants, and it is reasonably concluded that the requirements of product control equipment have contributed to this. The goal of the analysis is to identify the mechanism by which a flame can impinge on a container (which can result in a BLEVE), and to prevent container failure.

If the application of cooling water to a container is needed, it must be recognized that a fire department is well suited to apply water to LP-Gas containers. Water in the proper form can disperse gas clouds and cool containers and equipment. Aside from its rescue function, a fire department is essentially a person-machine system designed to apply water. The adequacy of a fire department should be determined, and frequently the best way to do so is to run a drill.

A discussion of the many aspects of the six fire analysis factors given in 3-10.2.3 is covered in the supplement, "Guidelines for Conducting a Firesafety Analysis."

Exception No. 1: If the analysis specified in 3-10.2.3 indicates a serious hazard does not exist, the fire protection provisions of 3-10.2.3 shall not apply.

There may be situations in which no fire protection is needed beyond what is provided in the facility in compliance with specific provisions in NFPA 58. An example of such a situation would be a well-isolated facility.

If an adequate water supply is not available, agreement should be reached with public safety authorities to limit the role of emergency forces to the control of onlookers.

Exception No. 2: If the analysis specified in 3-10.2.3 indicates that a serious hazard exists and the provisions of 3-10.2.3 cannot be met, special protection (see definition) shall be provided in accordance with 3-10.3.

Special protection addresses only the BLEVE hazard, because this hazard usually represents the greatest exposure to the facility, its neighbors, and emergency personnel. It must be recognized that special protection only simplifies, and thus enhances the effectiveness of, fire department activities such as rescue and controlling unignited leaks.

The provision of special protection does not eliminate the need to consider other fire protection measures, as the BLEVE of a container is not the only fire hazard. Frequently flammables other than LP-Gases, such as methanol and flammable paints, are stored at LP-Gas facilities. In addition, vehicle fires and structure fires must be considered.

3-10.2.4 Suitable roadways or other means of access for emergency equipment, such as fire department apparatus, shall be provided.

3-10.2.5 Each industrial plant, bulk plant, and distributing point shall be provided with at least one approved portable fire extinguisher having a minimum capacity of 18 lb (8.2 kg) of dry chemical with a B:C rating. (*Also see NFPA 10.*)

See commentary on 5-5.

3-10.2.6 LP-Gas fires shall not normally be extinguished until the source of the burning gas has been shut off or can be shut off.

The danger of ignition of a cloud of LP-Gas that has escaped into the atmosphere can be greater than that of LP-Gas burning as it escapes; therefore, this requirement is important for emergency personnel involved with LP-Gas.

3-10.2.7 Emergency controls shall be conspicuously marked, and the controls shall be located so as to be readily accessible in emergencies.

3-10.3 Special Protection.

This provision describes five modes of special protection to minimize the chances of a BLEVE from fire exposure. Special protection is not limited to these five modes, as noted in the definition in Section 1-6 which also permits "any means listed for this purpose." There has been some interest in designs for water application systems other than those given in 3-10.3.4 and 3-10.3.5, but the editor is unaware of any such systems listed for this purpose.

Although not specifically cited in this provision, any combination of these five modes (or another listed for the purpose) is also acceptable. For example, a common combination consists of earth mounding for the portion of a container not fitted with appurtenances or connections (which is customarily at least 75 percent of the container surface area) and water spray or monitor protection for the remainder. Because of the rather high water demand for water spray and many monitor nozzle systems, such a combination can preclude expensive water supply improvements.

Figure 3.39 Storage containers with special protection. Installation of two insulated 90,000-gal (341-m^3) containers with steel saddles. In addition, monitor nozzles, automatically activated by heat sensors, are installed at the tank openings. (Photo courtesy of Plant Systems, Inc.)

Figure 3.40 Storage container with water spray fixed protection. (Photo courtesy of Ransome Manufacturing Company.)

4

LP-Gas Liquid Transfer

4-1 Scope.

4-1.1 Application.

4-1.1.1 This chapter covers transfers of liquid LP-Gas from one container to another wherever this transfer involves connections and disconnections in the transfer system, or the venting of LP-Gas to the atmosphere. Included are provisions covering operational safety and methods for determining the quantity of LP-Gas permitted in containers.

It is believed by the editor that the chances of LP-Gas release (especially liquid) are greatest during liquid transfer operations (release is an operational necessity to a degree), and that accurate container filling is necessary to avoid release through pressure relief devices after the container has been filled. The standard, therefore, has always thoroughly addressed these hazards.

Prior to the 1972 edition, the provisions were scattered throughout the standard, leading to inconsistent treatment of the same basic hazard. By placing them in a single chapter, the inconsistencies were reduced and a focus on this major hazard was provided.

In the 1992 edition, parts of Chapter 4 were relocated to Chapter 3 because they were related to installation of containers, container appurtenances, and piping; provisions relating to transfer of liquid were relocated to Chapter 4 from Chapter 3.

4-1.1.2 Provisions for ignition source control at transfer locations are covered in Section 3-7. Fire protection shall be in accordance with Section 3-10.

4-2 Operational Safety.

4-2.1 Transfer Personnel.

4-2.1.1 Transfer operations shall be conducted by qualified personnel meeting the provisions of Section 1-5. At least one qualified person shall remain in attendance at the transfer operation from the time connections are made until the transfer is completed, shutoff valves are closed, and lines are disconnected.

The individual performing the transfer should not only be fully qualified in such work, but should also be an individual who will not panic under emergency conditions. The person must be sufficiently familiar with the operation of equipment to stop the transfer operation and minimize loss of product when an emergency arises. The individual must also be alert to the transfer operation and not just assume that everything is going to move along smoothly without incident. Knowledge of what is happening is most important in the liquid transfer.

The words "in attendance" clarify the Technical Committee's frequently misinterpreted intent, which is that transfer operations be continuously attended by a person who can view the transfer operation and take action if needed.

4-2.1.2 Transfer personnel shall exercise precaution to assure that the LP-Gases transferred are those for which the transfer system and the containers to be filled are designed.

Accidents can occur if propane or a mixture of propane and butane are placed into containers suitable only for butane. The lower setting for the pressure relief device has resulted in its operation, leading to injury and property damage. LP-Gas containers are marked (on a nameplate or stamped in the cylinder's collar or body) with the design pressure of a container.

4-2.2 Containers to Be Filled or Evacuated.

4-2.2.1 In the interest of safety, transfer of LP-Gas to and from a container shall be accomplished only by qualified persons trained in proper handling and operating procedures meeting the requirements of Section 1-6 and in emergency response procedures. Such persons shall notify the container owner and user in writing when noncompliance with Sections 2-2 and 2-3 is found.

The importance of qualified personnel is restated here, in addition to 4-2.1.1, and the need for training in emergency response procedures is added so that in the event of an incident, the person there is trained and prepared to take prompt action to minimize the extent of loss, damage, or fire. Personnel filling containers can be the only ones regularly viewing or inspecting containers, and they are required to notify the owner of the container and the user, in writing, of any condition in which the container no longer complies with the requirements of Sections 2-2, Containers and 2-3, Container Appurtenances. This is especially important with the trend in some areas toward greater ownership of fixed containers by the user, who may not be aware of the requirements of Sections 2-2 and 2-3.

4-2.2.2 Valve outlets on portable containers of 108 lb (49 kg) water capacity [nominal 45 lb (20 kg) propane capacity] or less shall be equipped with an effective seal such as a plug, cap, listed quick-closing coupling, or a listed quick-connect coupling. This seal shall be in place whenever the container is not connected for use.

The number of portable LP-Gas containers [especially the nominal 20-lb (9.1-kg) propane cylinder] has been increasing. The Barbecue Industry Association reports that gas grill sales have increased from 3,173,000 in 1985, to 5,022,065 in 1993. Most of these are believed to be propane grills, each with a cylinder. With the increasing number of grill cylinders, fires associated with these cylinders have probably increased also. Consistent with this, NFPA statistical analysis shows a rising trend in LP-Gas fires outdoors, mostly in yards outside homes, but the available data cannot provide a more detailed analysis. One type of incident has resulted from the cylinder valve being partially (and unintentionally) opened — possibly during transportation in passenger vehicles. This provision was added in the 1986 edition and is intended to minimize such accidents, and was the direct result of an incident of this type in Milford, Connecticut, in 1982. The sealing device is usually a plastic plug threaded into the cylinder opening, whenever the cylinder is not connected for use. Cylinders equipped with "quick coupling" devices are not provided with a plug, as the device shuts off the flow when not connected for use.

Exception: Nonrefillable (disposable) and new unused containers shall not be required to comply.

4-2.2.3 Containers shall be filled only after determination that they comply with the design, fabrication, inspection, marking, and requalification provisions of this standard. (*See 2-2.1.3 and 2-2.1.4.*)

In most cases the individual filling a container is more qualified than the owner to make this determination. This person is frequently the only one in contact with the container who can be expected to possess this ability.

This provision places a great deal of responsibility upon the filler, which may not always be adequately recognized. The filler must be familiar with the marking and inspection requirements, and the latter, especially, require considerable training. With the proliferation of ownership of 20-lb (9.1-kg) DOT specification cylinders by the general public, the necessary requalification of 12 years after the date of manufacture and every 5 to 12 years thereafter (*see Appendix C*), is often overlooked. When filling DOT cylinders, the filler must be careful to verify that the cylinder requalification period has not been exceeded. If a cylinder is out of date, the owner must be notified of his or her responsibility to have the cylinder requalified prior to filling. The filler must also check for obvious corrosion, and should look in the area within the foot ring at the bottom of the cylinder.

The responsibility also includes fillers of ASME tanks. The filler is not, however, expected to be superhuman, and cannot identify internal corrosion, worn valve seats, or other conditions not visible. The filler can and should be alert to visible corrosion, missing piping components, and the presence of a nameplate and report this to the owner, as required in 4-2.2.1. Many propane suppliers will have company policies that do not offer subsequent refilling, if the out-of-compliance condition is not corrected.

4-2.2.4 DOT specification cylinders authorized as "single trip," "nonrefillable," or "disposable" containers shall not be refilled with LP-Gas.

These containers are not designed for extended service. Their small size increases their chances of being overfilled using commonly available filling equipment and procedures. They are designed for weight filling, and are equipped with filling connections that common filling equipment will not fit, in an effort to control these hazards.

Accidents have resulted from the refilling of "nonrefillable" cylinders from a 20-lb (9.1-kg) cylinder. The practice is prohibited both by federal law and by NFPA 58. Adapters for this purpose are widely available, as there are refillable containers in the same sizes as nonrefillable containers.

4-2.2.5 Containers into which LP-Gas is to be transferred shall comply with the following with regard to service or design pressure in relation to the vapor pressure of the LP-Gas:

(a) For DOT specification cylinders, the service pressure marked on the container shall be not less than 80 percent of the vapor pressure of the LP-Gas at 130°F (54.4°C). For example, if the vapor pressure of a commercial propane is 300 psi (2.0 MPa) at 130°F (54.4°C), the service pressure must be at least 80 percent of 300, or 240 psi (1.6 MPa).

(b) For ASME containers, the minimum design pressure shall comply with Table 2-2.2.2 in relation to the vapor pressure of the LP-Gas.

This expands on the requirements of 4-2.1.2. The temperature of 130°F (54.4°C) is the design basis of the DOT requirements, as it is the highest temperature that a cylinder is expected to reach in the United States.

4-2.3 Arrangement and Operation of Transfer Systems.

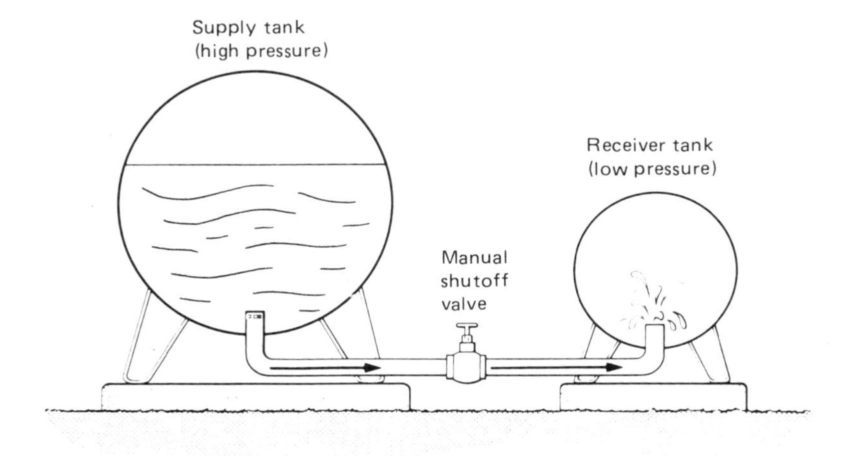

Figure 4.1 *Liquid transfer by pressure differential or gravity. (Drawing courtesy of National Propane Gas Association.)*

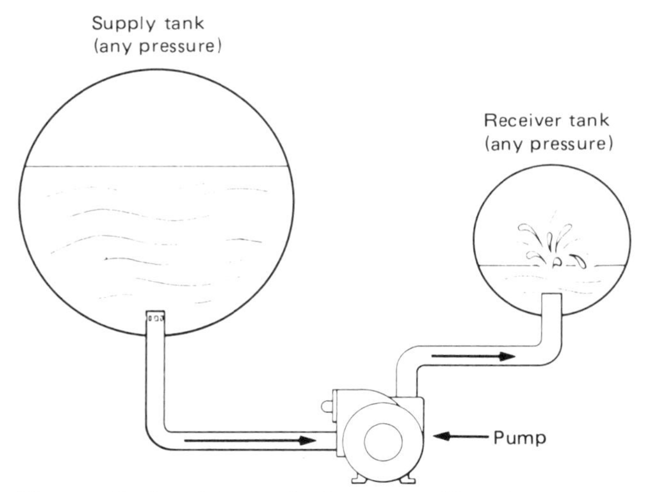

Figure 4.2 *Liquid transfer by pump. (Drawing courtesy of National Propane Gas Association.)*

LP-Gas can be transferred by pressure differential using gravity, pumps, or compressors to provide the force to accomplish the transfer.

Compressors are frequently used to unload tank cars and, occasionally, to transport trucks. Vapor is taken by a compressor from the container that will be receiving the product and is transferred to the container from which the liquid is being unloaded. This creates a pressure in the container being

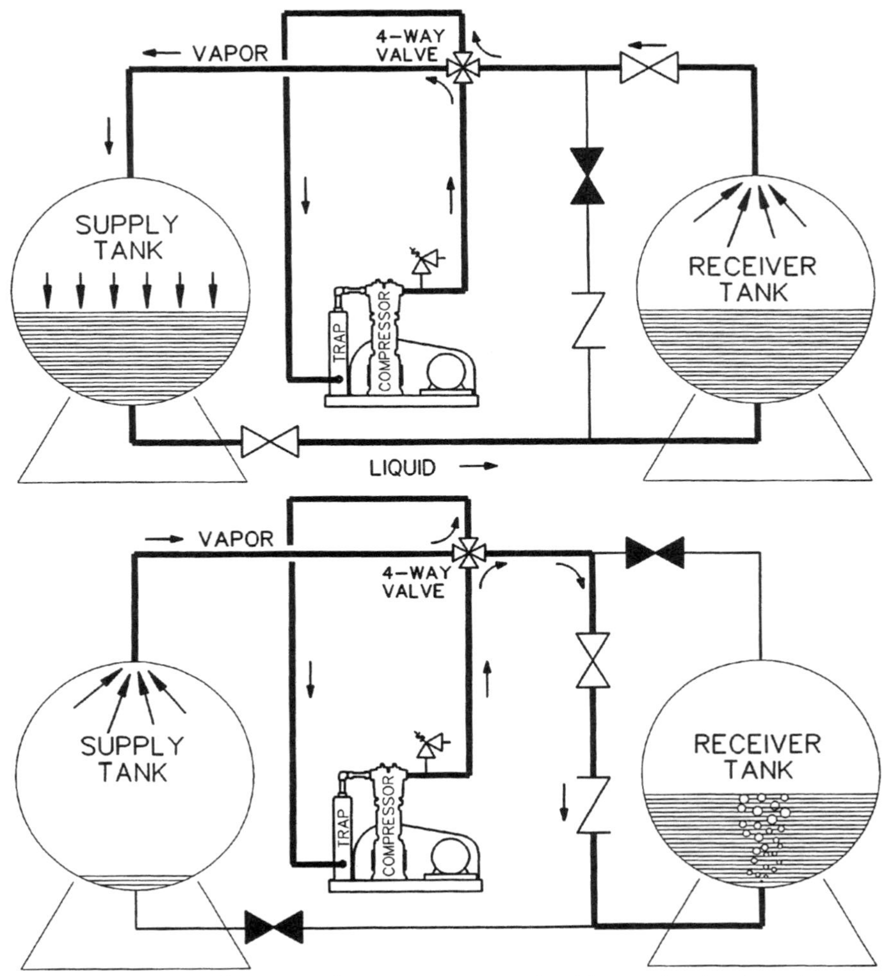

Figure 4.3 Transferring with a compressor—liquid transfer (top) and vapor recovery (bottom). (Drawing courtesy of Blackmer.)

unloaded that is higher than the pressure in the receiving container, causing liquid to flow into the receiving container. When the liquid has all been transferred, the residual vapors in the railroad or transport vehicle being unloaded can be recovered and placed into the receiving vessel by reversing the flow through the compressor.

It is advantageous to have vapor enter at the bottom of the receiving container so that it will bubble up through the liquid and be cooled, and more easily condensed. If the vapor is put into the top of the receiving container, heat of compression will create a higher pressure in the receiving container, slowing the transfer operation.

Figure 4.4 *Cylinder filling room with automatic cylinder filling scales. When the cylinder (not shown) is filled to its correct filling weight, the automatic shutoff valve on the hose inlet will shut off and automatically stop the flow. (Photo courtesy of Schultz Gas Service.)*

Care also has to be taken not to open valves too rapidly because the closing ("slugging") of any excess-flow check valves present can result.

Pumping of the liquid is the usual method of liquid transfer in operations such as delivery to a fixed container, transfer from bulk storage into delivery truck containers, and unloading of transport vehicles. With both methods, it is important to have proper correlation of the design size of the piping and an excess-flow check valve so that there will not be premature closing. This is also important from a safety standpoint because if there is a break or rupture in the line, the size of the line must permit the excess-flow valve to operate.

If unloading by gravity, the receiving container must be at a lower level than the container from which the liquid is being removed. Care must be taken to prevent closing of an excess-flow check valve.

4-2.3.1 Public access to areas where LP-Gas is stored and transferred shall be prohibited except where necessary for the conduct of normal business activities.

Areas where LP-Gas is stored or handled present hazards that are not apparent to those untrained in the properties of LP-Gas and the equipment used at the site. Therefore, the public is normally excluded from these areas.

Figure 4.5 Portable container filling. (Photo courtesy of Alternate Energy Installations.)

4-2.3.2 Sources of ignition shall be controlled during transfer operations, while connections or disconnections are made, or while LP-Gas is being vented to the atmosphere.

(a) Internal combustion engines within 15 ft (4.6 m) of a point of transfer shall be shut down while such transfer operations are in progress, except as follows:

1. Engines of LP-Gas cargo vehicles constructed and operated in compliance with Chapter 6 while such engines are driving transfer pumps or compressors on these vehicles to load containers as provided in 3-2.3.2.
2. Engines installed in buildings as provided in Section 8-3.

This practical provision acknowledges that vehicle engine operation must be allowed when entering or leaving filling stations. An operating internal combustion engine does not normally create sparks or have surfaces hot enough to ignite propane and, in fact, will stall in a propane rich environment due to lack of air for proper combustion.

(b) Smoking, open flame metal cutting or welding, portable electrical tools, and extension lights capable of igniting LP-Gas shall not be permitted within 25 ft (7.6 m) of a point of transfer while filling operations are in progress. Care shall be taken to assure that materials that have been heated have cooled before that transfer is started.

The spacing of sources of ignition and transfer operations was changed from 15 ft to 25 ft in the 1995 edition, for consistency with DOT regulations for separation distances for smoking.

(c) Sources of ignition, such as pilot lights, electric ignition devices, burners, electrical appliances, and engines located on the vehicle being refueled shall be turned off during the filling of any LP-Gas container on the vehicle.

This was revised in the 1995 edition to add control of ignition sources on vehicles waiting to be refueled. Fires have been reported to the editor that have resulted from recreational vehicles that were located too close to filling operations. The source of ignition in these fires could have been the electric ignition device of an appliance in the RV.

4-2.3.3 Cargo vehicles (*see Section 6-3*) unloading into storage containers shall be at least 10 ft (3.0 m) from the container and so positioned that the shutoff valves on both the truck and the container are readily accessible. The cargo vehicle shall not transfer LP-Gas into dispensing station storage while parked on a public way.

While this is a location matter, it is more importantly an operating issue. Dispensing stations are frequently used to fill engine fuel containers, so traffic is expected.

The unloading cargo vehicle should be a distance from the container receiving the product so that if something happens at either point, the other will not be involved to the extent that it would be if it were in close proximity. Also, it is important to have the cargo vehicle so located that it is easy to get to the valves on both the truck and the container so that they can quickly be shut off if there is an emergency need to do so. The prohibition of street parking is intended to avoid the risks associated with accidental street vehicle collisions when there is the added vulnerability associated with a transfer operation.

4-2.3.4 Transfers to containers serving agricultural or industrial equipment requiring refueling in the field shall comply with the following:

(a) Air-moving equipment, such as large blowers on crop dryers or on space heaters, shall be shut down while containers are being refilled, unless the point of transfer is at least 50 ft (15 m) from the air intake of the blower.

(b) Equipment employing open flames, or equipment with integral containers such as flame cultivators, weed burners, tractors, large blower-type space heaters, or tar kettles shall be shut down while refueling.

Agricultural and industrial installations present different hazards than residential fuel applications. The safety precautions required here recognize common factors that can contribute to accidents during transfer.

Figure 4.6 Filling a storage container. (Photo courtesy of Ace Gas Company.)

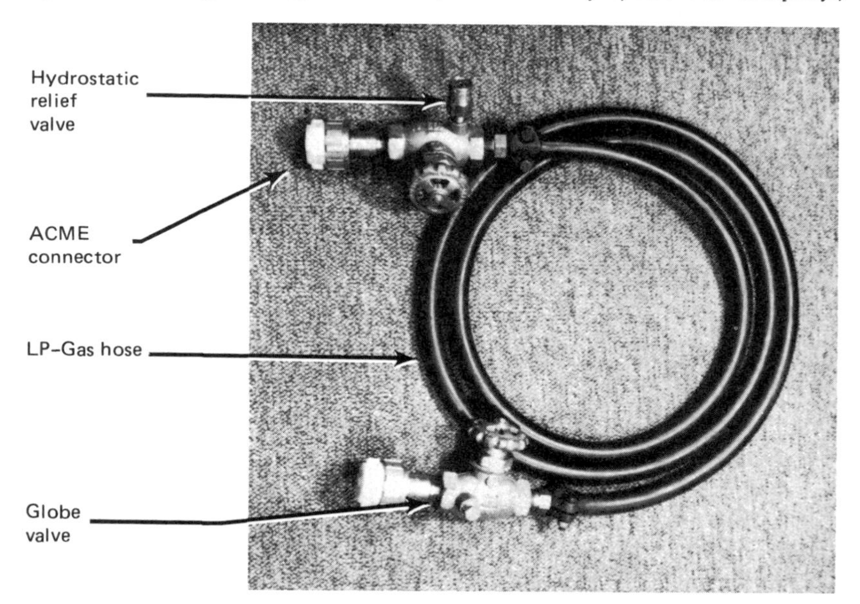

Figure 4.7 "Wet" liquid transfer hose is recommended to avoid having to evacuate the liquid from the hose when the delivery has been completed, which would create a hazard in the atmosphere. (Photo courtesy of National Propane Gas Association.)

Figure 4.8 Truck unloading station. This station includes a complete bulkhead to prevent valve breakage in the event of a truck pulling away with the hoses connected. In addition, the installation includes a shutoff valve, strainer, sight flow, and backflow check valve in the liquid fill line. The vapor equalizing line includes a shutoff valve, emergency shutoff valve with remote cable release, and excess-flow valve. Also installed are two crash posts for vehicular protection. (Photo courtesy of Schultz Gas Service.)

4-2.3.5 During the time tank cars are on sidings for loading or unloading, the following shall apply:

(a) A caution sign, with wording such as "STOP. TANK CAR CONNECTED," shall be placed at the active end(s) of the siding while the car is connected as required by DOT regulations.

(b) Wheel chocks shall be placed to prevent movement of the car in either direction.

This provision reiterates DOT regulations. There have been cases where tank cars have been moved when the unloading hose or connections were in place and accidents have occurred. There have also been cases where a tank car has been partially unloaded and the hoses disconnected, but the railroad personnel moved the car thinking that it was empty. Therefore, the warning sign should always be in place until the car has been completely unloaded and the railroad personnel advised to move the car. The value of chock blocks has been illustrated by incidents where trespassers have released the brake on the car and it has moved from its unloading point. In one such instance, a string of such cars ran down a grade and collided with a locomotive, killing two railroad workers.[1]

[1]From a report by Fire Division Chief Donald L. Elise, Ventura County, California Fire Protection District. "Propane Cloud and a Lot of Luck!" *Fire Command!* February 1974; page 19.

Figure 4.9 Tank car unloading connections. Note the small plastic tubing strapped to the hoses. This carries pressurized gas which opens the emergency shutoff valves used to connect to the tank car. In case of fire, the tubing will fail, and the emergency shutoff valves will close. (Photo courtesy of Ace Gas Company.)

4-2.3.6 Where a hose or swivel-type piping is used for loading or unloading railroad tank cars, an emergency shutoff valve complying with 2-4.5.4 shall be used at the tank car end of the hose or swivel-type piping.

See 3-2.8.10 for the requirements for emergency shutoff valves for stationary installations, which apply to installations over 4,000 gal. Note that 4-2.3.6 requires the use of an emergency shutoff valve regardless of any of the qualifications in 3-2.8.10.

4-2.3.7 Transfer hoses larger than 1/2-in. (12-mm) internal diameter shall not be used for making connections to individual containers being filled indoors.

This provision limits leakage rates.

4-2.3.8 Cargo tanks shall be permitted to be filled directly from railroad tank cars on a private track with non-stationary storage tanks involved provided the following requirements are met. Such operation shall be considered a bulk plant. (*See definition.*)

(a) Transfer protection shall be provided in accordance with 3-2.8.10.

(b) Compressors, if used, shall comply with 2-5.3 and 3-2.13.2.

Figure 4.10 *Liquid transfer chocking and warning sign for tank car loading and unloading. (Photo courtesy of National Propane Gas Association.)*

Figure 4.11 *Rail car unloading tower. This system utilizes a vapor compressor to withdraw vapor from the stationary storage tank and discharge the higher-pressure vapor into the tank car. This provides a higher vapor pressure in the tank car, and thereby liquid is discharged from the tank car to the stationary storage tank via a separate liquid fill line. In addition, the track must be grounded, bonded, and insulated to meet current American Railway Engineering Association requirements. A derailer, a "Stop. Tank Car Connected" sign, and wheel blocks are also required. (Photo courtesy of Plant Systems, Inc.)*

(c) Installations of liquid transfer facilities shall comply with 3-3.3.

(d) Protection against tampering for the compressor, fixed piping, and hose shall be provided in accordance with 3-3.6.

(e) Lighting shall be provided in accordance with 3-3.7, if operations are conducted during other than daylight hours.

(f) Ignition source control shall be in accordance with Section 3-7.

(g) Control of ignition sources during transfer shall be provided in accordance with 4-2.3.

(h) Fire extinguishers shall be provided in accordance with 3-10.2.5.

(i) Transfer personnel shall meet the provisions of 4-2.1.

(j) Cargo tanks shall meet the requirements of 4-2.2.3.

(k) Arrangement and operation of the transfer system shall be in accordance with 4-2.3.

(l) The points of transfer (*see definition*) shall be located in accordance with Table 3-2.3.3 with respect to exposures.

4-2.4 **Inspection.** Hose assemblies shall be visually inspected for leakage or damage that will impair their integrity. Such hose shall be immediately repaired or removed from service.

This requirement, added in 1992, recognizes that guidelines for hose inspection are needed in NFPA 58, as hoses are subject to deterioration in a short time if abused. When the requirement was added, consideration was given to requiring periodic hydrostatic testing of hose, but it was rejected, because the test can be detrimental to the hose.

4-3 Venting LP-Gas to the Atmosphere.

4-3.1 **General.** LP-Gas, in either liquid or vapor form, shall not be vented to the atmosphere.

Exception No. 1: Venting for the operation of fixed liquid level, rotary or slip tube gauges, provided the maximum flow does not exceed that from a No. 54 drill orifice.

Exception No. 2: Venting of LP-Gas between shutoff valves before disconnecting the liquid transfer line from the container. Where necessary, suitable bleeder valves shall be used.

Exception No. 3: LP-Gas shall be permitted to be vented for the purposes described in Exceptions No. 1 and 2 within structures designed for container filling as provided in 3-2.3.1 and Chapter 7.

Exception No. 4: Venting vapor from listed liquid transfer pumps using such vapor as a source of energy, provided the rate of discharge does not exceed that from a No. 31 drill size orifice. (See 3-2.3.3 for location of such transfer operations.)

Exception No. 5: Purging as permitted in 4-3.2.

Exception No. 6: Emergency venting as permitted in 4-3.3.

These exceptions were added in the 1992 edition, replacing four subparagraphs and adding the last two exceptions for completeness and ease of use.

While it is necessary to vent both vapor and a small amount of liquid from a fixed liquid level, rotary, or slip tube gauge, it should be done with consideration for the surroundings. For instance, discharge should not be close to or under a window to prevent LP-Gas from entering a building.

If it is necessary to bleed off liquid from a pipe or hose, the bleeding off must be done slowly and carefully. If this is not done, and the disconnection is made, a potential hazard is created because the rapid discharge could result in liquid propane being squirted into a person's eyes; the area of vapor ignition hazard is increased as well.

There are a few applications where propane is used as the source of energy in pumps and the vapors are vented to the air. Even though the amount of vented vapor is restricted by a No. 31 drill size opening, the extent of the vapor ignition hazard is sizable as indicated by the requirement in 3-2.3.3(b) that the distances specified in Table 3-2.3.3 be doubled.

4-3.2 Purging.

4-3.2.1 Venting of gas from containers for purging or for other purposes shall be accomplished as follows:

(a) If indoors, containers shall be permitted to be vented only in structures designed and constructed for container filling in accordance with 3-2.3.1 and Chapter 7 and with the following provisions:

1. Piping shall be provided to carry the vented product outside and to a point at least 3 ft (1 m) above the highest point of any building within 25 ft (7.6 m).
2. Only vapors shall be exhausted to the atmosphere.
3. If a vent manifold is used to allow for the venting of more than one container at a time, each connection to the vent manifold shall be equipped with a backflow check valve.

(b) Where located outdoors, container venting shall be performed under conditions that will result in rapid dispersion of the product being released. Consideration shall be given to such factors as distance to buildings, terrain, wind direction and velocity, and use of a vent stack so that a flammable mixture will not reach a point of ignition.

(c) If conditions are such that venting into the atmosphere cannot be accomplished safely, LP-Gas shall be permitted to be burned off providing such burning is done under controlled conditions at a distance of 25 ft (7.6 m) from combustibles or a hazardous atmosphere.

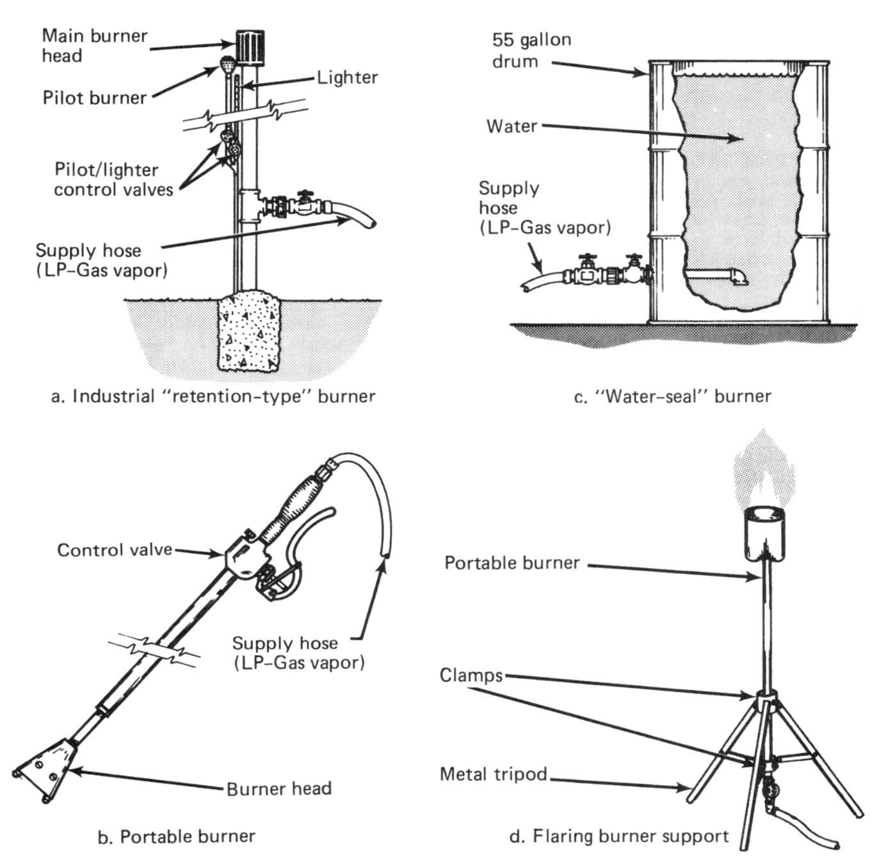

Figure 4.12 *Some burner arrangements for flaring LP-Gas. (Drawing courtesy of National Propane Gas Association.)*

The term "purging" is generally applied to the process of changing the contents of a container, either to completely empty a container of both LP-Gas liquid and vapor or to replace the air in a container with LP-Gas liquid and vapor. The most common other purpose referred to in 4-3.2.1 is the removal of LP-Gas from an overfilled container. Whatever the purpose, the release should be only vapor whenever it is feasible to do so. This not only considers the liquid-to-vapor expansion ratio but recognizes that liquid discharge can present a greater static ignition hazard than vapor. (*See the Note to 3-7.1.3, which refers the reader to NFPA 77, Recommended Practice on Static Electricity, for additional information on static electricity.*)

If a container has been grossly overfilled, it is impossible to prevent liquid discharge from the container outlet initially when correcting an overfill condition. In these circumstances, compliance with 4-3.2.1(a)(1) is especially vital because the piping provides a safe place for vaporization.

The backflow check valve required in each connection by 4-3.2.1(a)(3) is needed to minimize the possibility of a flammable mixture being created in any one of the connections. When a container is almost completely empty of vapor and the pressure is practically at atmospheric, there is a possibility of a flammable mixture being created. While the container could withstand the combustion explosion pressure developed, the piping could be subjected to detonation pressures it could not withstand.

When venting off the vapor, it must be remembered that the evaporation of the liquid vapor results in refrigeration of the remaining liquid, and the temperature of the liquid may approach the normal atmospheric pressure boiling point [about –44°F (–42°C) for propane]. Because there will be no pressure, the operator may erroneously assume that the container is empty. When the liquid warms up, vaporization will begin again and there will be a pressure generated, or if the valves have been left open there will be a flow of vapor into the atmosphere.

When venting is being done and the vapors are being discharged, it is important to check on conditions such as direction of wind, lack of wind, speed of wind, and the general layout of the area and structures. A given location might be perfectly satisfactory for venting the vapors under certain wind conditions, but be unsafe under other atmospheric conditions.

In some situations it is safer to burn off the vapors that are being vented rather than disperse them into the atmosphere. However, when this is done a properly engineered flare stack should be used.

4-3.3 Emergency Venting. The procedure to be followed for the disposal of LP-Gas in an emergency will be dictated by the conditions present, requiring individual judgment in each case and using, where practical, the provisions of this standard.

The need for the emergency venting cited here is most common in cargo vehicle and railroad tank car accidents, where a tank is damaged and failure is considered possible. These often require the services of experts.

In recent years, industry has increased the availability of such experts and there are teams available that have been organized on a state or regional basis. One group is Chemtrek, which is operated by the Chemical Manufacturers Association, for chemical cargoes. Frequently local propane distributors assist in accidents involving propane cargoes through state or regional associations.

4-4 Quantity of LP-Gas in Containers.

Filling containers to a proper capacity is important because LP-Gas liquids expand when heated and contract when cooled. Also, the pressure increases as the temperature of the liquid rises. However, the degree of expansion varies for the LP-Gases. Table 4-4.2.1 was developed to stipulate the filling densities for different liquids, as shown by the different specific gravities in the left-hand column. Propane generally falls within the 0.504–0.510 range of specific gravity. (*See Appendix B.*)

Determination of the maximum quantity that can be placed in a container is expressed as a percent of the water weight capacity of the container. As shown in the second column of Table 4-4.2.1, propane may be 42 percent of the water weight capacity for containers of up to 1,200-U.S. gal (4.5-m^3) capacity. For containers of more than 1,200-U.S. gal (4.5-m^3) capacity, the amount will be 45 percent of the water weight capacity. For example, a "100-lb" (45-kg) propane cylinder with a marked water capacity of 239 lb (108 kg) may contain 239 × 0.42, or 100.38 lb (45 kg) of propane.

A higher filling density is allowed for the containers of more than 1,200-U.S. gal (4.5-m^3) capacity because the larger quantity of LP-Gas is less subject to temperature rise. Two factors affect this: first, a larger container has less surface area per unit of volume, so that heat is absorbed more slowly and, second, there are only a limited number of hours in a day when the heat of the sun (the common source of heat causing container temperature rise) shines on a container. Buried containers are insulated from atmospheric and solar heating, and a higher percent of the water capacity is allowed for such containers. This is shown in the right-hand column of Table 4-4.2.1. Appendix F, Liquid Volume Tables, Computations, and Graphs, can be of material assistance in determining how to fill containers properly and should be included in the training of all personnel filling containers.

Proper filling is especially important for containers that are going to be used inside a building, e.g., industrial truck containers. The cylinder usually will be filled outdoors and the temperature might be extremely cold, even below zero. When it is taken inside the building, the temperatures may be in the 60°F (15.5°C) range. When these extreme differential temperatures are involved, the container must be very carefully filled to avoid having it become liquid-full and create a hydrostatic pressure situation within the building, which would cause the pressure relief device to open. The qualifications of personnel handling such containers should include knowledge of the importance of the liquid expansion factor.

The most accurate method of filling small portable containers is by weight, provided the scales are accurate. A second method is by volume, using a fixed liquid level dip tube gauge. If properly designed and installed, this gauge is accurate. Other volume gauges, such as a rotary gauge, should not be used for filling because they are not accurate enough. The rotary gauge is used to check on gas remaining in the tank, and not for specific filling accuracy gauging.

When using the volumetric method of filling, it is necessary to correct the gauge reading by referencing the liquid volume temperature correction table in Appendix F to determine the true amount of liquid at 60°F (16°C). If the volumetric method of filling is used, utilizing the fixed liquid level gauge, or a variable liquid level gauge without liquid volume temperature correction, the liquid level indicated by such gauges must be computed on the basis of the maximum permitted filling density when the liquid is at 40°F (4.4°C) for aboveground containers and at 50°F (10°C) for underground containers. [*See 2-3.4.3(a).*]

It is also suggested that when a variable-type gauging device such as the rotary gauge is used, its accuracy be checked periodically with the fixed liquid level gauge.

Many small portable cylinders built to DOT specifications incorporate only a fixed outage gauge and are not equipped with a means of measuring the temperature of the liquid propane. The fixed outage gauge is selected to provide a filling density of 80 percent, as required by 4-4.3.3(a), which states that the maximum filling density permitted must be based on a filling temperature of 40°F (4.4°C). Table 4-4.2.2(a), for aboveground containers up to 1,200 gal (4.5 m^3), has an asterisk at the 40°F (4.4°C) line to note this, and for the typical propane density of 0.504 to 0.510, permits an 80 percent maximum filling density. Note that for underground containers the filling temperature used is 50°F (10°C) when only a fixed outage gauge is used (which corresponds to a maximum filling density of 81 percent).

4-4.1 **Application.** This section includes provisions covering the maximum permissible LP-Gas content of containers and the methods of verifying this quantity.

4-4.2 **LP-Gas Capacity of Containers** (*see Appendix F*).

4-4.2.1* The maximum LP-Gas content of any container shall be that quantity that equals the maximum permitted filling limit provided in Table 4-4.2.1. (*See Appendix F.*)

A-4-4.2.1 The maximum permitted filling limit in percent by weight should be as shown in Table 4-4.2.1.

4-4.2.2 The compliance of the LP-Gas content of a container with Table 4-4.2.1 shall be determined either by weight or by volume in accordance with 4-4.3. If by volume, the volume having a weight equal to the maximum permitted filling limit shall be calculated using the formula in 4-4.2.2(b). These equivalent volumes are shown in Tables 4-4.2.2(a), (b), and (c).

(a) The maximum liquid LP-Gas content of any container depends upon the size of the container, whether it is installed aboveground or underground, the maximum permitted filling limit and the temperature of the liquid [*see Tables 4-4.2.2(a), (b), and (c)*].

Table 4-4.2.1 Maximum Permitted Filling Limit (Percent of marked water capacity in lb)

Specific Gravity at 60°F (15.6°C)	Aboveground Containers: 0 to 1,200 U.S. gal (0 to 4.5 m^3) Total Water Cap.	Aboveground Containers: Over 1,200 U.S. gal (0 to 4.5 m^3) Total Water Cap.	Underground Containers All Capacities
.496–.503	41%	44%	45%
.504–.510	42	45	46
.511–.519	43	46	47
.520–.527	44	47	48
.528–.536	45	48	49
.537–.544	46	49	50
.545–.552	47	50	51
.553–.560	48	51	52
.561–.568	49	52	53
.569–.576	50	53	54
.577–.584	51	54	55
.585–.592	52	55	56
.593–.600	53	56	57

(b) The maximum volume "V_t" (in percent of container capacity) of an LP-Gas at temperature "t," having a specific gravity "G" and a filling density of "L," shall be computed by use of the formula (*see Appendix F-4.1.2 for example*):

$$V_t = \frac{L}{G} \div F, \text{ or } V_t = \frac{L}{G \times F} \quad \text{where:}$$

V_t = percent of container capacity that may be filled with liquid.
L = filling limit.
G = specific gravity of particular LP-Gas.
F = correction factor to correct volume at temperature "t" to 60°F (16°C).

4-4.3 Compliance with Maximum Permitted Filling Limit Provisions.

4-4.3.1 The maximum permitted filling limit for any container, where practical, shall be determined by weight.

4-4.3.2 The volumetric method shall be permitted to be used for the following containers if designed and equipped for filling by volume:

Table 4-4.2.2(a) Maximum Permitted Liquid Volume (Percent of Total Water Capacity)

	Aboveground Containers 0 to 1,200 Gal (0 to 4.5 m³)												
	Specific Gravity												
Liquid Temperature °F (°C)	.496 to .503	.504 to .510	.511 to .519	.520 to .527	.528 to .536	.537 to .544	.545 to .552	.553 to .560	.561 to .568	.569 to .576	.577 to .584	.585 to .592	.593 to .600
−50 (−45.6)	70	71	72	73	74	75	75	76	77	78	79	79	80
−45 (−42.8)	71	72	73	73	74	75	76	77	77	78	79	80	80
−40 (−40)	71	72	73	74	75	75	76	77	78	79	79	80	81
−35 (−37.2)	71	72	73	74	75	76	77	77	78	79	80	80	81
−30 (−34.4)	72	73	74	75	76	76	77	78	78	79	80	81	81
−25 (−31.5)	72	73	74	75	76	77	77	78	79	80	80	81	82
−20 (−28.9)	73	74	75	76	76	77	78	79	79	80	81	81	82
−15 (−26.1)	73	74	75	76	77	77	78	79	80	80	81	82	83
−10 (−23.3)	74	75	76	76	77	78	79	79	80	81	81	82	83
− 5 (−20.6)	74	75	76	77	78	78	79	80	80	81	82	82	83
0 (−17.8)	75	76	76	77	78	79	79	80	81	81	82	83	84
5 (−15)	75	76	77	78	78	79	80	81	81	82	83	83	84
10 (−12.2)	76	77	77	78	79	80	80	81	82	82	83	84	84
15 (−9.4)	76	77	78	79	80	80	81	81	82	83	83	84	85
20 (−6.7)	77	78	78	79	80	80	81	82	83	84	84	84	85
25 (−3.9)	77	78	79	80	80	81	82	82	83	84	84	85	85
30 (−1.1)	78	79	79	80	81	81	82	83	83	84	85	85	86
35 (1.7)	78	79	80	81	81	82	83	83	84	85	85	86	86
*40 (4.4)	79	80	81	81	82	82	83	84	84	85	86	86	87
45 (7.8)	80	80	81	82	82	83	84	84	85	85	86	87	87
50 (10)	80	81	82	82	83	83	84	85	85	86	86	87	88
55 (12.8)	81	82	82	83	84	84	85	85	86	86	87	87	88
60 (15.6)	82	82	83	84	84	85	85	86	86	87	87	88	88
65 (18.3)	82	83	84	84	85	85	86	86	87	87	88	88	89
70 (21.1)	83	84	84	85	85	86	86	87	87	88	88	89	89
75 (23.9)	84	85	85	85	86	86	87	87	88	88	89	89	90
80 (26.7)	85	85	86	86	87	87	87	88	88	89	89	90	90
85 (29.4)	85	86	87	87	88	88	88	89	89	89	90	90	91
90 (32.2)	86	87	87	88	88	88	89	89	90	90	90	91	91
95 (35)	87	88	88	88	89	89	89	90	90	91	91	91	92
100 (37.8)	88	89	89	89	89	90	90	90	91	91	92	92	92
105 (40.4)	89	89	90	90	90	90	91	91	91	92	92	92	93
110 (43)	90	90	91	91	91	91	92	92	92	92	93	93	93
115 (46)	91	91	92	92	92	92	92	92	93	93	93	94	94
120 (49)	92	92	93	93	93	93	93	93	93	94	94	94	94
125 (51.5)	93	94	94	94	94	94	94	94	94	94	94	95	95
130 (54)	94	95	95	95	95	95	95	95	95	95	95	95	95

*See 4-4.3.3(a).

Table 4-4.2.2(b) Maximum Permitted Liquid Volume (Percent of Total Water Capacity)

Aboveground Containers Over 1,200 Gal (0 to 4.5 m^3)

Liquid Temperature °F (°C)	Specific Gravity												
	.496 to .503	.504 to .510	.511 to .519	.520 to .527	.528 to .536	.537 to .544	.545 to .552	.553 to .560	.561 to .568	.569 to .576	.577 to .584	.585 to .592	.593 to .600
–50 (–45.6)	75	76	77	78	79	80	80	81	82	83	83	84	85
–45 (–42.8)	76	77	78	78	79	80	81	81	82	83	84	84	85
–40 (–40)	76	77	78	79	80	80	81	82	83	83	84	85	85
–35 (–37.2)	77	78	78	79	80	81	82	82	83	84	84	85	86
–30 (–34.4)	77	78	79	80	80	81	82	83	83	84	85	85	86
–25 (–31.5)	78	79	79	80	81	82	82	83	84	84	85	86	86
–20 (–28.9)	78	79	80	81	81	82	83	83	84	85	85	86	87
–15 (–26.1)	79	79	80	81	82	82	83	84	85	85	86	87	87
–10 (–23.3)	79	80	81	82	82	83	84	84	85	86	86	87	87
– 5 (–20.6)	80	81	81	82	83	83	84	85	85	86	87	87	88
0 (–17.8)	80	81	82	82	83	84	84	85	86	86	87	88	88
5 (–15)	81	82	82	83	84	84	85	86	86	87	87	88	89
10 (–12.2)	81	82	83	83	84	85	85	86	87	87	88	88	89
15 (–9.4)	82	83	83	84	85	85	86	87	87	88	88	89	90
20 (–6.7)	82	83	84	85	85	86	86	87	88	88	89	89	90
25 (–3.9)	83	84	84	85	86	86	87	88	88	89	89	90	90
30(–1.1)	83	84	85	86	86	87	87	88	89	89	90	90	91
35 (1.7)	84	85	86	86	87	87	88	89	89	90	90	91	91
*40 (4.4)	85	86	86	87	87	88	88	89	90	90	91	91	92
45 (7.8)	85	86	87	87	88	88	89	89	90	91	91	92	92
50 (10)	86	87	87	88	88	89	90	90	91	91	92	92	92
55 (12.8)	87	88	88	89	89	90	90	91	91	92	92	92	93
60 (15.6)	88	88	89	89	90	90	91	91	92	92	93	93	93
65 (18.3)	88	89	90	90	91	91	91	92	92	93	93	93	94
70 (21.1)	89	90	90	91	91	91	92	92	93	93	94	94	94
75 (23.9)	90	91	91	91	92	92	92	93	93	94	94	94	95
80 (26.7)	91	91	92	92	92	93	93	93	94	94	95	95	95
85 (29.4)	92	92	93	93	93	93	94	94	95	95	95	96	96
90 (32.2)	93	93	93	94	94	94	95	95	95	95	96	96	96
95 (35)	94	94	94	95	95	95	95	96	96	96	96	97	97
100 (37.8)	94	95	95	95	95	96	96	96	96	97	97	97	98
105 (40.4)	96	96	96	96	96	97	97	97	97	97	98	98	98
110 (43)	97	97	97	97	97	97	97	98	98	98	98	98	99
115 (46)	98	98	98	98	98	98	98	98	98	99	99	99	99

*See 4-4.3.3(a)

Table 4-4.2.2(c) Maximum Permitted Liquid Volume (Percent of Total Water Capacity)

All Underground Containers													
	Specific Gravity												
Liquid Temperature °F (°C)	.496 to .503	.504 to .510	.511 to .519	.520 to .527	.528 to .536	.537 to .544	.545 to .552	.553 to .560	.561 to .568	.569 to .576	.577 to .584	.585 to .592	.593 to .600
–50 (–45.6)	77	78	79	80	80	81	82	83	83	84	85	85	86
–45 (–42.8)	77	78	79	80	81	82	82	83	84	84	85	86	87
–40 (–40)	78	79	80	81	81	82	83	83	84	85	86	86	87
–35 (–37.2)	78	79	80	81	82	82	83	84	85	85	86	87	87
–30 (–34.4)	79	80	81	81	82	83	84	84	85	86	86	87	88
–25 (–31.5)	79	80	81	82	83	83	84	85	85	86	87	87	88
–20 (–28.9)	80	81	82	82	83	84	84	85	86	86	87	88	88
–15 (–26.1)	80	81	82	83	84	84	85	86	86	87	87	88	89
–10 (–23.3)	81	82	83	83	84	85	85	86	87	87	88	88	89
– 5 (–20.6)	81	82	83	84	84	85	86	86	87	88	88	89	89
0 (–17.8)	82	83	84	84	85	85	86	87	87	88	89	89	90
5 (–15)	82	83	84	85	85	86	87	87	88	88	89	90	90
10 (–12.2)	83	84	85	85	86	86	87	88	88	89	90	90	91
15 (–9.4)	84	84	85	86	86	87	88	88	89	89	90	91	91
20 (–6.7)	84	85	86	86	87	88	88	89	89	90	90	91	91
25 (–3.9)	85	86	86	87	87	88	89	89	90	90	91	91	92
30 (–1.1)	85	86	87	87	88	89	89	90	90	91	91	92	92
35 (1.7)	86	87	87	88	88	89	90	90	91	91	92	92	93
40 (4.4)	87	87	88	88	89	90	90	91	91	92	92	93	93
45 (7.8)	87	88	89	89	90	90	91	91	92	92	93	93	94
*50 (10)	88	89	89	90	90	91	91	92	92	93	93	94	94
55 (12.8)	89	89	90	91	91	91	92	92	93	93	94	94	95
60 (15.6)	90	90	91	91	92	92	92	93	93	94	94	95	95
65 (18.3)	90	91	91	92	92	93	93	94	94	94	95	95	96
70 (21.1)	91	91	92	93	93	93	94	94	94	95	95	96	96
75 (23.9)	92	93	93	93	94	94	94	95	95	95	96	96	97
80 (26.7)	93	93	94	94	94	95	95	95	96	96	96	97	97
85 (29.4)	94	94	95	95	95	95	96	96	96	97	97	97	98
90 (32.2)	95	95	95	95	96	96	96	97	97	97	98	98	98
95 (35)	96	96	96	96	97	97	97	97	98	98	98	98	99
100 (37.8)	97	97	97	97	97	98	98	98	98	99	99	99	99
105 (40.4)	98	98	98	98	98	98	98	99	99	99	99	99	99

*See 4-4.3.3(a).

(a) DOT specification cylinders of less than 200 lb (91 kg) water capacity that are not subject to DOT jurisdiction (such as, but not limited to, motor fuel containers on vehicles not in interstate commerce or cylinders filled at the installation).

(b) DOT specification cylinders of 200 lb (91 kg) water capacity or more. (*See DOT regulations requiring spot weight checks.*)

(c) Cargo tanks or portable tank containers complying with DOT Specifications MC-330, MC-331 or DOT 51.

(d) ASME and API-ASME containers complying with 2-2.1.3 or 2-2.2.2.

4-4.3.3 Where used, the volumetric method shall be in accordance with the following:

(a) If a maximum fixed liquid level gauge or a variable liquid level gauge without liquid volume temperature correction is used, the liquid level indicated by these gauges shall be computed based on the maximum permitted filling limit when the liquid is at 40°F (4.4°C) for aboveground containers or at 50°F (10°C) for underground containers.

(b) When a variable liquid level gauge is used and the liquid volume is corrected for temperature, the maximum permitted liquid level shall be in accordance with Tables 4-4.2.2(a), (b), and (c).

(c) Containers with a water capacity of 2,000 gal (7.6 m) or less filled at consumer sites shall be gauged in accordance with 4-4.3.3(a), utilizing the fixed liquid level gauge.

Exception: Containers fabricated on or before December 31, 1965, shall be exempt from this provision.

4-4.3.4 Where containers are to be filled volumetrically by a variable liquid level gauge in accordance with 4-4.3.3(b), provisions shall be made for determining the liquid temperature (*see F-3.1.2*).

5

Storage of Portable Containers Awaiting Use, Resale, or Exchange

5-1 Scope.

The title of this chapter was revised in the 1995 edition to add the word "Exchange," recognizing that the exchange of cylinders, full or empty, is becoming a more popular alternate to filling portable cylinders, especially gas grill cylinders.

5-1.1 Application.

5-1.1.1 The provisions of this chapter are applicable to the storage of portable containers of 1,000 lb (454 kg) water capacity, or less, whether filled, partially filled, or empty (if they have been in LP-Gas service) as follows:

(a) At consumer sites or dispensing stations, where not connected for use.
(b) In storage for resale or exchange by dealer or reseller.

5-1.1.2 The provisions of this chapter shall not apply to containers stored at bulk plants.

Provisions for storing portable containers on users' premises and at locations for resale first appeared in the standard in the early 1940s. By that time, many filled cylinders were being resold at locations such as hardware stores and it was deemed necessary to regulate the manner in which these containers should be stored. Similarly, many users, particularly industrial users, stored cylinders on their premises awaiting use, and standards for their storage were also needed. The current provisions of NFPA 58 do not differentiate between storage at a resale or user location, but stipulate what is to be done if portable containers are stored in conventional buildings (whether frequented by the public or not), in special buildings or rooms, or outside of buildings. The provisions of this chapter apply only to the storage of portable containers on the premises of consumers, at dispensing stations, and at locations for resale or exchange by the cylinder dealer, or reseller. They do not apply to storage of portable containers at bulk plants.

5-2 General Provisions.

5-2.1 General Location of Containers.

5-2.1.1 Containers in storage shall be located to minimize exposure to excessive temperature rise, physical damage, or tampering.

Because they are infrequently occupied by persons, the temperatures of storage locations are often not rigorously controlled. Because of the smaller size of the containers covered in Chapter 5, their contents' temperatures tend to fluctuate more directly with ambient air temperatures or solar radiation. Even though [as noted in Appendix F, F-2.1(a)] these containers will not relieve LP-Gas through their pressure relief devices until the temperature of their contents exceeds 130°F (54°C), such temperatures could be reached in some poorly located, constructed, or ventilated storage locations. Some storage locations have considerable vehicular traffic (e.g., forklift trucks), and experience has shown the need to consider physical damage protection.

5-2.1.2 Containers in storage having individual water capacity greater than 2½ lb (1 kg) [nominal 1 lb (0.45 kg)] LP-Gas capacity, and filled with not more than 16.8 oz (0.522 kg) LP-Gas, shall be positioned such that the pressure relief valve is in direct communication with the vapor space of the container.

This is to ensure that, in case of overpressure, discharge is largely vapor rather than liquid. This is important to both the quantity of potential flammable vapor-air mixture and the relieving capacity of the pressure relief devices. Nominal 1-lb containers are used for handheld soldering torches; portable stoves; camping equipment; refillable portable appliances, such as cigarette lighters; and other uses. They are normally stored in cardboard shipping containers, and the proper storage orientation should be indicated on the shipping container.

Figure 5.1 *Storage rack for engine fuel (industrial truck) containers. Containers are so-called "universal" type in which a curved tube connects the relief valve inlet to the vapor space, thus permitting them to be stored on their sides. All enclosure surfaces, including the roof, are open mesh both for ventilation and so that cooling water can be applied in the event of leakage or fire. (Photo courtesy of H & H Equipment Company, Inc.)*

5-2.1.3 Containers stored in buildings in accordance with Section 5-3 shall not be located near exits, stairways, or in areas normally used, or intended to be used, for the safe egress of occupants.

This provision is consistent with NFPA *101*®, *Life Safety Code*®. In that code, an exit is defined as that portion of a means of egress separated from all other building spaces, such as an enclosed exit stairway. Exit access is the path that leads to an exit. Storage of LP-Gas containers is not permitted in an exit path or an exit access.

5-2.1.4 Empty containers that have been in LP-Gas service shall preferably be stored outdoors. If stored indoors, they shall be considered as full containers for the purposes of determining the maximum quantities of LP-Gas permitted in 5-3.1, 5-3.2.1, and 5-3.3.1.

Once filled, an LP-Gas container is seldom ever truly empty. At the least, it will usually be full of vapor and may contain some liquid or a residue that could contain the flammable odorant. If empty containers were not counted as full containers it would be impossible for an enforcing authority to determine whether the storage limits were being exceeded without lifting or tilting the cylinders.

5-2.1.5 Containers that are not connected for use shall not be stored on roofs.

Rooftops are largely "out of sight and out of mind," and are often places where combustible materials accumulate. Even an "empty" container can complicate fire control activities in a location that is difficult to handle.

5-2.2 Protection of Valves on Containers in Storage.

5-2.2.1 Container valves shall be protected as required by 2-2.4.1. Screw-on type caps or collars shall be securely in place on all containers stored, regardless of whether they are full, partially full, or empty, and container outlet valves shall be closed and plugged or capped. The provisions of 4-2.2.2 for valve outlet plugs and caps shall apply.

See commentary on 4-2.2.2.

5-3 Storage within Buildings.

5-3.1 Storage within Buildings Frequented by the Public.

DOT specification cylinders with a maximum water capacity of 2½ lb (1 kg) [nominal 1 lb (0.45 kg)] LP-Gas capacity, and filled with no more than 16.8 oz (0.522 kg) LP-Gas, used with completely self-contained hand torches and similar applications, shall be permitted to be stored or displayed in a building frequented by the public. The quantity of LP-Gas shall not exceed 200 lb (91 kg).

Earlier editions of the standard could be interpreted to mean that 20-lb (9.1-kg) LP-Gas cylinders for wallpaper removal (steamers) could be stored at paint and wallpaper stores. A formal interpretation was issued to indicate that the intent of these provisions was that the size of the cylinder be limited to 2½ lb (1 kg) water capacity or about 1 lb (0.5 kg) LP-Gas capacity torch-type cylinders. Storage of containers larger than this is covered in 5-3.2, 5-3.3, or Section 5-4. For storage at hardware or paint stores, see 5-3.3 or Section 5-4.

In earlier days, the storage of 2½ lb (1 kg) water capacity torch-type cylinders was appropriately covered by limiting such storage to 24 units of each brand of display with a maximum of 200 lb (91 kg) of LP-Gas. With the popularity of discount stores and mass merchandising, however, this limitation was not practical. These types of retail establishments needed more inventory on hand; consequently, the current standards continue to limit the amount on display.

The exceptions permit storing up to 48 containers each holding no more than 10 oz of butane in restaurants and food service locations. This was added in the 1992 edition along with a provision permitting portable butane-fired cooking appliances (*see 3-4.8.4*). The addition here recognizes that spare cylinders must be on hand for these appliances and limits them to a reasonable quantity. These cylinders are constructed to DOT specification 2P or 2Q and to UL Standards 147A or 147B, and are similar to aerosol containers. [*See Figure 3.31.*]

Exception No. 1: Storage in restaurants and at food service locations of 10-oz (283-gr) butane nonrefillable containers shall be limited to no more than 24 containers.

Exception No. 2: An additional twenty-four 10-oz (283-gr) butane nonrefillable containers shall be permitted to be stored in another location within the building provided that the storage area is constructed with at least a 2-hr fire wall protection.

5-3.2 Storage within Buildings Not Frequented by the Public (Such as Industrial Buildings).

5-3.2.1 The maximum quantity allowed in one storage location shall not exceed 735 lb (334 kg) water capacity [nominal 300 lb (136 kg) LP-Gas]. If additional storage locations are required on the same floor within the same building, they shall be separated by a minimum of 300 ft (91.4 m). Storage beyond these limitations shall comply with 5-3.3.

The standard originally limited storage in an industrial building to 300 lb (136 kg) LP-Gas. It became apparent that this limitation was not realistic for large industrial complexes where buildings may be as large as 300,000 sq ft (28,000 m^2). As a consequence, the standards were revised to permit more than one storage location with a maximum of 300 lb (136 kg) LP-Gas, if such locations had a separation of 300 ft (91 m). If additional storage was needed, a facility described by 5-3.3 would have to be used, or the storage would have to be outdoors.

5-3.2.2 Containers carried as part of the service equipment on highway mobile vehicles shall not be considered part of the total storage capacity in the requirements of 5-3.2.1 provided such vehicles are stored in private garages and carry no more than 3 LP-Gas containers with a total aggregate capacity per vehicle not exceeding 100 lb (45.4 kg) of LP-Gas. Container valves shall be closed when not in use.

The provision was developed originally to recognize the carrying of portable containers on telephone and electric utility service vehicles and the garaging of such vehicles in buildings owned and occupied by such firms. The current limit of up to three containers on a vehicle while maintaining the total quantity at 100 lb (45 kg) LP-Gas recognizes that many utility vehicles carry multiple containers to meet service requirements and the fire experience is satisfactory.

5-3.3 Storage within Special Buildings or Rooms.

5-3.3.1 The maximum quantity of LP-Gas that may be stored in special buildings or rooms shall be 10,000 lb (4540 kg).

5-3.3.2 Special buildings or rooms for storing LP-Gas containers shall not be located adjoining the line of property occupied by schools, churches, hospitals, athletic fields, or other points of public gathering.

5-3.3.3 The construction of all such special buildings, and rooms within, or attached to, other buildings, shall comply with Chapter 7 and the following:

(a) Adequate vents, to the outside only, shall be provided at both top and bottom and shall be located at least 5 ft (1.5 m) from any building opening.
(b) The entire area shall be classified for purposes of ignition source control in accordance with Section 3-7.

5-3.4 Storage within Residential Buildings. Storage of containers within a residential building, including the basement or any storage area in a common basement storage area in multiple family buildings and attached garages, shall be limited to 2 containers each with a maximum water capacity of 2.7 lb (1.2 kg) and shall not exceed 5.4 lb (2.4 kg) total water capacity for smaller containers per each living space unit. Each container shall meet DOT specifications.

This provision clearly states the Committee's intent on the restriction of LP-Gas container storage in residential buildings. It limits storage to two 1-lb (0.5-kg) (net) containers. Previously, some users of the standard have interpreted that a 20-lb (9.1-kg) barbecue grill cylinder could be stored in a building if it were not connected for use. There have been many incidents resulting in injury, death, and property damage due to fire and explosion caused by leakage from 20-lb (9.1-kg) barbecue grill and recreational vehicle cylinders stored in buildings.[1] It has been the Committee's long-standing position that LP-Gas cylinder storage in buildings be limited to small containers, necessary attended uses, and larger cylinders in industrial buildings only.

5-4 Storage Outside of Buildings.

5-4.1* **Location of Storage Outside of Buildings.** Storage outside of buildings for containers awaiting use, exchange, or resale shall be located at least 5 ft (1.5 m) from any doorway in a building frequented by the public in accordance with Table 5-4.1 with respect to:

A-5-4.1 The filling process in (e) refers to the time period beginning when a cylinder or cylinders are brought to a dispensing station to be filled and ending when the last cylinder is filled and all the cylinders are removed from the filling area. This is meant to define a continuous process with the cylinders being unattended for only brief periods, such as operator breaks or lunch.

Table 5-4.1

Quantity of LP-Gas Stored	Horizontal Distance to: (a) and (b)	(c) and (d)	(e)
720 lb (227 kg) or less	0	0	5 ft (1.5 m)
721 (227 + kg) to 2,500 lb (1134 kg)	0	10 ft (3 m)	10 ft (3 m)
2,501 (1134 + kg) to 6,000 lb (2721 kg)	10 ft (3 m)	10 ft (3 m)	10 ft (3 m)
6,001 (2721 + kg) to 10,000 lb (4540 kg)	20 ft (6 m)	20 ft (6 m)	20 ft (6 m)
Over 10,000 lb (4540 kg)	25 ft (7.6 m)	25 ft (7.6 m)	25 ft (7.6 m)

(a) The nearest important building or group of buildings.

(b) The line of adjoining property that can be built upon.

(c) Busy thoroughfares or sidewalks.

(d) The line of adjoining property occupied by schools, churches, hospitals, athletic fields, or other points of public gathering.

(e) A dispensing station.

[1] "Special Data Information Package: LP-Gas Cylinders and Tanks." Fire Analysis and Research Division, National Fire Protection Association, Quincy, MA, 1993.

Prior to the 1967 edition of the standard, the provisions for outside storage could be interpreted to mean that any amount less than 10,000 lb (4,536 kg) of LP-Gas could be stored next to a building or located adjoining a line of property occupied by schools, churches, hospitals, public gatherings, or thoroughfares. Also, many years ago it was a custom for hardware stores in small communities to store cylinders on the sidewalk adjacent to the store. Consequently, the standards were revised to delineate clearly how much storage could be located with respect to certain types of exposures. Also, in the case of construction work it is sometimes impractical to limit the amount of storage necessary for operations according to NFPA 58 provisions. In that case, the authority having jurisdiction must determine the conditions for storage (*See 5-4.3*).

In the 1992 edition paragraph (e) was added along with column (e) in the table covering distances to dispensing stations. This recognizes the practice of offering filled cylinders to the public on an exchange basis, rather than filling cylinders at a retail site. The new column (e) requires a minimum 5-ft (1.5-m) spacing from dispensing stations if on the same site. The cylinders, which are usually stored in ventilated cabinets, can be located adjacent to buildings but are required to be 5 ft (1.5 m) from a doorway frequented by the public.

Exception: Location of cylinders in the filling process shall not be considered to be in storage.

The exception clarifies that cylinders in the process of being filled are not covered by the 5-ft (1.5-m) spacing requirement from dispensing stations. While this is redundant to 5.1.1.1 it is placed here due to much misinterpretation. The term "in the process of being filled" is a broad one that recognizes efficient filling practices, and means not only the cylinder being filled, but the group of cylinders being filled. Thus, a group of cylinders can be located within 5 ft (1.5 m) of a dispensing station and all can be filled prior to moving any of them away to storage. No time limit is specified, but it is usually limited to the reasonable time it takes to perform the filling operation, allowing for other duties and lunches and breaks. Cylinders should be removed by the end of the work shift, unless responsibility is assigned to workers on the next shift.

5-4.2 Protection of Containers.

5-4.2.1 Containers at a location open to the public shall be protected by either:

(a) An enclosure in accordance with 3-3.6(a), or

(b) A lockable ventilated metal locker or rack that prevents tampering with valves and pilferage of the cylinder.

This section recognizes cabinets used for exchange cylinders at retail locations. Note that the cabinet must be ventilated and lockable. Ventilation can be accomplished by using expanded metal or screening for the sides of the locker. The lockability requirement is intended to prevent tampering with the cylinders when the location is not attended.

5-4.2.2 Protection against vehicle impact shall be provided in accordance with good engineering practice where vehicle traffic normally is expected at the location.

This new requirement recognizes that full containers awaiting sale or use must be protected from vehicle damage, when vehicles are operated in areas where cylinders are stored. The requirement that good engineering practice be used recognizes that the protection needed varies with the size and anticipated speed of vehicles operated in the storage area, and allows flexibility in design and materials.

5-4.3 **Alternate Location and Protection of Storage.** Where the provisions of 5-4.1 and 5-4.2.1 are impractical at construction sites, or at buildings or structures undergoing major renovation or repairs, the storage of containers shall be acceptable to the authority having jurisdiction.

5-5 Fire Protection.

Storage locations, other than supply depots at separate locations apart from those of the dealer, reseller, or user's establishments, shall be provided with at least one approved portable fire extinguisher having a minimum capacity of 18 lb (8.2 kg) dry chemical with a B:C rating. (*Also see NFPA 10, Standard for Portable Fire Extinguishers.*)

Prior to the 1983 edition, any type of extinguisher with a B:C rating was acceptable. It could have had as little as 2 lb (0.9 kg) of agent and a very short discharge time. Considering that it may not always be possible to stop the flow of escaping gas after extinguishing the fire, it was felt to be advisable for the safety of the operator to provide enough agent and a longer discharge time to enable him to recognize a problem. It also recognizes that "B" ratings are not derived from tests on gas fires but are based upon flammable liquid pan fires, and that some extinguishers have been optimized to fight these flammable liquid pan fires rather than gas fires.

The minimum capacity was changed from 20 lb to 18 lb in the 1992 edition to recognize extinguishers using agents of lower density that provide equivalent extinguishment.

References Cited in Commentary

The following publication is available from the National Fire Protection Association, 1 Batterymarch Park, P.O. Box 9101, Quincy, MA 02269-9101.

NFPA *101,® Life Safety Code®*.

6

Vehicular Transportation of LP-Gas

Little or no information on the transportation of LP-Gas existed in NFPA standards during the early years of the LP-Gas industry, since bulk transportation was generally by rail and delivery to the consumer site was usually by cylinder delivery. Good practice requirements for the construction and operation of tank trucks for bulk transportation on highways were adopted in 1935 and published by the National Board of Fire Underwriters as *Regulations for the Design, Construction, and Operation of Automobile Tank Trucks and Tank Trailers for the Transportation of Liquefied Petroleum Gases*, NBFU 59. These provisions were incorporated into the 1940 edition of NFPA 58 as Division III. Division III related solely to tank trucks and trailers until expanded in the 1961 edition to cover all types of truck transportation of LP-Gas, including transportation in portable containers, bulk tanks to the consumer site, movable fuel storage tenders, and farm carts. Also included were the first provisions for parking and garaging LP-Gas tank vehicles.

The current Chapter 6 has further modified these earlier provisions to take into account the increased transportation of cylinders by general public consumers from the filling facility to their location, and also to recognize that DOT and CTC regulations are being used increasingly at the local level, whether or not the transportation is in interstate commerce.

6-1 Scope.

6-1.1 Application.

6-1.1.1 This chapter includes provisions that apply to containers, container appurtenances, piping, valves, equipment, and vehicles used in the transportation of LP-Gas, as follows:

(a) Transportation of portable containers.

The primary purpose of Chapter 6 is to set forth provisions for transporting LP-Gas for subsequent use. As noted in the exception to 6-1.1.1(a) and in 6-1.1.2, this chapter does not apply to those installations where the gas is being used on transportation equipment for tar kettles, catering vehicles, engine fuel, etc. Those installations are covered in Chapters 3 and 8.

Exception: The provisions of this chapter shall not apply to LP-Gas containers and related equipment incident to their use on vehicles as covered in Section 3-8 and Chapter 8.

(b) Transportation in cargo vehicles, whether fabricated by mounting cargo tanks on conventional truck or trailer chassis, or constructed as integral cargo units in which the container constitutes in whole, or in part, the stress member of the vehicle frame. Transfer equipment and piping, and the protection of such equipment and the container appurtenances against overturn, collision, or other vehicular accidents, are also included.

(c) Vehicles and procedures under the jurisdiction of DOT shall comply with DOT regulations.

NOTE: Most truck transportation of LP-Gas is subject to regulation by the U.S. Department of Transportation. Many of the provisions of this chapter are identical or similar to DOT regulations and are intended to extend these provisions to areas not subject to DOT regulation.

Until the late 1960s, the provisions on "tank trucks" (now "cargo vehicles") were quite detailed and were essentially specifications. After World War II, the U.S. and Canadian government agencies concerned with interstate and interprovence transportation of hazardous materials expanded their regulations. Eventually, treatment in these regulations and in Chapter 6 of NFPA 58 grew very similar. For a time, this dual coverage was justified on the basis that some transportation did not involve interstate or interprovence commerce and that NFPA 58 was needed to cover such vehicles. In spite of efforts to keep NFPA 58 consonant with DOT/CTC regulations, however, some differences remained. These led to confusion and, because it makes little economic sense to build a cargo vehicle that cannot be used in a different state or that cannot cross a state line, the federal regulations became dominant.

Many of the earlier provisions on cargo vehicle construction have been deleted and references made to the federal regulations, since the vast majority of state laws are based on adoption of these regulations. However, certain provisions have been retained that are not included in the federal regulations, such as those for emergency shutoff valves, flexible connectors, chock blocks, and fire extinguishers. They may be incorporated in federal regulations in the future.

Today, cargo vehicles in the U.S. are constructed to Specification MC-331 of the *Code of Federal Regulations*, Title 49, Part 178.337. Older units constructed to Specification MC-330 have undergone some retrofitting and are also in operation today. No further construction to the MC-330 specification is permitted. Additionally, the U.S. Department of Transportation (DOT) has issued a ruling to recognize those vehicles constructed to NFPA 58 provisions for continued operation as long as certain minimum requirements are met. Compliance with pertinent regulations promulgated by the Bureau of Motor Carrier Safety, in addition to regulations issued by the Office of Hazardous Materials of DOT, is mandatory.

6-1.1.2 The provisions of this chapter shall not be applicable to the transportation of LP-Gas on vehicles incident to its use on these vehicles as covered in Sections 3-8, 8-5, 8-6, and 8-7.

6-1.1.3 If LP-Gas is used for engine fuel, the supply piping and regulating, vaporizing, gas-air mixing, and carburetion equipment shall be designed, constructed, and installed in accordance with Chapter 8. Fuel systems (including fuel containers) shall be constructed and installed in accordance with Section 3-8. Fuel shall be used from the cargo tank of tank trucks or motor fuel containers installed in accordance with 8-2.6. Fuel shall not be used from cargo tanks on trailers or semitrailers.

Many retail bulk delivery trucks (referred to as "bobtails" in the propane distribution industry) utilize propane as an engine fuel, although some newer vehicles use diesel fuel for fuel economy. Although provisions for the installation of the engine fuel system are covered in Chapter 8, it is important to note that fuel for the engine may be obtained from the cargo container on this type of truck with one chassis, but cannot be obtained from cargo containers on trailers or semitrailers. An LP-Gas–fueled trailer or semitrailer requires a separate LP-Gas fuel container.

6-1.1.4 No artificial light other than electrical shall be used with the vehicles covered by this chapter. Wiring used shall have adequate mechanical strength and current-carrying capacity with suitable overcurrent protection (fuses or automatic circuit breakers) and shall be properly insulated and protected against physical damage.

6-2 Transportation in Portable Containers.

6-2.1 Application. This section applies to the vehicular transportation of portable containers filled with LP-Gas delivered as "packages," including containers built to DOT cylinder specifications and other portable containers (such as DOT portable tank containers and skid tanks). The design and construction of these containers is covered in Chapter 2.

6-2.2 Transportation of DOT Specification Cylinders or Portable ASME Containers.

6-2.2.1 Portable containers having an individual water capacity not exceeding 1,000 lb (454 kg) [nominal 420 lb (191 kg) LP-Gas capacity], when filled with LP-Gas, shall be transported in accordance with the requirements of this section.

The maximum size of an individual cylinder permitted under the DOT regulations is 1,000 lb (454 kg) water capacity [nominal 420 lb (191 kg) of LP-Gas]. Portable ASME containers, which generally serve the same purpose as DOT cylinders, are also limited to this size for the purpose of this section.

6-2.2.2 Containers shall be constructed as provided in Section 2-2 and equipped in accordance with Section 2-3 for transportation as portable containers.

The referenced sections require construction to the ASME or DOT codes, and specify minimum appurtenances and appurtenance protection.

6-2.2.3 The quantity of LP-Gas in containers shall be in accordance with Chapter 4.

6-2.2.4 Valves of containers shall be protected in accordance with 2-2.4.1. Screw-on type protecting caps or collars shall be secured in place. The provisions of 4-2.2.2 shall apply.

The reference to 2-2.4.1 provides details on protection of container valves. Paragraph 4-2.2.2 mandates the sealing of valve outlets with a plug, cap, or approved quick-closing coupling on cylinders with capacities of up to 45 lb (20 kg) of LP-Gas (except single-trip nonrefillable and new, unused containers). The reference here is also to ensure that this seal is in place while the cylinder is being transported.

Note that cylinders of more than 45-lb (20-kg) LP-Gas capacity are not required to be plugged, as they are almost exclusively handled by employees of LP-Gas distribution firms who are trained in proper practice. The requirement for plugs was added in the 1986 edition following a tragic accident involving a consumer-transported cylinder.

6-2.2.5 The cargo space of the vehicle shall be isolated from the driver's compartment, the engine, and its exhaust system. Open-bodied vehicles shall be considered to be in compliance with this provision. Closed-bodied vehicles having separate cargo, driver's, and engine compartments shall also be considered to be in compliance with this provision.

Exception: Closed-bodied vehicles such as passenger cars, vans, and station wagons shall not be used for transporting more than 215-lb (98-kg) water capacity [nominal 90 lb (41 kg) LP-Gas capacity] but not more than 108 lb (49 kg) water capacity [nominal 45 lb (20 kg) LP-Gas capacity] per container (see 6-2.2.6 and 6-2.2.7), unless the driver's and engine compartments are separated from the cargo space by a vaportight partition that contains no means of access to the cargo space.

These container capacity provisions rule out transportation of the commonly used 100-lb (45-kg) LP-Gas cylinder in this type of vehicle.

6-2.2.6 Containers and their appurtenances shall be determined to be leak-free before being loaded into vehicles. Containers shall be loaded into vehicles with substantially flat floors or equipped with suitable racks for holding containers. Containers shall be securely fastened in position to minimize the possibility of movement, tipping, or physical damage.

Figure 6.1 *Open portable container cargo vehicle (delivery truck). (Photo courtesy of Eastern Propane.)*

Racks that hold the cylinder in a position tilted upward slightly from horizontal are commonly used for the delivery of industrial truck cylinders. The safety experience with this type of transportation has been good and is the basis for the treatment in 6-2.2.7 and 6-2.2.8, as far as cylinders with a maximum size of 45 lb (20 kg) of LP-Gas are concerned. Generally, this is the maximum size of a portable cylinder for industrial truck use.

6-2.2.7 Containers having an individual water capacity not exceeding 108 lb (49 kg) [nominal 45 lb (20 kg) LP-Gas] capacity transported in open vehicles and containers having an individual water capacity not exceeding 10 lb (4.5 kg) [nominal 4.2 lb (2 kg) LP-Gas] capacity transported in enclosed spaces of the vehicle shall be permitted to be transported in other than the upright position. Containers having an individual water capacity exceeding 108 lb (49 kg) [nominal 45 lb (20 kg) LP-Gas] capacity transported in open vehicles and containers having an individual water capacity exceeding 10 lb (45 kg) [nominal 4.2 lb (1.9 kg) LP-Gas] capacity transported in enclosed spaces shall be transported with the relief device in direct communication with the vapor space.

Transporting containers upright and with the relief valve in communication with the vapor space minimizes the quantity of LP-Gas released under abnormal conditions. The exception for lift truck cylinders [maximum 45 lb (20 kg) LP-Gas] in open vehicles stipulates that 20-lb (9.1-kg) cylinders—e.g., for gas grills—cannot be transported inside a vehicle on their sides.

6-2.2.8 Vehicles transporting more than 1,000 lb (454 kg) of LP-Gas, including the weight of the containers, shall be placarded as required by DOT regulations or state law.

6-2.3 Transportation of Portable Containers of More than 1,000 lb (454 kg) Water Capacity.

Portable containers of this size are used when a need exists to transport LP-Gas in bulk to a point of use where the container is not part of a vehicle, such as in cargo vehicles (Section 6-3) and trailer types (Section 6-4). If these portable containers are permanently mounted on a vehicle and used for making deliveries, they become cargo tanks and must comply with Section 6-3. They are definitely not storage containers used for a permanent installation, and such a storage container should never be used for this purpose. The primary difference between the two is that portable containers of this size are designed, and the fittings are of a type and so protected, such that the container may be transported filled to its maximum quantity of LP-Gas. Storage containers do not have such features.

DOT regulations provide for a portable container designed for transporting quantities of LP-Gas greater than 420 lb (191 kg). This container is a Specification 51 container, and its construction and fitting arrangement are

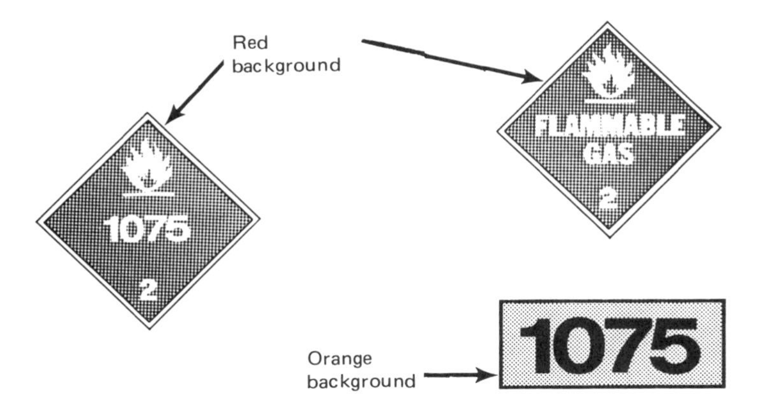

Figure 6.2 *DOT placards. Note that the weight of the container must be considered. A load of six typical 100-lb (45-kg) propane capacity DOT cylinders would weigh more than 1,000 lb (454 kg) and require placarding even though the LP-Gas weight is only 600 lb (272 kg). (Drawing courtesy of National Propane Gas Association.)*

outlined in the *Code of Federal Regulations* (CFR), Title 49, Parts 100–199. A counterpart constructed in accordance with Section 2-2 of NFPA 58 and equipped for portable use according to Section 2-3 may be used. Vehicles transporting portable containers of this size must be placarded as required by DOT regulations.

Portable containers of this size are used temporarily at construction sites, for emergency power generation, and for delivery in bulk to remote areas. Certain applications would dictate the use of a Specification 51 container, particularly if transported by vessel under the jurisdiction of the U.S. Coast Guard, as this is the only type this DOT agency recognizes.

6-2.3.1 Portable containers having an individual water capacity exceeding 1,000 lb (454 kg) [nominal 420 lb (191 kg) LP-Gas capacity] when filled with LP-Gas shall be transported in compliance with the requirements of this section.

6-2.3.2 Containers shall be constructed in accordance with Section 2-2 and equipped in accordance with Section 2-3 for portable use or shall comply with DOT portable tank container specifications for LP-Gas service.

6-2.3.3 The quantity of LP-Gas put into containers shall be in accordance with Chapter 4.

6-2.3.4 Valves and other container appurtenances shall be protected in accordance with 2-2.4.2.

6-2.3.5 Containers and their appurtenances shall be determined to be leak-free before being loaded into vehicles. Containers shall be transported in a suitable rack or frame or on a flat surface. Containers shall be fastened securely in a position to minimize the possibility of movement, tipping, or physical damage, relative to each other or to the supporting structure, while in transit.

6-2.3.6 Containers shall be transported with relief devices in communication with the vapor space.

6-2.3.7 Vehicles carrying more than 1,000 lb (454 kg) of LP-Gas, including the weight of the containers, shall be placarded as required by DOT regulations or state law.

6-2.3.8 Where portable containers complying with the requirements of this section are mounted permanently or semipermanently on vehicles to serve as cargo tanks, so that the assembled vehicular unit can be used for making liquid deliveries to other containers at points of use, the provisions of Section 6-3 shall apply.

6-2.4 **Fire Extinguishers.** Each truck or trailer transporting portable containers as provided by 6-2.2 or 6-2.3 shall be equipped with at least one approved portable fire extinguisher having a minimum capacity of 18 lb (8.2 kg) dry chemical with a B:C rating. (*Also see NFPA 10, Standard for Portable Fire Extinguishers.*)

See commentary on 5-5.

6-3 Transportation in Cargo Vehicles.

Cargo vehicles must comply with DOT regulations, CFR 49, Parts 100–199, and the Federal Motor Carrier Safety Regulations. Added requirements are contained in NFPA 58 as noted in the commentary following the Note to 6-1.1.1.

6-3.1 Application.

6-3.1.1 This section includes provisions for cargo vehicles used for the transportation of LP-Gas as liquid cargo, normally loaded into the cargo container at the bulk plant or manufacturing point and transferred into other containers at the point of delivery. Transfer shall be permitted to be made by a pump or compressor mounted on the vehicle or by a transfer means at the delivery point.

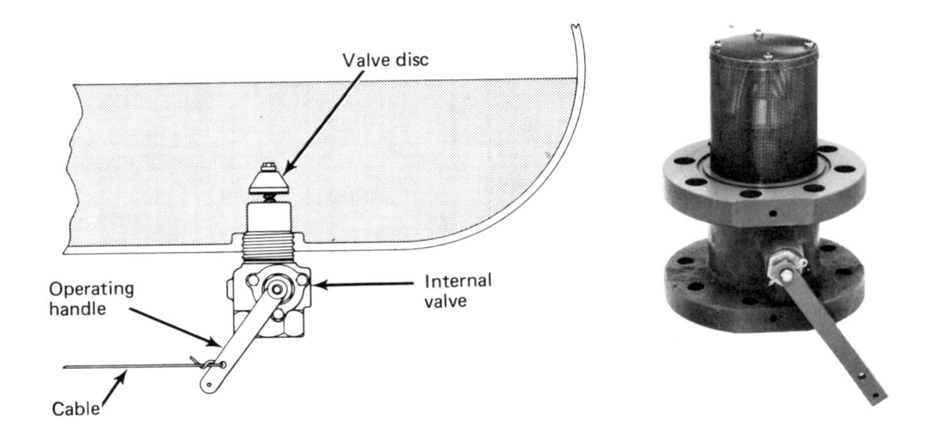

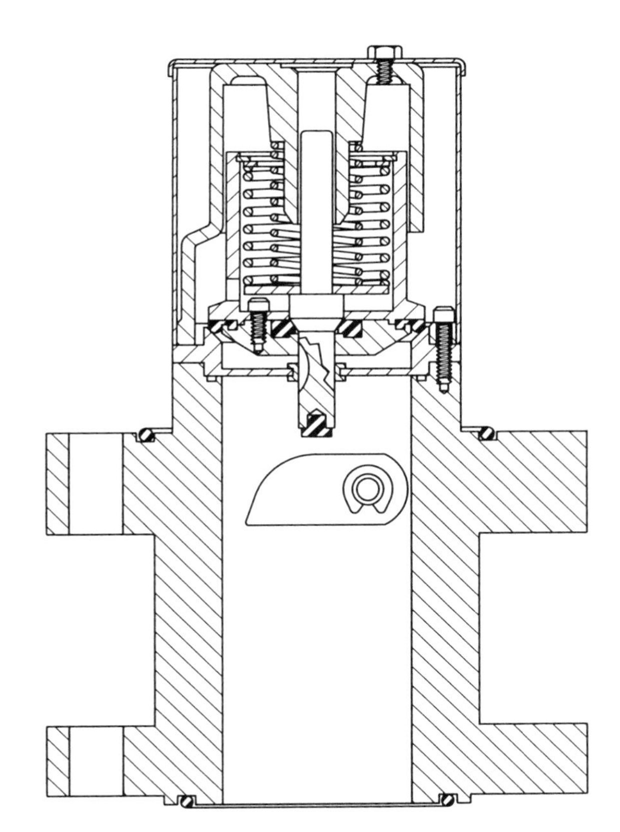

Figure 6.8 *Internal valve in cargo vehicle container. (Photo and drawing courtesy of Engineered Controls, Inc.)*

6-3.3 Piping (Including Hose), Fittings, and Valves.

6-3.3.1 Pipe, tubing, pipe and tubing fittings, valves, hose, and flexible connectors shall comply with Section 2-4, with the provisions of DOT cargo tank specifications for LP-Gas, and shall be suitable for the working pressure specified in 2-5.1.2. In addition, the following shall apply:

(a) Pipe shall be wrought iron, steel, brass, or copper in accordance with 2-4.2.
(b) Tubing shall be steel, brass, or copper in accordance with 2-4.3(a), (b), or (c).
(c) Pipe and tubing fittings shall be steel, brass, copper, malleable iron, or ductile (nodular) iron suitable for use with the pipe or tubing used as specified in 6-3.3.1(a) or (b).
(d) Pipe joints shall be threaded, flanged, welded, or brazed. Fittings, where used, shall comply with 6-3.3.1(c).

1. Where joints are threaded, or threaded and back welded, pipe and nipples shall be Schedule 80 or heavier. Copper or brass pipe and nipples shall be of equivalent strength.
2. Where joints are welded or brazed, the pipe and nipples shall be Schedule 40 or heavier. Fittings or flanges shall be suitable for the service. (*See 6-3.3.2.*)
3. Brazed joints shall be made with a brazing material having a melting point exceeding 1,000°F (538°C).

(e) Tubing joints shall be brazed, using a brazing material having a melting point of at least 1,000°F (538°C).

These provisions for piping, fittings, and valves are more detailed than those in the DOT regulations. They also specify the working pressure of such materials, particularly the 350-psi (2.4-MPa) minimum working pressure on the discharge of pumps required in 2-5.1.2.

6-3.3.2 Pipe, tubing, pipe and tubing fittings, valves, hose, and flexible connectors, and complete cargo vehicle piping systems including connections to equipment (*see 6-3.4*), after assembly, shall comply with 2-5.1.2.

The operating pressure requirements for piping and equipment in LP-Gas service are established in 2-5.1.2, and this makes them applicable to cargo vehicles. These requirements apply to piping on all vehicles used for transportation of LP-Gas. They complement the provisions of 6-3.3.1.

6-3.3.3 Valves, including shutoff valves, excess-flow valves, backflow check valves, and remotely controlled valves, used in piping shall comply with the applicable provisions of DOT cargo tank specifications for LP-Gas service, and with 2-4.5, and shall comply with the minimum design pressure requirements of 2-5.1.2.

The requirements for valves are established in 2-4.5, and this makes those requirements applicable to cargo vehicles. It specifies materials and sets pressure limits, similar to 2-5.1.2 for piping.

6-3.3.4 Hose, hose connections, and flexible connectors shall comply with 2-4.6 and 6-3.3.1. Flexible connectors used in the piping system to compensate for stresses and vibration shall be limited to 3 ft (1 m) in overall length. Flexible connectors on existing LP-Gas cargo units replaced after December 1, 1967, shall comply with 2-4.6.

(a) Flexible connectors assembled from rubber hose and couplings installed after December 31, 1974, shall be permanently marked to indicate the date of assembly of the flexible connector, and the flexible portion of the connector shall be replaced within 6 years of the indicated date of assembly of the connector.

(b) The rubber hose portion of flexible connectors shall be replaced whenever a cargo unit is remounted on a different chassis, or whenever the cargo unit is repiped, if such repiping encompasses that portion of piping in which the connector is located.

Exception: This shall not be required if the remounting or repiping is performed within 1 year of the date of assembly of the connector.

6-3.3.5 All threaded primary valves and fittings used in liquid filling or vapor equalization directly on the cargo container of transportation equipment shall be of steel, malleable, or ductile iron construction. All existing equipment shall be so equipped not later than the scheduled requalification date of the container.

Experience with brass valves in liquid filling or vapor equalization lines (which are permitted on stationary containers) where the threads were stripped and the valves pulled out in drive-aways with hose connected necessitated this provision prohibiting their further use on cargo vehicles. This provision is more restrictive than the requirement for valves and fittings used with stationary containers, where brass is permitted and its use is common.

6-3.4 Equipment.

These provisions are also more detailed than DOT regulations.

6-3.4.1 LP-Gas equipment, such as pumps, compressors, meters, dispensers, regulators, and strainers, shall comply with Section 2-5 for design and construction and shall be installed in accordance with the applicable provisions of 3-2.13. Equipment on vehicles shall be securely mounted in place and connected to the piping system in accordance with the manufacturer's instructions, taking into account the greater (compared with stationary service) jarring and vibration problems incident to vehicular use.

6-3.4.2 Pumps or compressors used for LP-Gas transfer shall be permitted to be mounted on tank trucks, trailers, semitrailers, or tractors. If an electric drive is used, obtaining energy from the electrical installation at the delivery point, the installation on the vehicle (and at the delivery point) shall comply with 3-7.2.

6-3.4.3 The installation of compressors shall comply with the applicable provisions of 3-2.13.2 and 6-3.4.1.

6-3.4.4 The installation of liquid meters shall be in accordance with 3-2.13.4(a). If venting of LP-Gas to the air is necessary, provision shall be made to vent it at a safe location.

6-3.4.5 Where wet hose is carried while connected to the truck's liquid pump discharge piping, an automatic device, such as a differential regulator, shall be installed between the pump discharge and the hose connection to prevent liquid discharge while the pump is not operating. Where a meter or dispenser is used, this device shall be installed between the meter outlet and the hose connection. If an excess-flow valve is used, it shall not be the exclusive means of complying with this provision.

Generally, transport trucks carry "dry" hose, which is completely disconnected following a liquid transfer operation, and carried in protective containers during travel. The smaller retail bulk delivery tank trucks (bobtails) carry "wet" hose, which contains liquid LP-Gas at all times. This latter type of hose is protected by a differential regulator, which is open only when the pump is running.

6-3.5 Protection of Container Appurtenances, Piping System, and Equipment. Container appurtenances, piping, and equipment comprising the complete LP-Gas system on the cargo vehicle shall be securely mounted in position (*see 6-3.2.1 for container mounting*), shall be protected against damage to the extent it is practical, and in accordance with DOT regulations.

6-3.6 Painting and Marking Liquid Cargo Vehicles. Painting of cargo vehicles shall comply with *Code of Federal Regulations*, Title 49, Part 195. Placarding and marking shall comply with CFR 49.

6-3.7 Fire Extinguishers. Each tank truck or tractor shall be provided with at least one approved portable fire extinguisher having a minimum capacity of 18 lb dry chemical with a B:C rating. (*Also see NFPA 10, Standard for Portable Fire Extinguishers.*)

See commentary on 5-5.

6-3.8 Chock Blocks for Liquid Cargo Vehicles. Each tank truck and trailer shall carry chock blocks, which shall be used to prevent rolling of the vehicle whenever it is being loaded or unloaded, or is parked.

6-3.9 Exhaust Systems. The truck engine exhaust system shall comply with *Federal Motor Carrier Safety Regulations*.

6-3.10 Smoking Prohibition. No person shall smoke or carry lighted smoking material on or within 25 ft (7.6 m) of a vehicle required to be placarded in accordance with DOT regulations containing LP-Gas liquid or vapor. This requirement shall also apply at points of liquid transfer and while delivering or connecting to containers.

This requirement is consistent with federal regulations (49 CFR 397.13). Note that the separation distance for smoking of 25 ft (7.6 m) is greater than the separation distance for internal combustion engines during transfer operations specified in 4-2.3.2(a) [15 ft (5 m]. The Technical Committee agrees that 25 ft (7.6 m) is the correct separation distance for smoking, and the separation distance of other sources of ignition may be changed in a future edition of the standard.

6-4 Trailers, Semitrailers, and Movable Fuel Storage Tenders, Including Farm Carts.

6-4.1 Application. This section applies to all tank vehicles, other than trucks, that are parked at locations away from bulk plants.

6-4.2 Trailers or Semitrailers Comprising Parts of Vehicles in Accordance With Section 6-3. When parked, tank trailers or semitrailers covered by Section 6-3 shall be positioned so that the pressure relief valves shall communicate with the vapor space of the container.

Trailers or semitrailers are used where large volumes are needed on a temporary basis, such as at a temporary asphalt mixing plant for road construction or repair. They are not to be used as permanent storage at a small bulk plant for filling cylinders or as retail delivery cargo vehicles. If this were done, the system would have to meet the provisions of Sections 3-2 and 3-3, which would be impractical.

6-4.3 Fuel Storage Tenders Including Farm Carts.

6-4.3.1 Movable fuel storage tenders including farm carts (*see definition*) shall comply with this section. Where used over public ways, they shall comply with applicable state regulations.

Figure 6.9 Farm cart. (Photo courtesy of Schultz Gas Service, Inc.)

6-4.3.2 Such tenders shall be constructed in accordance with Section 2-2 and equipped with appurtenances as provided in Section 2-3. Mounting of containers shall be adequate for the service involved.

6-4.3.3 Threaded piping shall be not less than Schedule 80, and fittings shall be designed for not less than 250 psi (1.7 MPa).

6-4.3.4 Piping, hoses, and equipment, including valves, fittings, pressure relief valves, and container accessories, shall be adequately protected against collision or upset.

6-4.3.5 Tenders shall be so positioned that container safety relief valves communicate with the vapor space.

6-4.3.6 Such tenders shall not be filled on a public way.

6-4.3.7 Such tenders shall contain no more than 5 percent of their water capacity in liquid form during transportation to or from the bulk plant.

6-4.3.8 The shortest practical route, consistent with safety, shall be used when transporting such tenders between points of utilization.

Movable fuel storage tenders and farm carts are limited in size to 1,200 gal (4,543 L) water capacity. They are nonhighway vehicles but occasionally are moved over roads for short distances to be used as a temporary fuel supply for farm tractors, construction equipment, etc. When transported over public roads, they must comply with state laws and DOT requirements where they are enforced as state law. Paragraphs 6-4.3.7 and 6-4.3.8 were added in the 1992 edition to recognize that the regulations of the Department of Transportation 49 CFR 173.315(j) apply to tenders (6-4.3 through 6-6) and that routes must be as short as possible (6-4.3.8).

6-5 Transportation of Stationary Containers to and from Point of Installation.

6-5.1 Application. This section applies to the transportation of containers designed for stationary service at the point of use and secured to the vehicle only for transportation. Such containers shall be transported in accordance with 6-5.2.1.

Storage containers are moved to the consumer site with a small amount of LP-Gas (5 percent maximum) for initial startup. It is not possible to measure tank contents below 5 percent with the types of gauges used in LP-Gas containers. Section 173.315(j) of Title 49, CFR also sets forth rules consistent with NFPA 58 provisions for the shipment of storage containers to consumers' premises by private motor carrier. Occasionally, it may be necessary to move the container with more than 5 percent LP-Gas—for example, when the liquid evacuation valve does not function and the container cannot be pumped out at its installed site; or where it will be safer to transfer the contents at the propane dealers facility than in a residential neighborhood. This may be done subject to the limitations specified by the authority having jurisdiction.

6-5.2 Transportation of Containers.

6-5.2.1 ASME containers of 125 gal (0.5 m^3) or more water capacity shall contain no more than 5 percent of their water capacity in liquid form during transportation.

Exception: Containers shall be permitted to be transported with more LP-Gas than 5 percent of their water capacity in a liquid form, but less than the maximum permitted by Section 4-4, provided that:

(a) Such transportation shall be permitted only to move containers from a stationary or temporary installation to a bulk plant; and

(b) The owner of the container or the owner's designated representative authorizes its transportation; and

(c) Valves and fittings shall be protected to minimize the possibility of damage by a method approved by the authority having jurisdiction; and

(d) Lifting lugs shall not bear more than the empty weight of the container plus 5 percent liquid volume. Additional means for lifting, securing, and supporting the container shall be provided.

Containers smaller than 125 gal (0.5 m^3) are not required to be drained for two reasons. First, they are not required to be equipped with a connection for liquid evacuation (*see 2-2.3.3*), so they can be difficult to drain, and their weight, even when full, is not so great that they cannot be safely handled.

The limit of 5 percent water capacity for transportation of containers is a practical one in that it represents a low weight and recognizes that the gauges on containers cannot accurately measure contents below 5 percent.

This exception was modified in the 1995 edition to add criteria for safe transportation of containers with more than 5 percent of their maximum capacity. The four requirements permit transportation only to a bulk plant, require valve and fitting protection to prevent damage during transportation, and require that lifting of the container is done in a safe manner. These requirements recognize that it is sometimes required to move a container to safely empty it, and that the equipment usually available at a bulk plant can expedite the procedure, enhancing safety.

Tanks built prior to 1961 do not have an actuated liquid withdrawal excess-flow valve and may not have any bottom fitting, or the actuated liquid withdrawal excess-flow valve may be inoperable. In such a case at the installed location, the only option is to roll the container on its side to withdraw liquid through the vapor withdrawal valve. When this is done, there is no excess-flow protection as it is not required on vapor withdrawal connections. To perform this operation as safely as possible, it is often preferable to do so away from a residential location, and where the bulk plant has personnel better equipped to handle the procedure.

The lifting lugs on a container are normally designed for the weight of the container and 5 percent of its maximum propane capacity only. When it is necessary to lift a container with more than 5 percent of its capacity, alternate lifting means must be used.

6-5.2.2 Containers shall be safely secured to minimize movement relative to each other or to the carrying vehicle while in transit, giving consideration to the sudden stops, starts, and changes of direction normal to vehicular operation.

6-5.2.3 Valves, regulators, and other container appurtenances shall be adequately protected against physical damage during transportation.

6-5.2.4 Pressure relief valves shall be in direct communication with the vapor space of the container.

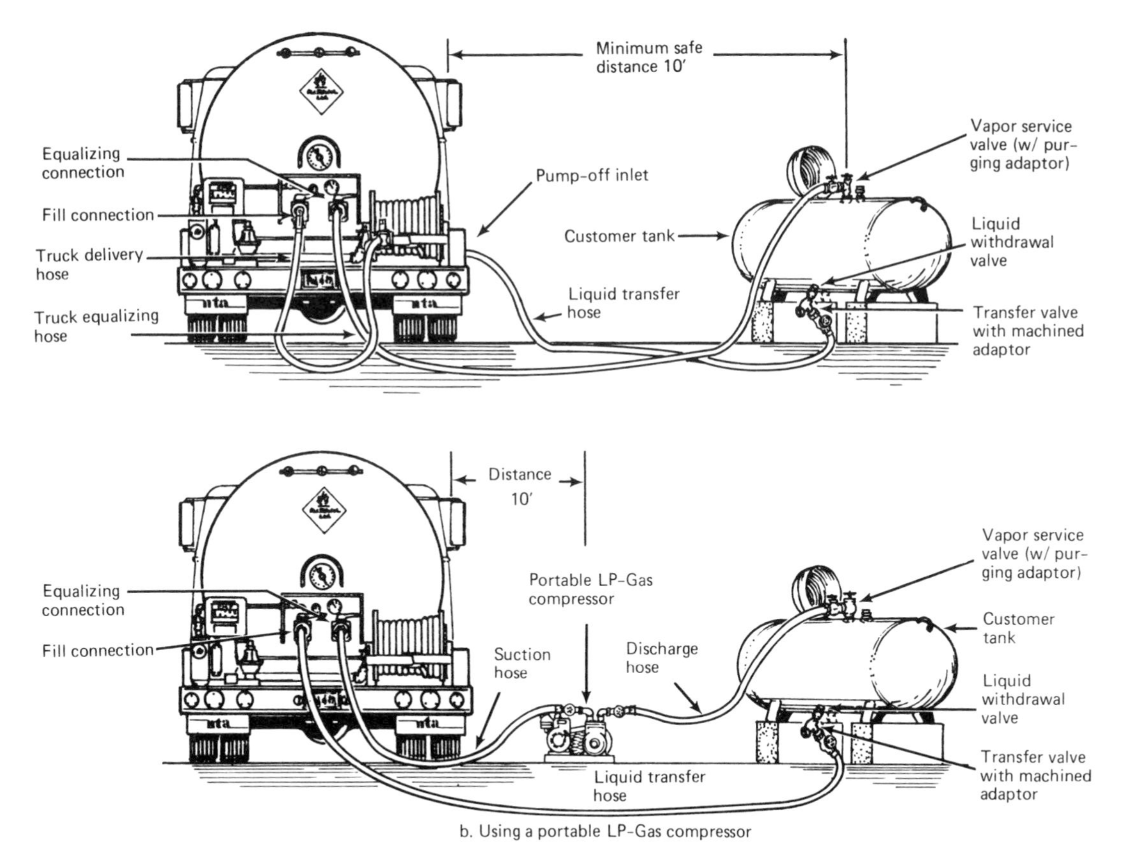

Figure 6.10 *Removing liquid from container to comply with 5 percent provision. (Drawing courtesy of National Propane Gas Association.)*

Figure 6.11 Transporting ASME stationary storage containers. (Top) Flatbed truck with crane. (Bottom) Saddle trailer. [Photos courtesy of Schultz Gas Service (top) and the National Propane Gas Association (bottom).]

Figure 6.12 Lifting ASME stationary storage container using lifting lugs. (Photo courtesy of the National Propane Gas Association.)

6-6 Parking and Garaging Vehicles Used to Carry LP-Gas Cargo.

6-6.1 Application. This section applies to the parking (except parking associated with a liquid transfer operation) and garaging of vehicles used for the transportation of LP-Gas. Such vehicles include those used to carry portable containers (*see Section 6-2*) and those used to carry LP-Gas in cargo tanks (*see Section 6-3*).

This section sets out provisions for parking cargo vehicles and cylinder delivery trucks outdoors in congested and uncongested areas, and indoors within public buildings or at the owner's premises for temporary parking or for service and repair.

6-6.2 Parking.

6-6.2.1 Vehicles carrying or containing LP-Gas parked outdoors shall comply with the following:

(a) Vehicles shall not be left unattended on any street, highway, avenue, or alley, provided that drivers are not prevented from those necessary absences from the vehicle connected with their normal duties, nor shall this requirement prevent stops for meals or rest stops during the day or night.

Exception No. 1: This shall not apply in an emergency.

Exception No. 2: Where parked in accordance with 6-6.2.1(b).

(b) Vehicles shall not be parked in congested areas. Such vehicles shall be permitted to be parked off the street in uncongested areas if at least 50 ft (15 m) from any building used for assembly, institutional, or multiple residential occupancy. This requirement shall not prohibit the parking of vehicles carrying portable containers or cargo vehicles of 3,500 gal (13 m^3) water capacity or less on streets adjacent to the driver's residence in uncongested residential areas, provided such parking locations are at least 50 ft (15 m) from a building used for assembly, institutional, or multiple residential occupancy.

In remote areas where the driver's residence may be a considerable distance from bulk plant facilities and consumers are in the area of the driver's home, it is permissible to park the retail bulk delivery tank truck or cylinder delivery truck near the driver's home under the conditions stipulated.

6-6.2.2 Vehicles parked indoors shall comply with the following:

(a) Cargo vehicles parked in any public garage or building shall have LP-Gas liquid removed from the cargo container, piping, pump, meter, hoses, and related equipment, and the pressure in the delivery hose and related equipment shall be reduced to approximately atmospheric, and all valves shall be closed before the vehicle is moved indoors. Delivery hose or valve outlets shall be plugged or capped before the vehicle is moved indoors.

(b) Vehicles used to carry portable containers shall not be moved into any public garage or building for parking until all portable containers have been removed from the vehicle.

(c) Vehicles carrying or containing LP-Gas shall be permitted to be parked in buildings complying with Chapter 7 and located on premises owned or under the control of the operator of such vehicles, provided:

1. The public is excluded from such buildings.
2. There is adequate floor level ventilation in all parts of the building where such vehicles are parked.
3. Leaks in the vehicle LP-Gas systems are repaired before the vehicle is moved indoors.
4. Primary shutoff valves on cargo tanks and other LP-Gas containers on the vehicle (except propulsion engine fuel containers) are closed and delivery hose outlets plugged or capped to contain system pressure before the vehicle is moved indoors. Primary shutoff valves on LP-Gas propulsion engine fuel containers shall be closed while the vehicle is parked.
5. No LP-Gas container is located near a source of heat or within the direct path of hot air being blown from a blower-type heater.
6. LP-Gas containers are gauged or weighed to determine that they are not filled beyond the maximum filling limit according to Section 4-4.

If it is necessary to park cargo vehicles in public buildings, all liquid LP-Gas must be removed from the cargo tank and equipment on tank trucks and cylinders removed from cylinder trucks. In very cold climates it may be necessary to park these vehicles in garages on the owner's premises. The type of building construction required by 6-6.2.2(c) must not be overlooked.

6-6.2.3 Vehicles shall be permitted to be serviced or repaired indoors as follows:

(a) When it is necessary to move a vehicle into any building located on premises owned or operated by the operator of such vehicle for service on engine or chassis, the provisions of 6-6.2.2(a) or (c) shall apply.

(b) When it is necessary to move a vehicle carrying or containing LP-Gas into any public garage or repair facility for service on the engine or chassis, the provisions of 6-6.2.2(a) or (b) shall apply, unless the driver or a qualified representative of an LP-Gas operator is in attendance at all times while the vehicle is indoors. In this case, the following provisions shall apply under the supervision of such qualified persons:

1. Leaks in the vehicle LP-Gas systems shall be repaired before the vehicle is moved indoors.
2. Primary shutoff valves on cargo tanks, portable containers, and other LP-Gas containers installed on the vehicle (except propulsion engine fuel containers) shall be closed. LP-Gas liquid shall be removed from the piping, pump, meter, delivery hose, and related equipment and the pressure therein reduced to approximately atmospheric before the vehicle is moved inside. Delivery hose or valve outlets shall be plugged or capped before the vehicle is moved indoors.
3. No container shall be located near a source of heat or within the direct path of hot air blown from a blower or from a blower-type heater.
4. LP-Gas containers shall be gauged or weighed to determine that they are not filled beyond the maximum filling capacity according to Section 4-4.

(c) If repair work or servicing is to be performed on a cargo tank system, all LP-Gas shall be removed from the cargo tank and piping, and the system shall be thoroughly purged before the vehicle is moved indoors.

All LP-Gas is to be removed when service or repairs are done on the vehicle in public garages, unless a qualified person is present at all times and certain precautions are taken. At the owner's premises the above measures are to be taken for servicing, which would not make it necessary to remove LP-Gas from the vehicle. However, if work is to be done on the cargo tank system, whether at a public garage or owner's premises or not, all LP-Gas must be removed and the system purged beforehand.

References Cited in Commentary

The following publication is available from the U.S. Government Printing Office, Washington, DC.

Code of Federal Regulations, Title 49, Parts 100–199, and 393, 396, and 397.

7

Buildings or Structures Housing LP-Gas Distribution Facilities

7-1 Scope.

7-1.1 Application.

7-1.1.1 This chapter covers the construction, ventilation, and heating of structures housing certain types of LP-Gas systems as referenced in this standard. Such structures can be separate buildings used exclusively for the purpose (or for other purposes having similar hazards), or they can be rooms attached to, or located within, buildings used for other purposes.

Chapter 7 was introduced in the 1972 edition because certain systems or operations present combustion explosion hazards when they are located or conducted inside structures; where accidental escape of liquid or vapor could result in the formation of a quantity of flammable vapor-air mixture large enough that its ignition could produce destructive pressure in the structure. In such instances, the construction can limit the damage to the structure and its surroundings. For an in-depth discussion of this concept (commonly called "explosion venting"), see NFPA 68, *Guide for Venting of Deflagrations.*

Generally, Chapter 7 applies to structures where liquid LP-Gas is used in volume or where high-pressure gas is present; however, the application of Chapter 7 is stipulated in specific provisions throughout the standard, e.g., 3-2.3.1(b).

The aerosol packaging industry, which uses LP-Gas propellants, has used Chapter 7 extensively in the design of filling rooms. It is noted, however, that NFPA 58 was not developed with such facilities in mind [they are considered chemical plants and are excluded by 1-1.3.1(d)]. Therefore, Chapter 7 should be used with considerable judgment in such applications and may need to be augmented by other safeguards not cited in NFPA 58, such as higher ventilation rates and stricter electrical area classification requirements. NFPA 30B, *Code for the Manufacture and Storage of Aerosol Products*, may provide additional information.

7-1.1.2 The provisions of this chapter apply only to buildings constructed or converted after December 31, 1972.

Exception: Buildings previously constructed under the provisions of 5-3.3. (Also, see 1-2.5.)

This applies Chapter 7 retroactively to storage within special buildings or rooms covered by 5-3.3, which includes all such structures not in a bulk plant. Most of these locations are in occupancies having considerable public exposure, such as mercantile occupancies.

7-2 Separate Structures or Buildings.

Chapter 7 recognizes three basic locations for structures: separate (no physical connection with another structure); attached (connected to another structure with the common walls having a perimeter not exceeding 50 percent of the perimeter of the attached structure); and within (wholly or partly inside another structure). These are presented in order of decreasing degree of overall safety to the facility; this order should be considered in the facility design. There should be good reasons, for example, to use a room within a structure if one of the other options will do the job.

7-2.1 Construction of Structures or Buildings.

7-2.1.1 Separate buildings or structures shall be one story in height and shall have walls, floors, ceilings, and roofs constructed of noncombustible materials. Exterior walls, ceilings, and roofs shall be constructed as follows:

(a) Of lightweight material designed for explosion venting, or

(b) If of heavy construction, such as solid brick masonry, concrete block, or reinforced concrete construction, explosion venting windows or panels in walls or roofs shall be provided having an explosion venting area of at least 1 sq ft (0.1 m^2) for each 50 cu ft (1.4 m^3) of the enclosed volume.

From functional and economic considerations, the lightweight material designed for explosion venting type of construction in 7-2.1.1(a) has definite advantages over that in 7-2.1.1(b). A simple steel framing, to which roof and wall panels are affixed by fastenings just strong enough to withstand wind loadings, allows for the maximum possible explosion venting area. If these panels are of nonfrangible material, such as steel or aluminum, they stay more or less in one piece when they blow off, are lightweight, and do not travel very far, or can be hinged or restrained with cables. While also reasonably light in weight, cement-asbestos panels tend to fragment and present more of a flying missile hazard. When such a structure explodes, the steel framing usually survives with little damage. The roof and walls can be replaced quickly and the structure can be returned to use promptly.

Aside from being much more costly, the heavy construction described in 7-2.2.1(b) does not have as much explosion venting area, and the masonry types present flying missile hazards and take much longer to rebuild. They do, however, have advantages where external security is a problem and aesthetics need to be considered. This type of construction is referred to a damage limiting construction by some insurance companies who recommend it in some applications.

Additional information on explosion venting can be found in NFPA 68, *Guide to Venting of Deflagrations*.

7-2.1.2 The floor of such structures shall not be below ground level. Any space beneath the floor shall preferably be of solid fill. If not so filled, the perimeter of the space shall be left entirely unenclosed.

This provision recognizes that LP-Gas is heavier than air.

7-2.2 Structure or Building Ventilation.

7-2.2.1 The structure shall be ventilated using air inlets and outlets, the bottom of which shall be not more than 6 in. (150 mm) above the floor, and shall be arranged to provide air movement across the floor as uniformly as practical and in accordance with the following:

(a) Where mechanical ventilation is used, air circulation shall be at least 1 cu ft per min per sq ft (0.3 $m^3/min/m^2$) of floor area. Outlets shall discharge at least 5 ft (1.5 m) from any opening into the structure or any other structure.

(b) Where natural ventilation is used, each exterior wall [up to 20 ft (6.1 m) in length] shall be provided with at least one opening, with an additional opening for each 20 ft (6.1 m) of length or fraction thereof. Each opening shall have a minimum size of 50 sq in. (32 250 mm^2) and the total of all openings shall be at least 1 sq in. per sq ft (720 mm^2/m^2) of floor area.

Mechanical ventilation is more predictable than natural ventilation and is preferred, especially in buildings used for liquid transfer. Where the occupancy is used for storage only, natural ventilation has generally proven to be adequate.

The ventilation provisions reflect the fact that LP-Gas is heavier than air. The criteria for mechanical ventilation are based on the practice of continuously removing a stratum of air 1-ft (0.3-m) deep along the floor. It is important that the air discharged be made up by incoming air. The most common ventilation defect is blockage of inlets for mechanical systems and blockage of both inlets and outlets for natural arrangements in cold weather for comfort reasons. The inlet air should be heated to solve this problem. It is considerably easier to do this with a mechanical system because the inlets can be manifolded and a single heater installed in the common duct.

In large buildings many smaller ventilation openings are preferable. Prior to 1989 the standard could be met with two openings, each opening having an area of 1 sq in. (600 mm^2) per sq ft (0.1 m^2) of floor area, and with larger buildings being constructed the openings were very large, creating heating problems in colder regions.

7-2.3 Structure or Building Heating. Heating shall be by steam or hot water radiation or other heating transfer medium with the heat source located outside of the building or structure (*see Section 3-7, Ignition Source Control*), or by electrical appliances listed for Class I, Group D, Division 2 locations, in accordance with NFPA 70, *National Electrical Code* (*see Table 3-7.2.2*).

7-3 Attached Structures or Rooms within Structures.

This requirement maintains the explosion venting function of the attached structure while minimizing the possibility of damage (explosion and fire) to the structure to which it is attached. The common wall(s) is built to be strong and with a degree of fire resistance, and the presence and character of openings is controlled.

While a wall designed for a static pressure of 100 lb/ft^2 (4.79 kPa/m^2) is a substantial wall [even a wall of minimal height would be at least 12 in. (0.3 m) of masonry or steel-reinforced concrete block], such a wall should be as strong as is feasible to build.

7-3.1 Construction of Attached Structures. Attached structures shall comply with 7-2.1 (attachment shall be limited to 50 percent of the perimeter of the space enclosed; otherwise such space shall be considered as a room within a structure—*see 7-3.2*), and shall comply with the following:

(a) Common walls at points at which structures are to be attached shall:

1. Have, as erected, a fire resistance rating of at least 1 hr.

NOTE: For information on fire resistance of building materials, see NFPA 251, *Standard Methods of Fire Tests of Building Construction and Materials.*

2. Have no openings. Common walls for attached structures used only for storage of LP-Gas shall be permitted to have doorways that shall be equipped with 1½-hr (B) fire doors.

NOTE: For information of fire doors, see NFPA 80, *Standard for Fire Doors and Fire Windows.*

3. Be designed to withstand a static pressure of at least 100 lb (0.7 MPa) per sq ft (0.1 m^2).

(b) The provisions of 7-3.1(a) shall be permitted to be waived if the building to which the structure is attached is occupied by operations or processes having a similar hazard.

(c) Ventilation and heating shall comply with 7-2.2.1 and 7-2.3.

7-3.2 Construction of Rooms within Structures.

The principle here is similar to that of 7-3.1, and the provisions are also similar. A first-story location is required in order to remove the floor as an area of exposure to the remainder of the building and to facilitate fire control activities by fire departments.

The reference to NFPA 68 was added in the 1992 edition to provide a source of information for explosion venting. NFPA 68 provides a method of calculating the area required to vent ignition of an LP-Gas/air mixture.

7-3.2.1 Rooms within structures shall be located in the first story and shall have at least one exterior wall with unobstructed free vents for freely relieving explosion pressures.

NOTE: For information on explosion venting, see NFPA 68, *Guide for Venting of Deflagrations.*

(a) Walls, floors, ceilings, or roofs of such rooms shall be constructed of noncombustible materials. Exterior walls and ceilings shall either be of lightweight material designed for explosion venting, or, if of heavy construction (such as solid brick masonry, concrete block, or reinforced concrete construction), shall be provided with explosion venting windows or panels in the walls or roofs having an explosion venting area of at least 1 sq ft (0.1 m^2) for each 50 cu ft (1.4 m^3) of the enclosed volume.

(b) Walls and ceilings common to the room and to the building within which it is located shall:

1. Have, as erected, a fire resistance rating of at least 1 hr.

NOTE: For information on fire resistance rating of building materials, see NFPA 251, *Standard Methods of Fire Tests of Building Construction and Materials.*

2. Not have openings. Common walls for rooms used only for storage of LP-Gas shall be permitted to have doorways that shall be equipped with 1½-hr (B) fire doors.

NOTE: See NFPA 80, *Standard for Fire Doors and Fire Windows.*

3. Be designed to withstand a static pressure of at least 100 lb (0.7 MPa) per sq ft (0.1 m^2).

Exception: The provisions of 7-3.2.1(b) shall be permitted to be waived if the building within which the room is located is occupied by operations or processes having a similar hazard.

(c) Ventilation and heating shall comply with 7-2.2.1 and 7-2.3.

References Cited in Commentary

The following publications are available from the National Fire Protection Association, 1 Batterymarch Park, P.O. Box 9101, Quincy, MA 02269-9101.

NFPA 30B, *Code for the Manufacture and Storage of Aerosol Products*, 1990 edition.

NFPA 68, *Guide to Venting of Deflagrations*, 1988 edition.

8

Engine Fuel Systems

This chapter was created in the 1992 edition by relocation of the former Section 3-6. This was done for two reasons: first, to isolate this unique installation in its own chapter, and second, to remove some text from Chapter 3, which is very large.

Propane has been used as an engine fuel for stationary engines (ranch water pumps, pipeline compressors, etc.) since the 1920s and for vehicle propulsion in farm tractors, buses, etc., since shortly thereafter. It was the fuel used by the Graf Zeppelin in the world voyage of 1928. Standards for LP-Gas engine fuel installations first appeared in the 1937 edition of NFPA 58 in a separate chapter.

Coverage of specific applications can make the standard as a whole hard to understand and can lead to inconsistencies that are difficult to explain. Engine fuel installations are a special field, however. It became apparent in the early 1980s that a different approach was needed because it was confusing to search through the entire standard to find the requirements for a proper engine fuel installation. All provisions relating to LP-Gas engine installations of any sort are now located in Chapter 8 of NFPA 58. There is little need to look elsewhere in the standard for information on this subject.

The consolidation of the current Chapter 8 was made in the 1983 edition. In addition, changes were made to remove inconsistencies, clarify the provisions, and provide for newer methods and equipment. Many of the changes reflected a study of worldwide regulations, practices, trends, and experience. To allow for flexibility and innovation, the provisions are generally written in terms of performance rather than detailed product specifications.

Chapter 8 applies to fuel systems on vehicles for any purpose, stationary and portable engines, and the garaging of vehicles having propane-fueled engines. Also, anyone making an installation, repairing, servicing, or refueling is required to be properly trained in these procedures.

8-1 Application.

8-1.1 This chapter applies to fuel systems using LP-Gas as a fuel for internal combustion engines. Included are provisions for containers, container appurtenances, carburetion equipment, piping, hose and fittings, and provisions for their installation. This chapter covers engine fuel systems for engines installed on vehicles for any purpose, as well as fuel systems for stationary and portable engines. It also includes provisions for garaging of vehicles upon which such systems are installed.

NOTE: See Section 3-8 for systems on vehicles for purposes other than for engine fuel.

8-1.2 Containers supplying fuel to stationary engines, or to portable engines used in lieu of stationary engines, shall be installed in accordance with Section 3-2 (*see Section 3-4 for portable engines used in buildings or on roofs or exterior balconies under certain conditions*).

8-1.3 Containers supplying fuel to engines on vehicles, regardless of whether the engine is used to propel the vehicle or is mounted on it for other purposes, shall be constructed and installed in accordance with this section.

8-1.4 In the interest of safety, each person engaged in installing, repairing, filling, or otherwise servicing an LP-Gas engine fuel system shall be properly trained in the necessary procedures.

8-2 General Purpose Vehicle Engines Fueled by LP-Gas.

General purpose vehicles and industrial trucks are the two categories of vehicles using propane engines for their propulsion. General purpose vehicles covered in this section include practically any nonindustrial truck vehicles using internal combustion engines. Most general purpose vehicles are used in fleet operations serviced from central sources of supply not available to the public. Propane is not readily available in large enough quantities to accommodate masses of individual private vehicle owners. All over-the-road general purpose vehicles powered by LP-Gas, must be labeled by a diamond shaped marking. (*See 8-2.10.*)

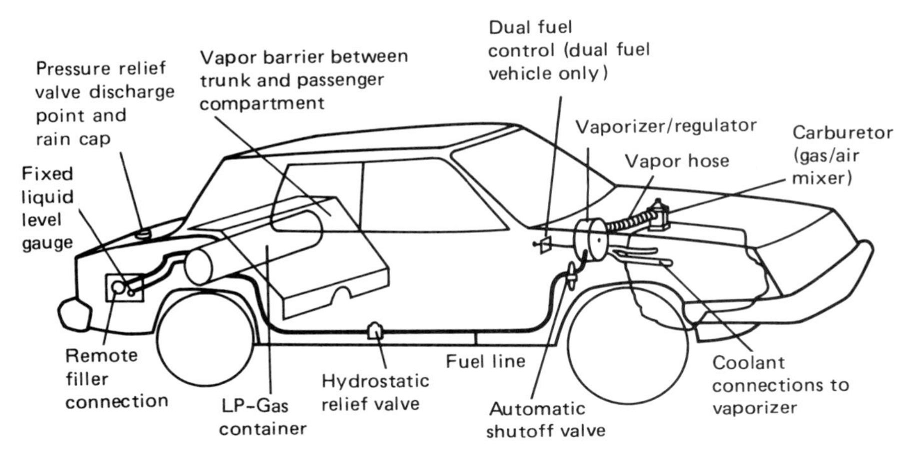

Figure 8.1 Typical propane fuel system on a passenger car. (Drawing courtesy of the National Propane Gas Association.)

8-2.1 This section covers the installation of fuel systems supplying engines used to propel vehicles such as passenger cars, taxicabs, multipurpose passenger vehicles, buses, recreational vehicles, vans, trucks (including tractors, tractor semi-trailer units, and truck trains), and farm tractors.

8-2.2 Containers.

The basic provisions for LP-Gas containers in Chapter 2 are repeated here for general purpose vehicle engine fuel containers. There are several special considerations for those constructed to the ASME Code, because of environmental factors unique to vehicles.

(a)* Containers designed, fabricated, tested, and marked (or stamped) in accordance with the regulations of the U.S. Department of Transportation (DOT); or the "Rules for Construction of Unfired Pressure Vessels," Section VIII, Division I, ASME *Boiler and Pressure Vessel Code*, applicable at the date of manufacture shall be used as follows:

A-8-2.2(a) Prior to April 1, 1967, these regulations were promulgated by the Interstate Commerce Commission. In Canada, the regulations of the Canadian Transport Commission apply, which are available from the Canadian Transport Commission, Union Station, Ottawa, Canada.

1. Adherence to applicable ASME Code Case Interpretations and Addenda shall be considered as compliance with the ASME Code.
2. Containers fabricated to earlier editions of regulations, rules, or codes shall be permitted to be continued in use in accordance with 1-1.5. (*See Appendices C and D.*)

3. Containers that have been involved in a fire and that show no distortion shall be requalified for continued service in accordance with the code under which they were constructed before being reused.

4. DOT containers shall be designed and constructed for at least 240 psi (1.6 MPa) service pressure.

5. DOT specification containers shall be requalified in accordance with DOT regulations. The owner of the container shall be responsible for such requalification. (*See Appendix C.*)

6. ASME containers covered in this section shall be constructed for a minimum 250-psi (1.7-MPa) design pressure.

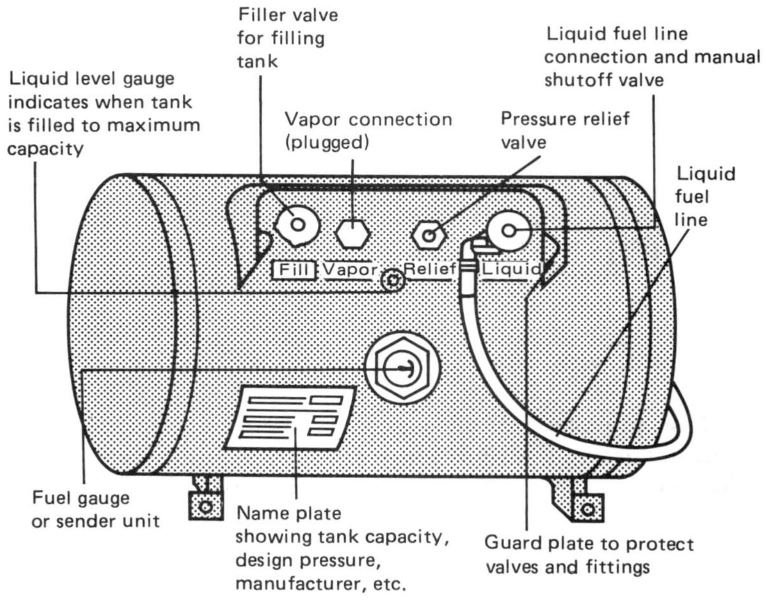

Figure 8.2 *Typical engine fuel container and appurtenances. (Drawing courtesy of the National Propane Gas Association.)*

ASME Code containers installed in enclosed spaces (e.g., trunks), buses (including school buses), multipurpose vehicles (ten persons or fewer), and industrial trucks must have a minimum design pressure of 312.5 psi (2.1 MPa). All other containers must have a minimum 250-psi (1.7-MPa) design pressure. Specifying a 312.5-psi (2.1-MPa) design pressure is done primarily to provide a higher pressure relief valve setting for these particular containers. Containers in this type of service may be subjected to higher temperatures or be on vehicles containing large numbers of persons (especially children); by having a higher setting, unnecessary premature discharge of the relief valve is minimized.

Exception: Containers installed in enclosed spaces on vehicles and all engine fuel containers for industrial trucks, buses (including school buses), and multipurpose passenger vehicles shall be constructed of design pressure of at least 312.5 psi (2.1 MPa).

7. Repair or alterations of containers shall comply with the regulations, rules, or code under which the container was fabricated. Field welding on containers shall be limited to attachments to nonpressure parts, such as saddle pads, wear plates, lugs, or brackets applied by the container manufacturer.
8. Containers showing serious denting, bulging, gouging, or excessive corrosion shall be removed from service.

(b) Containers shall comply with 8-2.2(a) or shall be designed, fabricated, tested, and marked using criteria that incorporate an investigation to determine that they are safe and suitable for the proposed service, are recommended for that service by the manufacturer, and are acceptable to the authority having jurisdiction.

(c) ASME containers shall be marked in accordance with 8-2.2(c)1. through 12. The markings specified shall be on a stainless steel metal nameplate attached to the container, which shall be located to remain visible after the container is installed. The nameplate shall be attached in such a way to minimize corrosion of the nameplate or its fastening means and not contribute to corrosion of the container.

1. Service for which the container is designed; i.e., aboveground.
2. Name and address of container manufacturer or trade name of container.
3. Water capacity of container in lb or U.S. gallons.
4. Design pressure in psi.
5. The wording "This container shall not contain a product having a vapor pressure in excess of 215 psi (1.5 MPa) at 100°F (37.8°C)."
6. Tare weight of container fitted for service in order for containers to be filled by weight.
7. Outside surface area in sq ft.
8. Year of manufacture.
9. Shell thickness ________head thickness ________.
10. OL ________OD ________HD ________.
11. Manufacturer's serial number.
12. ASME Code symbol.

The only difference in nameplate markings for ASME engine fuel containers is the obvious deletion of the reference to underground service. The nameplate markings must be visible after the container is installed. In some instances, a lamp and mirror may have to be used.

(d) LP-Gas fuel containers used on passenger-carrying vehicles shall not exceed 200 gal (0.8 m^3) aggregate water capacity.

(e) Individual LP-Gas containers used on other than passenger carrying vehicles normally operating on the highway shall not exceed 300 gal (1 m^3) water capacity.

Total fuel capacity for passenger carrying vehicles is limited to 200 gal (0.8 m^3) water capacity [approximately 160 gal (0.6 m^3) of LP-Gas]. More than one tank may be used and manifolded together, but the total volume cannot exceed this limit. For nonpassenger vehicles the maximum size of individual containers that may be used is 300 gal (1.1 m^3) water capacity [approximately 240 gal (0.9 m^3) of LP-Gas].

(f) Containers covered in this section shall be equipped for filling into the vapor space.

Exception: Containers having a water capacity of 30 gal (0.1 m^3) or less shall be permitted to be filled into the liquid space.

Prior to the 1989 edition, engine fuel containers smaller than 30 gal (0.1 m^3) had not been allowed to be filled in the liquid space to ensure equilibrium between liquid and vapor in the container, with corresponding pressure reduction on filling (*see commentary following 2-2.3.2*). This was changed because of problems experienced in the field with some automatic stop-fill valves [required by paragraph 8-2.3(a)(8)].

1. The connections for pressure relief valves shall be located and installed in such a way as to have direct communication with the vapor space of the container and shall not reduce the relieving capacity of the relief device.
2. If the connection is located at any position other than the uppermost point of the container, it shall be internally piped to the uppermost point practical in the vapor space of the container.

See commentary following 2-2.3.5.

(g) The container openings, except those for pressure relief valves and gauging devices, shall be labeled to designate whether they communicate with the vapor or liquid space. Labels shall be permitted to be on valves.

(h) Engine fuel containers constructed of steel shall be painted to retard corrosion. (*See Appendix A-3-2.4.1(f).*)

8-2.3 Container Appurtenances.

Basic provisions for container appurtenances in Chapter 2 are also repeated here. Requirements for container appurtenances are located in Table 2-3.3.2 (a). Additional special considerations recognize factors unique to these applications.

(a) Container appurtenances (such as valves and fittings) shall comply with Section 2-3 and the following. Container appurtenances subject to working pressures in excess of 125 psi (0.9 MPa) but not to exceed 250 psi (1.7 MPa) shall be suitable for a working pressure of at least 250 psi (1.7 MPa).

1. Manual shutoff valves shall be designed to provide positive closure under service conditions and shall be equipped with an internal excess-flow check valve designed to close automatically at the rated flows of vapor or liquid specified by the manufacturers.

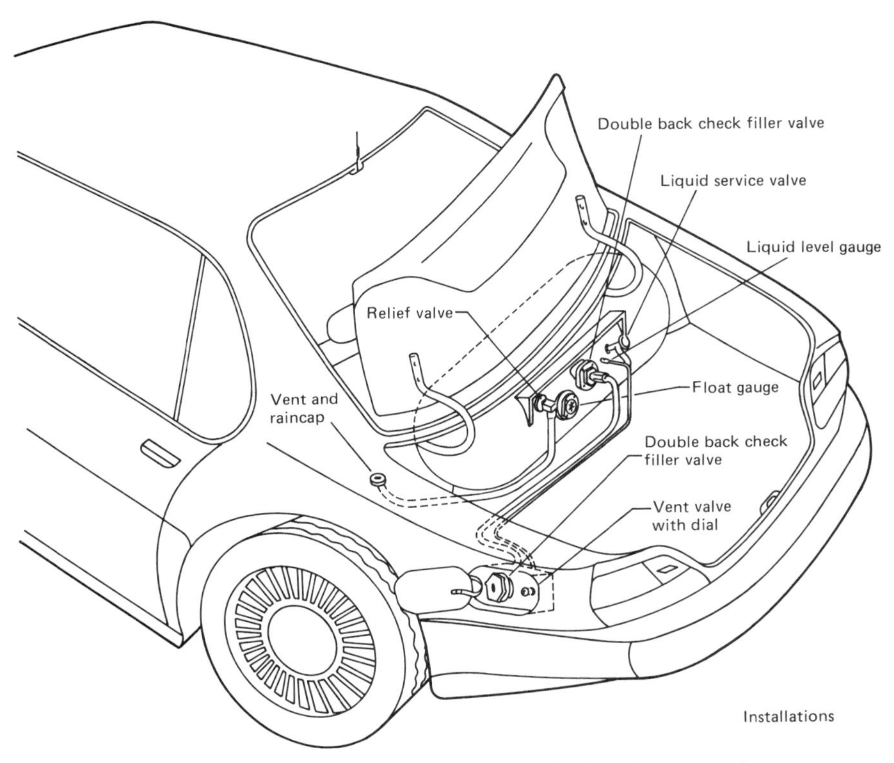

Figure 8.3 *Container shown in Figure 8.2 installed and arranged for remote filling and for external pressure relief valve discharge. (Drawing courtesy of Engineered Controls, Inc.)*

In the event the service manual valve is broken off at the container, the excess-flow check valve is designed to close and keep the gas from escaping. This excess-flow check valve will not close in the event of partial breakage of the service manual valve or leakage in the system downstream of this valve. The term "internal excess-flow check valve" is not defined; however both "internal valve" and "excess-flow check valve" are defined, and the internal excess-flow check valve is defined by those two definitions.

2. Double backflow check valves shall be of the spring loaded type and shall close when flow is either stopped or reversed. This valve shall be installed in the fill opening of the container for either remote or direct filling.

3. Containers shall be fabricated so they can be equipped with a fixed liquid level gauge capable of indicating the maximum permitted filling level in accordance with 4-4.2.2. Fixed liquid level gauges in the container shall be designed so the bleeder valve maximum opening to the atmosphere is not larger than a No. 54 drill size. If the bleeder valve is installed at a location remote from the container, the container fixed liquid level gauge opening and the remote bleeder valve opening shall not be larger than a No. 54 drill size.

4. ASME containers shall be equipped with full internal or flush-type full internal pressure relief valves conforming with applicable requirements of UL 132, *Safety Relief Valves for Anhydrous Ammonia and LP-Gas*, or other equivalent pressure relief valve standards. The start-to-leak setting of such pressure relief valve, with relation to the design pressure of the container, shall be in accordance with Table 2-3.2.3. These relief valves shall be plainly and permanently marked with:

a. the pressure in psi (MPa) at which the valve is set to start to leak;

b. the rated relieving capacity in cu ft per minute of air at 60°F (15.6°C) and 14.7 psia (an absolute pressure of 0.1 MPa); and

c. the manufacturer's name and catalog number. Fusible plugs shall not be used.

5. DOT containers shall be equipped with full internal or flush-type full internal pressure relief valves in accordance with DOT regulations (*see Appendix E*). Fusible plugs shall not be used.

Pressure relief valves on ASME and DOT containers must be the spring-loaded internal type, as the working elements must remain intact within the container so they can still function in the event of an accident. A shear section is generally employed. Fusible plugs are permitted as the safety relief device for small LP-Gas containers that would not be used for engine fuel service. The prohibition of fusible plugs is stated here to prevent misinterpretation of the DOT rules.

6. A float gauge if used shall be designed and approved for use with LP-Gas.

7. A solid steel plug shall be installed in unused openings.

8. Containers fabricated after January 1, 1984, for use as engine fuel containers on vehicles shall be equipped or fitted with an automatic means to prevent filling in excess of the maximum permitted filling limit.

An overfilling prevention device shall be permitted to be installed on the container or exterior of the compartment where remote filling is used, provided that a double backflow-check valve is installed in the container fill valve opening.

Overfilled containers can be hazardous when liquid LP-Gas expands to the point where the container becomes liquid-full, causing the pressure relief valve to open and discharge LP-Gas. To minimize this hazard on general purpose vehicles, containers manufactured after January 1, 1984 must be fitted or equipped with an automatic means to prevent overfilling. Automatic stop-fill valves utilizing a float are listed for this purpose. There are other means available to accomplish this objective, e.g., externally operated solenoid valves.

8-2.4 Carburetion Equipment.

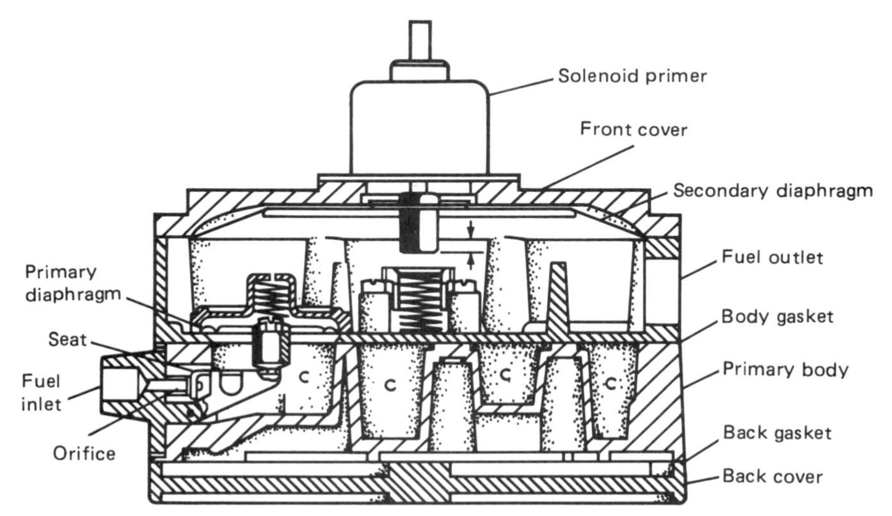

Figure 8.4 Vaporizer-regulator. Engine coolant circulating through chambers (C) vaporizes liquid LP-Gas. (Drawing courtesy of the National Propane Gas Association.)

Carburetion equipment is specially designed and tested for this application. An important feature of LP-Gas carburetion systems is the automatic shutoff valve. This is either an approved electric solenoid valve controlled by vacuum or oil pressure, or a vacuum lockoff that will not permit fuel flow even if the ignition is in the "on" position (it must not provide fuel when the engine is not running). This valve is located as close to the regulator as possible to minimize the volume involved. A primer valve is used for starting.

(a) *General.* Carburetion equipment shall comply with (b) through (e) of this section or shall be designed, fabricated, tested, and marked using criteria that incorporate an investigation to determine that they are safe and suitable for the proposed service, are recommended for that service by the manufacturer, and are acceptable to the authority having jurisdiction. Carburetion equipment subject to working pressures in excess of 125 psi (0.9 MPa) but not to exceed 250 psi (1.7 MPa) shall be suitable for a working pressure of at least 250 psi (1.7 MPa).

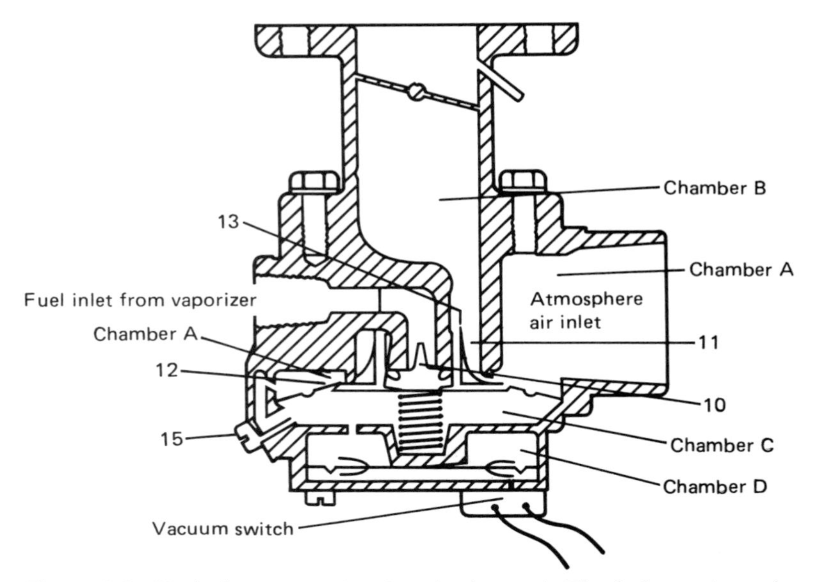

Figure 8.5 Updraft-type gas-air mixer (carburetor). The fuel-metering valve (10) is connected to the air valve (11) and diaphragm assembly (12). Fuel enters the mixer from the vaporizer under slight pressure. It passes through the fuel-metering valve (12), which also acts as the secondary regulator valve, and is reduced to subatmospheric pressure. Opening of the air valve is determined by engine vacuum in chamber B, which is sensed through passage (13). The power-adjusting valve is (15). This unit also has a vacuum switch that serves as a control for an electric automatic shutoff valve [8-2.4(d)]. (Photo courtesy of the National Propane Gas Association.)

(b) *Vaporizer.*

1. Vaporizers shall be fabricated of materials suitable for LP-Gas service and shall be resistant to the action of LP-Gas under service conditions. Such vaporizers shall be designed and approved for engine fuel service and shall comply with the following:

a. The vaporizer proper, any of its parts, or any devices used with it that can be subjected to container pressure shall have a design pressure of at least 250 psi (1.7 MPa), where working pressures do not exceed 250 psi (1.7 MPa), and shall be plainly and permanently at a readily visible point with the marked design pressure of the fuel containing portion in psi (MPa).

2. The vaporizer shall not be equipped with a fusible plug.

3. Each vaporizer shall have a valve or suitable plug located at or near the lowest portion of the section occupied by the water or other heating liquid to allow substantially complete drainage. The engine cooling system drain or water hoses shall be permitted to serve this purpose, if effective.

4. Engine exhaust gases shall be permitted to be used as a direct source of heat to vaporize the fuel if the materials of construction of those parts of the vaporizer in contact with the exhaust gases are resistant to corrosion from these gases and if the vaporizer system is designed to prevent pressure in excess of 200 psi (1.4 MPa).

5. Devices that supply heat directly to the fuel container shall be equipped with an automatic device to cut off the supply of heat before the pressure in the container reaches 200 psi (1.4 MPa).

(c) *Regulator.* The regulator shall be approved and shall be permitted to be either part of the vaporizer unit or a separate unit.

(d) *Automatic Shutoff Valve.* An approved automatic shutoff valve shall be provided in the fuel system as close as practical to the inlet of the gas regulator. The valve shall prevent flow of fuel to the carburetor when the engine is not running even if the ignition switch is in the "on" position. Atmospheric-type regulators (zero governors) shall not be considered as automatic shutoff valves for this purpose.

(e) *Fuel Filter.* Fuel filters, if used, shall be approved and shall be permitted to be either a separate unit or part of a combination unit.

8-2.5 Piping, Hose, and Fittings.

Piping and hose requirements are identical to those specified in Chapter 2 with one exception with respect to hose. The requirements of 8-2.5(d)(1) specifies that hose used for vapor or liquid in excess of 5 psi (34.5 kPa) must meet certain requirements as specified in Chapter 2 and repeated here. This type of hose is tested and listed, and used for remote filling and for providing fuel from the container to the carburetion equipment. Hose used for vapor service at 5 psi (34.5 kPa) or less need be constructed only of materials resistant to the action of LP-Gas. This low-pressure hose is used between the regulator or vaporizer-regulator, sometimes called "converter," and the carburetor, sometimes called "gas-air mixer." It is generally either under a vacuum or slightly positive pressure, and there is no need to have hose in this service of the quality required for higher pressures.

(a) *Pipe.*

1. Pipe shall be wrought iron or steel (black or galvanized), brass, or copper and shall comply with the following:

a. Wrought-iron pipe: ASME B36.10M, *Welded and Seamless Wrought Steel Pipe.*

b. Steel pipe: ASTM A53, *Specification for Pipe, Steel, Black and Hot-Dipped, Zinc-Coated Welded and Seamless.*

c. Steel pipe: ASTM A106, *Specification for Seamless Carbon Steel Pipe for High-Temperature Service.*

d. Brass pipe: ASTM B43, *Specification for Seamless Red Brass Pipe, Standard Sizes.*

e. Copper pipe: ASTM B42, *Specification for Seamless Copper Pipe, Standard Sizes.*

2. For LP-Gas vapor in excess of 125 psi (0.9 MPa) or for LP-Gas liquid, the pipe shall be Schedule 80 or heavier. For LP-Gas vapor at pressures of 125 psi (0.9 MPa) or less, the pipe shall be Schedule 40 or heavier.

(b) *Tubing.*

1. Tubing shall be steel, brass, or copper and shall comply with the following:

a. Steel tubing: ASTM A539, *Specification for Electric-Resistance-Welded Coiled Steel Tubing for Gas Fuel Oil Lines*, with a minimum wall thickness of 0.049 in.

b. Copper tubing: Type K or L, ASTM B88, *Specification for Seamless Copper Water Tube.*

c. Copper tubing: ASTM B280, *Specification for Seamless Copper Tube for Air Conditioning and Refrigeration Field Service.*

d. Brass tubing: ASTM B135, *Specification for Seamless Brass Tube.*

(c) *Pipe and Tube Fittings.*

1. Cast-iron pipe fittings such as ells, tees, crosses, couplings, unions, flanges, or plugs shall not be used. Fittings shall be steel, brass, copper, malleable iron, or ductile iron and shall comply with the following:

a. Pipe joints in wrought iron, steel, brass, or copper pipe shall be permitted to be screwed, welded, or brazed. Tubing joints in steel, brass, or copper tubing shall be flared, brazed, or made up with approved gas tubing fittings.

See commentary following 2-4.4(a)(4).

(i) Fittings used with liquid LP-Gas, or with vapor LP-Gas at operating pressures over 125 psi (0.9 MPa), where working pressures do not exceed 250 psi (1.7 MPa), shall be suitable for a working pressure of at least 250 psi (1.7 MPa).

(ii) Fittings for use with vapor LP-Gas at pressures in excess of 5 psi (34.5 kPa) and not exceeding 125 psi (0.9 MPa) shall be suitable for a working pressure of 125 psi (0.9 MPa).

(iii) Brazing filler material shall have a melting point exceeding 1,000°F (538°C).

(d) *Hose, Hose Connections, and Flexible Connectors.*

1. Hose, hose connections, and flexible connectors (*see definition*) used for conveying LP-Gas liquid or vapor at pressures in excess of 5 psi (34.5 kPa) shall be fabricated of materials resistant to the action of LP-Gas both as liquid and

vapor and shall be of wire braid reinforced construction. The wire braid shall be stainless steel. The hose shall comply with the following:

a. Hose shall be designed for a working pressure of 350 psi (240 MPa) with a safety factor of 5 to 1 and shall be continuously marked "LP-GAS," "PROPANE," "350 PSI WORKING PRESSURE," and the manufacturer's name or trademark. Each installed piece of hose shall contain at least one such marking.

b. Hose assemblies after the application of connections shall have a design capability to withstand a pressure of not less than 700 psi (4.8 MPa). If a test is performed, such assemblies shall not be leak-tested at pressures higher than the working pressure [350 psi (2.4 MPa) minimum] of the hose.

2. Hose used for vapor service at 5 psi (34.5 kPa) or less shall be constructed of material resistant to the action of LP-Gas.

3. Hose in excess of 5 psi (34.5 kPa) service pressure and quick connectors shall be approved for this application by the authority having jurisdiction as specified in Section 1-3.

8-2.6 Installation of Containers and Container Appurtenances.

Particular attention is given to the protection of the container and its fittings from damage resulting from collision, overturn, running over objects (such as curbs), or brushing against objects. Protection from road material thrown up from the ground and exposure to heat from the engine or exhaust is also considered. There are limitations as to where the fuel container may be located. The container or its fittings are not to protrude beyond the widest or highest point of the vehicle: they cannot be ahead of the front axle or beyond the rear bumper, or mounted directly on vehicle roofs.

Specific provisions are given for clearances between the road and the container and its fittings. Figure 8-2.6(e) depicts shaded areas beneath the vehicle where the container must be installed. This procedure is primarily for those installations where a vehicle is being converted to LP-Gas from another fuel or to operate as a dual fuel vehicle. Original manufacturers can deviate from this procedure, as they have the facilities for individual design and road testing to verify their design. Installers may install a substitute container within the space used by the original manufacturer.

(a) Containers shall be located in a place and in a manner to minimize the possibility of damage to the container and its fittings. Containers located in the rear of the vehicles, where protected by substantial bumpers, shall be considered in conformance with this requirement. If the fuel container is installed within 8 in. (20 cm) of the engine or exhaust systems, it shall be shielded against direct heating.

The requirement for shielding of a container within 8 in. of any portion of an engine or exhaust system was added in the 1995 edition to provide specific guidance to the installer and vehicle designer. Previous text required shielding when the container was "near" an engine or exhaust system. The 8 in. distance was taken from NFPA 52, *Compressed Natural Gas Vehicular Fuel Systems*, and is consistent with Canadian requirements.

(b) Container markings shall be readable after a container is permanently installed on a vehicle. A portable lamp and mirror shall be permitted to be used when reading markings.

(c) Container valves, appurtenances, and connections shall be adequately protected to prevent damage due to accidental contacts with stationary objects or from stones, mud, or ice thrown up from the ground and from damage due to overturn or similar vehicular accident. Location on the container where parts of the vehicle furnish the necessary protection or the use of a fitting guard furnished by the manufacturer of the container shall be permitted to meet these requirements.

(d) Containers shall not be mounted directly on roofs or ahead of the front axle or beyond the rear bumper of the vehicles. In order to minimize the possibility of physical damage, no part of a container or its appurtenances shall protrude beyond the sides or top of the vehicle.

This provision was changed in the 1989 edition by the deletion of the words "at the point where it is installed" at the end of the paragraph. The revision permits a side-mounted container as long as it does not protrude beyond the widest point of the vehicle.

(e) Containers shall be installed with as much road clearance as practical. This clearance shall be measured to the bottom of the container or the lowest fitting, support, or attachment on the container or its housing, if any, whichever is lowest, as follows [*see Figure 8-2.6(e)*]:

1. Containers installed between axles shall comply with 8-2.6(e)3 or shall be not lower than the lowest point forward of the container on:

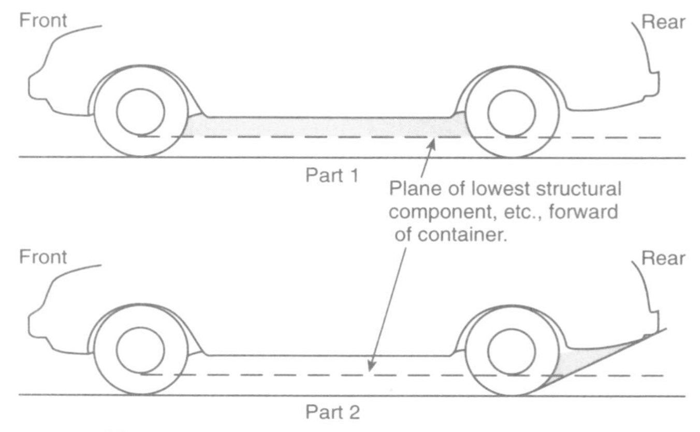

Figure 8-2.6(e) *Container installation clearances.*

a. The lowest structural component of the body;

b. The lowest structural component of the frame or subframe if any;

c. The lowest point on the engine;

d. The lowest point of the transmission (including the clutch housing or torque converter housing as applicable) [*see Part 1, Figure 8-2.6(e)*].

2. Containers installed behind the rear axle and extending below the frame shall comply with 8-2.6(e)3 or shall be not lower than the lowest of the following points and surfaces.

a. They shall not be lower than the lowest point of a structural component of the body, engine, transmission (including clutch housing or torque converter housing, as applicable), forward of the container. Also they shall not be lower than lines extending rearward from each wheel at the point where the wheels contact the ground directly below the center of the axle to the lowest and most rearward structural interference (e.g., bumper, frame, etc.). [*See Figure 8-2.6(e), Part 2.*]

b. Where there are two or more rear axles, the projections shall be made from the rearmost axle.

3. Where an LP-Gas container is substituted for the fuel container installed by the original manufacturer of the vehicle (whether or not that fuel container was intended for LP-Gas), the LP-Gas container either shall fit within the space in which the original fuel container was installed or shall comply with 8-2.6(e)1 or 2.

(f) Fuel containers shall be securely mounted to prevent jarring loose and slipping or rotating, and the fastenings shall be designed and constructed to withstand without permanent visible deformation static loading in any direction equal to four times the weight of the container filled with fuel.

One must consider not only the strength of fasteners for mounting the container, but also the strength of that portion of the vehicle to which the container is mounted. For example, reinforcement must be used when mounting containers to thin metal decking. A preferable method is to mount the container on the chassis frame, if one exists.

(g) Welding for the repair or alterations of containers shall comply with 8-2.2(a)7.

(h) Main shutoff valves on a container for liquid and vapor shall be readily accessible without the use of tools, or other means shall be provided to shut off the container valves.

The phrase "readily accessible without the use of tools" is used only in Chapter 8 [the only other use of the phrase "use of tools" is in 8-2.7(d)]. Its use here reflects the Committee's concern that vehicle manufacturers would locate the main shutoff valve behind a plate secured by screws. It is the Committee's intent that the valve be accessible in case of an emergency without compromising the need for the valve to be protected from accidental damage.

Figure 8.6 Pressure relief valve discharge line extended above the roof of the cab. (Photo courtesy of Eastern Propane.)

Formal Interpretation 89-1
Reference: 8-2.6(h)

Question: Is it the intent of the above referenced code provision to prohibit installation of an underbody van propane tank(s) which require the operator to reach underneath the vehicle to open or close the shut off valve, assuming the shut off valve does not require tools to gain access to it or to operate it?
Answer: No.

Issue Edition: 1989
Reference: 3-6.2.6(h)
Issue Date: May 4, 1990
Effective Date: May 24, 1990 ■

Figure 8.7 LP-Gas container inside an enclosed pickup truck bed. The pressure relief valve discharge line runs from the container to the roof of the topper. (Photo courtesy of Eastern Propane.)

(i) The pressure relief valve discharge system(s) shall be installed so that:

1. The relief valve discharge from fuel containers on vehicles other than industrial (and forklift) trucks (*see 8-3.3*) shall be directed upward or downward within 45 degrees of vertical; shall not directly impinge on the vehicle fuel container(s), the exhaust system, or any other part of the vehicle; and shall not be directed into the interior of the vehicle.

2. Where the relief valve discharge must be piped away, the pipeaway system shall consist of a breakaway adapter recommended by the relief valve manufacturer, a length of nonmetallic hose, and a protective cover to minimize the possibility of the entrance of water or dirt into either the relief valve or its discharge system. No portion of the system shall have an internal diameter less than the internal diameter of the recommended breakaway adapter.

a. The breakaway adapter shall be threaded for direct connection to the relief valve and shall not interfere with the operation of the relief valve. It shall break away without impairing the function of the relief valve and shall have a melting point of not less than 1,500°F (816°C).

Exception: The breakaway adapter shall be permitted to be an integral part of the pressure relief valve.

b. The nonmetallic hose shall be as short as practical and shall be able to withstand the downstream pressure from the relief valve in the full open position. The hose shall be fabricated of materials resistant to the action of LP-Gas.

c. Where hose is used to pipe away the relief valve discharge on containers installed on the outside of the vehicle, the breakaway adapter and any attached fitting shall deflect the relief valve discharge upward or downward within 45 degrees of vertical and shall meet the other requirements of 8-2.6(i)1 without the hose attached. If an additional fitting is necessary to meet this requirement, it shall have a melting point not less than 1,500°F (816°C) and shall meet the requirements of 8-2.6(i)2.

3. The pipeaway system connections shall be mechanically secured and shall not depend on adhesives or sealing compounds. The system shall not be routed between a bumper system and the vehicle body.

4. Where a pipeaway system is not required, the pressure relief valve shall have a protective cover in accordance with 8-2.6(i)2.

Pressure relief valve discharge location must be considered in the installation of engine fuel containers, as valves should not be directed in such a way as to create an unsafe condition. Most vehicles must have the relief directed upward or downward within 45 degrees of vertical, and located in such a way that it is not directed at any part of the vehicle. Prior to the 1992 edition, these requirements were more restrictive, permitting only vertical discharge. They were modified based on the fact that downward discharge could be safer than vertical discharge, as the potential of water freezing in the valve is eliminated, and a downward discharge might be less likely to impinge on people or vehicles. The Committee was told that many school buses in Canada have used downward discharge for several years without problems.

These restrictions do not apply to industrial (forklift) trucks, which are covered by Section 8-3.

8-2.7 Containers Mounted in the Interior of Vehicles.

(a) Containers mounted in the interior of vehicles shall be installed so that any LP-Gas released from container appurtenances due to operation, leakage, or connection of the appurtenances will not be in an area communicating directly with the driver or passenger compartment or with any space containing radio transmitters or other spark-producing equipment. This shall be permitted to be accomplished by the following means:

1.* Locating the container, including its appurtenances, in an enclosure that is securely mounted to the vehicle, is gastight with respect to driver or passenger compartments and to any space containing radio transmitters or other spark-producing equipment, and is vented outside the vehicle.

A-8-2.7(a)1 The luggage compartment (trunk) of a vehicle can constitute such an enclosure provided it meets all these requirements.

2. Enclosing the container appurtenances and their connections in a structure that is securely mounted on the container, is gastight with respect to the driver or passenger compartments or with any space carrying radio transmitters or other spark-producing equipment, and is vented outside the vehicle.

(b) Fuel containers shall be installed and fitted so that no gas from fueling and gauging operations can be released inside of the passenger or luggage compartments, by permanently installing the remote filling connections [double backflow check valve, see 8-2.3(a)(2)], and fixed liquid level gauging device to the outside of the vehicle.

(c) Container pressure relief valve installation shall comply with 8-2.6(i).

(d) Enclosures, structures, seals, and conduits used to vent enclosures shall be fabricated of durable materials and shall be designed to resist damage, blockage, or dislodgement through movement of articles carried in the vehicle or by the closing of luggage compartment enclosures or vehicle doors and shall require the use of tools for removal.

An alternative to mounting the fuel containers on the outside is installation in the interior of the vehicle (trunk compartment, etc.). Certain conditions are specified to ensure that no LP-Gas is released into the passenger compartment.

Figures 8.8, 8.9, and 8.10 *Containers mounted in the interior of vehicles.*

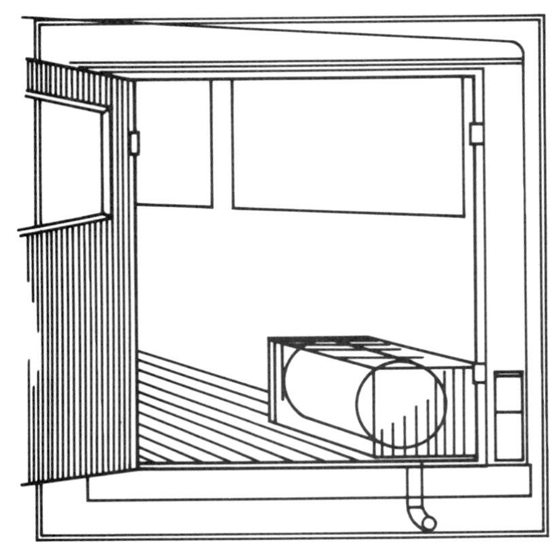

Figure 8.8 *Gastight box built around the container to provide a vapor barrier in van-type vehicles. (Drawing courtesy of the National Propane Gas Association.)*

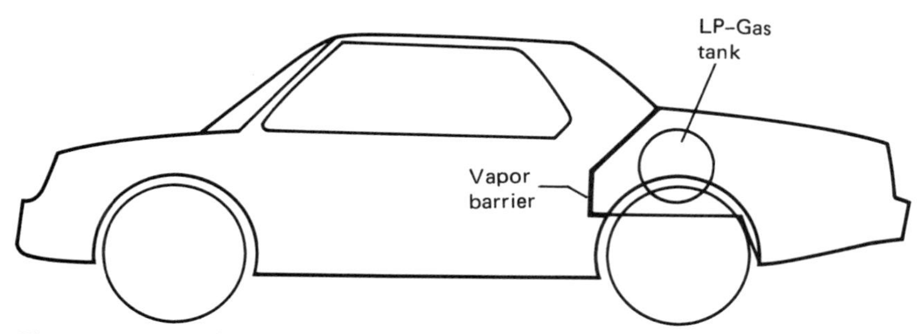

Figure 8.9 *Trunk-mounted fuel container. A vapor barrier must be created between the passenger and luggage compartments. The barrier must be 100 percent gastight so no propane vapor can enter the car. (Drawing courtesy of the National Propane Gas Association.)*

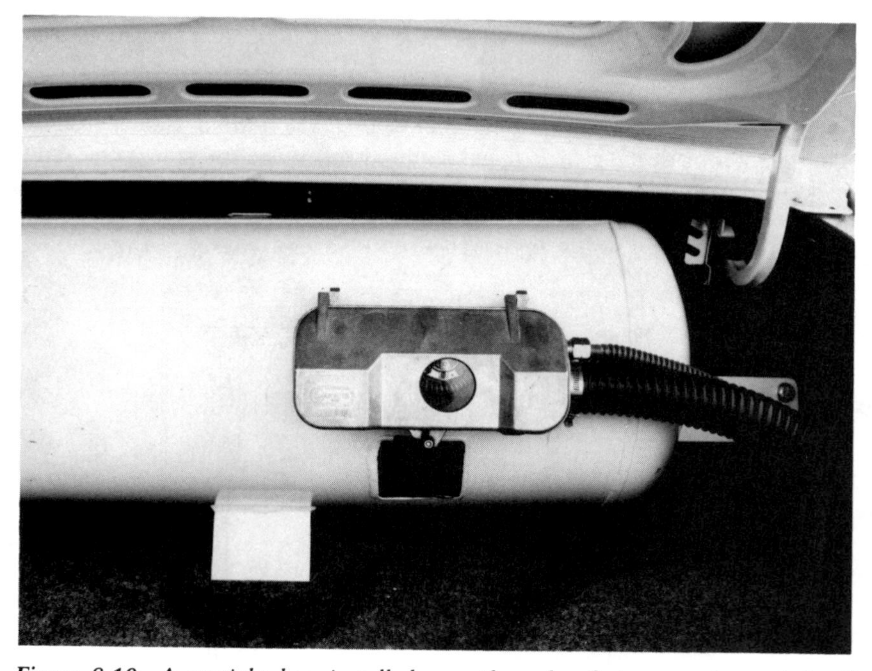

Figure 8.10 *A gastight box installed over the valve fittings on the tank itself can also provide a vapor barrier. (Photo courtesy of Manchester Tank Company.)*

Four options are given for installation of containers mounted in the interior as far as sealing the container appurtenances and their connections from the passenger space is concerned. These are:

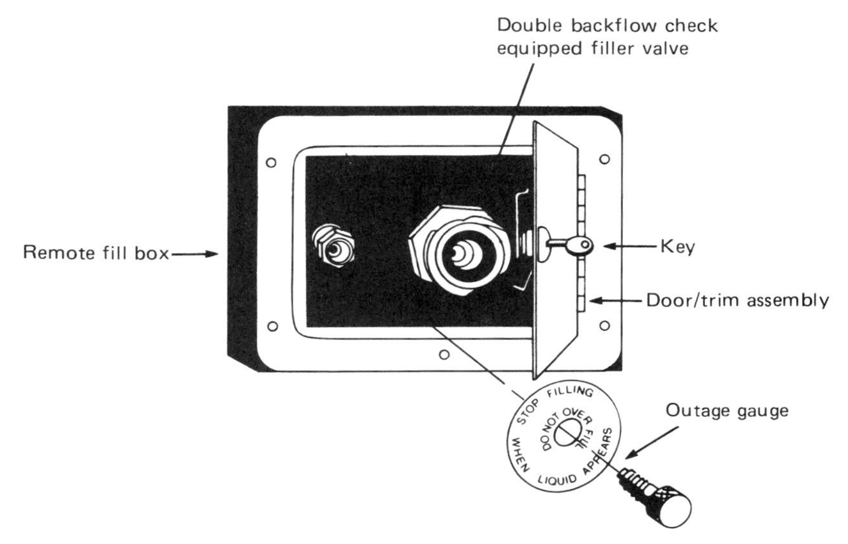

Figure 8.11 Remote filling connection with fixed liquid level gauge. (Photo courtesy of the National Propane Gas Association.)

1. Locating the container and its appurtenances in the luggage compartment (trunk) and sealing the trunk from the passenger carrying space
2. Putting the entire container and its appurtenances in a sealed compartment
3. Using a gastight box totally enclosing the container appurtenances
4. Mounting the container so appurtenances and their connections are outside of the passenger space. Spark-producing equipment (radio transmitters, truck lid motors) are to be isolated from container appurtenances and their connections.

Remote filling is to be accomplished by permanently installing a double backflow check valve on the exterior of the vehicle in addition to one installed in the container. Remote fixed liquid level gauges are also to be permanently mounted and orificed to a No. 54 drill size in addition to the orificed connection on the container. This clearly indicates that coiled filling hoses in trunks are not permitted.

8-2.8 Pipe and Hose Installation.

(a) The piping system shall be designed, installed, supported, and secured in such a manner to minimize the possibility of damage due to expansion, contraction, vibration, strains or wear and to preclude any loosening while in transit.

(b) Piping (including hose) shall be installed in a protected location. If outside, piping shall be under the vehicle and below any insulation or false bottom. Fastening or other protection shall be installed to prevent damage due to vibration or abrasion. At each point where piping passes through sheet metal or a structural member, a rubber grommet or equivalent protection shall be installed to prevent chafing.

(c) Fuel line piping that must pass through the floor of a vehicle shall be installed to enter the vehicle through the floor directly beneath, or adjacent to, the container. If a branch line is required, the tee connection shall be in the main fuel line under the floor and outside the vehicle.

(d) Where liquid service lines of two or more individual containers are connected together, a spring-loaded backflow check valve or equivalent shall be installed in each of the liquid lines prior to the point where the liquid lines tee together to prevent the transfer of LP-Gas from one container to another.

This requirement is intended to prevent transfer of LP-Gas between containers when more than one is used, and to prevent possible overfilling by transfer.

(e) Exposed parts of the piping system shall be of corrosion-resistant material or shall be adequately protected against exterior corrosion.

(f) Piping systems, including hose, shall be tested and proven free of leaks at not less than normal operating pressure.

(g) There shall be no fuel connection between a tractor and trailer or other vehicle units.

(h) A hydrostatic relief valve or device providing pressure-relieving protection shall be installed in each section of piping (including hose) in which liquid LP-Gas can be isolated between shutoff valves so as to relieve to a safe atmosphere the pressure that could develop from the trapped liquid. This hydrostatic relief valve shall have a pressure setting not less than 400 psi (2.8 MPa) or more than 500 psi (3.5 MPa).

8-2.9 Equipment Installation.

(a) Installation shall be made in accordance with the manufacturer's recommendations and, in the case of listed or approved equipment, it shall be installed in accordance with the listing or approval.

(b) Equipment installed on vehicles shall be considered a part of the LP-Gas system on the vehicle and shall be protected against vehicular damage in accordance with 8-2.6(a).

(c) The gas regulator and the approved automatic shutoff valve shall be installed as follows:

1. Approved automatic pressure reducing equipment, properly secured, shall be installed between the fuel supply container and the carburetor to regulate the pressure of the fuel delivered to the carburetor.
2. An approved automatic shutoff valve shall be provided in the fuel system in compliance with 8-2.4(d).

(d) Vaporizers shall be securely fastened in position.

8-2.10 Marking. Each over-the-road general purpose vehicle powered by LP-Gas shall be identified with a weather-resistant diamond-shaped label located on an exterior vertical or near vertical surface on the lower right rear of the vehicle (on the trunk lid of a vehicle so equipped, but not on the bumper of any vehicle) inboard from any other markings. The label shall be approximately 4¾ in. (120 mm) long by 3¼ in. (83 mm) high. The marking shall consist of a border and the word "PROPANE" [1 in. (25 mm) minimum height centered in the diamond] in silver or white reflective luminous material on a black background. (*See Figure 8-2.10.*)

Figure 8-2.10 *Example of vehicle identification marking.*

8-3 Industrial (and Forklift) Trucks Powered by LP-Gas.

8-3.1 This subsection applies to LP-Gas installation on industrial trucks (including forklift trucks), both to propel them and to provide the energy for their materials handling attachments. LP-Gas fueled industrial trucks shall comply with NFPA 505, *Fire Safety Standard for Powered Industrial Trucks Including Type Designations, Areas of Use, Maintenance and Operation.*

NFPA 505 is a companion standard to NFPA 58 with respect to LP-Gas fueled industrial trucks. While NFPA 58 sets forth installation provisions and certain conditions for their use, NFPA 505 sets forth type designations, where they are to be used, dual fuel trucks, truck maintenance and, to a certain extent, fuel handling and storage.

The basic provisions for engine fuel systems on general purpose vehicles are also applicable to those for industrial truck engines except for some differences peculiar to this type of service.

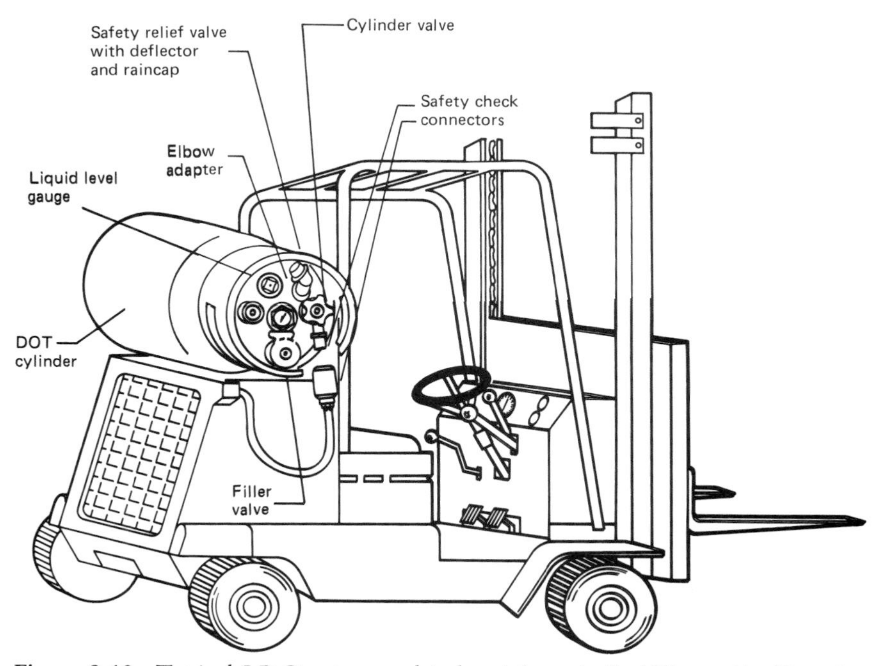

Figure 8.12 *Typical LP-Gas powered industrial truck (forklift truck). (Drawing courtesy of Engineered Controls, Inc.)*

8-3.2 ASME and DOT fuel containers shall comply with 8-2.2 and 8-2.3(a)1 through 7.

Fuel containers for industrial trucks need not comply with 8-2.3(a)(8) (that is, be equipped with an automatic means for overfill prevention) as required for general purpose vehicles. These containers are refilled under more closely controlled conditions, and experience has shown that overfilling has not been as much of a problem.

(a) Portable containers shall be permitted to be designed, constructed, and fitted for filling in either the vertical or horizontal position, or, if of the portable universal type [*see 2-3.4.2(c)2*], in either position. The container shall be in the appropriate position while being filled or, if of the portable universal type, shall be permitted to be filled in either position, provided:

1. The fixed level gauge indicates correctly the maximum permitted filling level in either position.
2. The pressure relief valves are located in, or connected to, the vapor space in either position.

In addition to portable industrial truck containers designed for refilling and used specifically in a horizontal or a vertical position, a universal type is available in which the same container can be filled and used in either position. With this type of container, the fixed liquid level gauge must indicate the maximum filling level in either the correct horizontal or vertical position (for horizontal position, the relief valve must be at 12 o'clock), and the pressure relief valve must communicate with the vapor space in either position.

8-3.3 The container relief valve shall be vented upward within 45 degrees of vertical and otherwise shall comply with 8-2.6(i).

This provision is similar to the requirements for other vehicles. Generally, industrial trucks do not require pressure relief valve pipeaways, and adapters can be used for deflecting the discharge in view of the configurations of these vehicles.

8-3.4 Gas regulating and vaporizing equipment shall comply with 8-2.4(b)1 through 5 and 8-2.4(c), (d), and (e).

8-3.5 Piping and hose shall comply with 8-2.5(a) through (d).

Exception: Hose 60 in. (1.5 m) in length or less shall not be required to be of stainless steel wire braid construction.

8-3.6 The operation of industrial trucks (including forklift trucks) powered by LP-Gas engine fuel systems shall comply with NFPA 505, *Fire Safety Standard for Powered Industrial Trucks Including Type Designations, Areas of Use, Maintenance and Operation*, and with the following:

(a) Refueling of such trucks shall be accomplished as follows:

1. Trucks with permanently mounted containers shall be refueled outdoors.

2. Exchange of removable fuel containers preferably should be done outdoors, but shall be permitted to be done indoors. If done indoors, means shall be provided in the fuel piping system to minimize the release of fuel when containers are exchanged, using one of the following methods:

a. Use of an approved quick-closing coupling (a type that closes in both directions when uncoupled) in the fuel line; or

b. Closing the shutoff valve at the fuel container and allowing the engine to run until the fuel in the line is exhausted.

(b) LP-Gas fueled industrial trucks shall be permitted to be used in buildings or structures as follows:

1. The number of fuel containers on such a truck shall not exceed two.

2. With the approval of the authority having jurisdiction, industrial trucks shall be permitted to be used in buildings frequented by the public, including those times when such buildings are occupied by the public. The total water capacity of the fuel containers on an individual truck shall not exceed 105 lb (48 kg) [nominal 45 lb (20 kg) LP-Gas capacity].

3. Trucks shall not be parked and left unattended in areas occupied by or frequented by the public except with the approval of the authority having jurisdiction. If left unattended with approval, the fuel system shall be checked to make certain there are no leaks and that the container shutoff valve is closed.

4. In no case shall industrial trucks be parked and left unattended in areas of excessive heat or near sources of ignition.

8-3.7 All containers used in industrial truck service (including forklift truck cylinders) shall have the container pressure relief valve replaced by a new or unused valve within 12 years of the date of manufacture of the container and every 10 years thereafter.

8-4 General Provisions for Vehicles Having Engines Mounted on Them (Including Floor Maintenance Machines).

This section includes coverage of floor maintenance machines. Use of these propane-powered floor buffers has grown since their introduction into the standard in 1989. This widespread use has resulted in specific coverage in NFPA 58 for the benefit of users of this equipment and enforcers of the standard.

8-4.1 This section includes provisions for the installation of equipment on vehicles that supplies LP-Gas as a fuel for engines mounted on these vehicles. The term "vehicles" includes floor maintenance and any other readily portable mobile unit, whether the engine is used to propel the vehicle or is mounted on it for other purposes.

8-4.2 Gas vaporizing, regulating, and carburetion equipment to provide LP-Gas as a fuel for engines shall be installed in accordance with 8-2.8 and 8-2.9.

(a) In the case of industrial trucks (including forklift trucks) and other engines on vehicles operating in buildings other than those used exclusively to house engines, an approved automatic shutoff valve shall be provided in the fuel system in compliance with 8-2.4(d).

(b) The source of air for combustion shall be completely isolated from the driver and passenger compartment, ventilating system, or air conditioning system on the vehicle.

Although approved automatic shutoff valves are required, this provision recognizes the use of atmospheric-type regulators (zero governors) for this purpose, with portable engines of 12 horsepower or less with magneto ignition used exclusively outdoors. An atmospheric-type regulator, as its name implies, depends on the atmospheric pressure for its control. It is not an approved vacuum lockoff, which is considered an approved automatic shutoff valve. It operates on the principle that when there is no vacuum in the carburetor venturi, the regulator shuts off the flow of fuel. Atmospheric-type regulators have been used for many years on outdoor engine applications.

8-4.3 **Piping and Hose Shall Comply with 8-3.5.**

8-4.4 Non-self-propelled floor maintenance machinery (floor polishers, scrubbers, buffers) and other similar portable equipment shall be listed and shall comply with the following.

(a) The provisions of 8-3.2 through 8-3.5 and 8-3.6(a) and (b) shall apply.

(b) The storage of LP-Gas containers mounted or used on such machinery or equipment shall comply with Chapter 5.

There is great concern that a propane-powered floor buffer, with its 20-lb (9-kg) LP-Gas container, will be stored inside buildings in concealed areas, such as closets. The Technical Committee, consistent with its long prohibition (with certain exceptions) of LP-Gas containers larger than 1 lb (0.5 kg) in buildings, wished to ensure that containers are stored properly in accordance with Chapter 5. The Technical Committee was especially concerned that the introduction of engine fuel containers would lead to improper storage and required specific labeling.

Propane-powered floor buffers and forklift trucks are both categorized as "industrial trucks." It is highly improbable, however, that a forklift truck will be left in a closet. If a floor buffer and its 20-lb (9-kg) LP-Gas container were stored in a closet and a fire occurred, the propane in the cylinder could be a threat to firefighters due to the potential for a BLEVE and the fuel that can accelerate a fire.

A label shall be affixed to the machinery or equipment, with the label facing the operator, denoting that the container or portion of the machinery or equipment containing the LP-Gas container must be stored in accordance with Chapter 5.

8-4.5 With approval of the authority having jurisdiction, floor maintenance machines shall be permitted to be used in buildings frequented by the public, including the times when such buildings are occupied by the public.

Figure 8.13 *Floor maintenance machine. (Photo courtesy of Swiss Clean.)*

8-5 Engine Installation Other than on Vehicle.

8-5.1 Stationary engines and gas turbines installed in buildings, including portable engines used in lieu of, or to supplement, stationary engines, shall comply with NFPA 37, *Standard for the Installation and Use of Stationary Combustion Engines and Gas Turbines*, and the applicable provisions of Chapters 1 and 2 and Section 3-2 of this standard.

8-5.2 Portable engines, except as provided in 8-4.1, shall be permitted to be used in buildings only for emergencies, and the following shall apply:

(a) The capacity of the LP-Gas containers used with such engines and the equipment used to provide fuel to them shall comply with the applicable provisions of Section 3-4.

(b) An approved automatic shutoff valve shall be provided in the fuel system in compliance with 8-2.4(d). Atmospheric-type regulators (zero governors) used for portable engines of 12 horsepower or less with magneto ignition and used exclusively outdoors shall be considered as in compliance with 8-2.4(d).

(c) Provision shall be made to supply sufficient air for combustion and cooling. Exhaust gases shall be discharged to a point outside the building or to an area in which they will not constitute a hazard.

8-5.3 Piping and hose shall comply with 8-2.5(a) through (d).

8-5.4 Gas regulating, vaporizing, and carburetion equipment shall comply with 8-2.4(b)1 through 5, 8-2.4(c), and 8-2.4(e).

8-5.5 Installation of piping, carburetion, vaporizing, and regulating equipment for the engine fuel system shall comply with 8-2.8 and 8-2.9.

8-5.6 Engines installed or operated exclusively outdoors shall comply with 8-5.3, 8-5.4, and 8-5.5.

Atmospheric-type regulators (zero governor) shall be considered as automatic shutoff valves only in the case of completely outdoor operations, such as farm tractors, construction equipment, or similar outdoor engine applications.

8-5.7 Engines used to drive portable pumps and compressors or pumps shall be equipped in accordance with 2-5.1.4.

8-6 Garaging of Vehicles.

Vehicles with LP-Gas engine fuel systems mounted on them and general purpose vehicles propelled by LP-Gas engines shall be permitted to be stored or serviced inside garages, provided:

(a) The fuel system is leak-free, and the container(s) is not filled beyond the limits specified in Chapter 4.

(b) The container shutoff valve is closed when vehicles or engines are under repair except when engine is operated.

(c) The vehicle is not parked near sources of heat, open flames, or similar sources of ignition, or near inadequately ventilated pits.

For additional information involving the provisions of Chapter 8, refer to the following publications of the National Propane Gas Association (formerly National LP-Gas Association), 1600 Eisenhower Lane, Lisle, IL 60532.

NPGA 602, *Safe Use of LP-Gas in Industrial Trucks* (Spanish version available).

NPGA 803, *Safety Considerations in Converting Passenger Carrying Vehicles from Gasoline to LP-Gas.*

NPGA 804, *Safety Bulletin for Drivers of LP-Gas Powered Vehicles.*

NPGA 4006, *Training Guidebook: LP-Gas Carburetion.*

NPGA 5825, *Preventive Maintenance and Filling Procedures.*

NPGA 5909, *Safety Considerations for Motor Fuel Systems.*

NPGA 5915, *Safety Factors in Converting School Buses to LP-Gas.*

9

Refrigerated Storage

9-1 Refrigerated Containers.

The chapter on refrigerated storage of LP-Gases was added in the 1989 edition. It was revised and relocated to Chapter 9 in the 1992 edition to enhance safety and make it consistent with API Standard 2510, *Design and Construction of Liquefied Petroleum Gas Installations (LPG)*.

Refrigerated LP-Gas storage systems are those in which the liquefied gas is stored at or near its boiling point [–44°F (–42°C) for propane] at atmospheric pressure. The temperature of the liquid is maintained by a refrigeration process. Storage containers are characterized by their large size [over 1 million gal (3800 m^3) is not unusual] and low design pressure [$^1/_2$ psi (3 kPa) is common]. They are thermally insulated as a matter of operational necessity and can withstand only very low internal pressures.

Such containers are not only very different from nonrefrigerated LP-Gas containers, they are few in number and found usually in specialized applications in LP-Gas production facilities, marine and pipeline terminals, and natural gas peak-shaving plants.

9-1.1 Refrigerated LP-Gas containers shall be designed and constructed in accordance with the applicable provisions of the following codes as appropriate for the design loadings including design pressure and temperature.

9-1.1.1 For pressures of 15 psi (103 kPa) or more, the ASME Code, Section VIII shall apply, except that construction using joint efficiencies in Table UW 12, Column C, Division 1, shall not be permitted. Material shall be selected from those recognized by ASME that meet the requirements of API 620, *Design and Construction of Large, Welded, Low-Pressure Storage Tanks,* including Appendix R.

Section VIII of the ASME Code would be used to design and fabricate shop-built vessels or semirefrigerated containers operating above 15 psi (103 kPa). While few of these vessels are anticipated for refrigerated LP-Gas service, their coverage is included to make the standard complete.

The exception of Table UL-12 of the ASME Code strengthens inspection requirements.

The material selection reference to Appendix R of API 620 reflects the somewhat unique problems associated with the design and construction of low-temperature vessels.

9-1.1.2 For pressures below 15 psi (103 kPa), API 620, *Design and Construction of Large, Welded, Low-Pressure Storage Tanks*, including Appendix R, shall apply.

API 620 is recognized worldwide as a standard for the construction of large welded storage tanks. Appendix R of that standard specifically addresses the design and material requirements of such tanks operating at refrigerated LP-Gas temperatures.

9-1.1.3 Where austenitic steels or nonferrous materials are used, API 620, Appendix Q, shall be used in the selection of materials.

Appendix Q of API 620 contains guidance for the use of austenitic steels and nonferrous materials that are not included in Appendix R.

9-1.2 The operator shall specify the maximum allowable working pressure, which includes a suitable margin above the operating pressure and the maximum allowable vacuum.

Because of the large size of LP-Gas refrigerated storage containers, small increments of design pressure and vacuum can carry large costs. Container pressure is significantly affected by small variations in filling rates, product composition, barometric pressure changes, and process upsets. The design pressure of these containers is usually quite low [$^1/_2$ to 2 psi (3 to 14 kPa)], and their ability to withstand vacuum is minimal [usually $^1/_4$ psi (2 kPa)]. The operator must take all these factors into account when specifying the container design pressure, and ensure that the plant is designed and operated in such a way that this design pressure is not exceeded.

9-1.2.1 For ASME vessels, the positive margin shall be at least 5 percent of the absolute vapor pressure of the LP-Gas at the design storage temperature. The margin (both positive and vacuum) for low-pressure API vessels shall include the control range of the boil-off handling system, the effects of flash or vapor collapse during filling operations, the flash that can result from withdrawal pump recirculation, and the normal range of barometric pressure changes.

9-1.2.2 The minimum design temperature for those parts of a refrigerated LP-Gas container that are in contact with the refrigerated liquid or vapor shall be the boiling point at atmospheric pressure of the product to be stored.

9-1.3 The design wind loading on refrigerated LP-Gas containers shall be based on the projected area at various height zones aboveground in accordance with ASCE 7, *Design Loads for Buildings and Other Structures*. Wind speeds shall be based on a Mean Occurrence Interval of 100 years.

Wind loading can be a significant factor in designing refrigerated storage vessels because of their larger size.

9-1.4 Field-erected containers for refrigerated storage shall be designed as an integral part of the storage system including tank insulation, compressors, condensers, controls, and piping. Allowance shall be made for the service temperature limits of the particular process and the products to be stored when determining material specifications and the design pressure. Welded construction shall be used.

This was added in the 1995 edition and as a Tentative Interim Amendment to the 1992 edition. It had been inadvertently deleted from the 1992 edition.

9-1.5 The design seismic loading on refrigerated LP-Gas containers shall be based on forces recommended in the ICBO *Uniform Building Code*. In those areas identified as zones 3 and 4 on the Seismic Risk Map of the United States, Figures 1, 2, and 3 of Chapter 23 of the *UBC*, a seismic analysis of the proposed installation shall be made that meets the approval of the authority having jurisdiction.

Because of the larger size of refrigerated storage vessels, a detailed seismic analysis may be required in the more seismically active zones. A competent geotechnical engineer and structural engineer should perform this analysis.

9-1.6 All piping that is part of a refrigerated LP-Gas container shall be in accordance with ASME B31.3, *Chemical Plant and Refinery Piping*. This container piping shall include all piping internal to the container, within the insulation spaces, and external piping attached or connected to the container up to the first circumferential external joint of the piping. Inert gas purge systems wholly within the insulation spaces are exempt from this provision.

This requirement is an exception to API 620 because the requirements of ANSI B31.3 are somewhat more stringent than the piping requirements of API 620.

9-1.7 Refrigerated LP-Gas containers shall be installed on foundations designed by an engineer experienced in foundations and soils and shall be constructed in accordance with recognized structural engineering practices. Prior to the start of design and construction of the foundation, a subsurface investigation shall be conducted by a soils engineer to determine the stratigraphy and physical properties of the soils underlying the site.

NOTE: See ASCE 56, *Sub-Surface Investigation for Design and Construction of Foundation for Buildings, and Appendix C, API 620, Design and Construction of Large, Welded, Low-Pressure Storage Tanks*, for further information.

These requirements emphasize the importance of the tank foundation in maintaining the integrity of the tank. If the foundation should shift, large stresses would be placed on the tank and failure could occur. The publications referenced in the Note can be used for guidance.

9-1.7.1 The refrigerated LP-Gas container foundation shall be periodically monitored for settlement during the life of the facility including construction, hydrostatic testing, commissioning, and operation. Any settlement in excess of that anticipated in the design shall be investigated and corrective action taken if appropriate.

9-1.7.2 The bottom of a refrigerated LP-Gas container, either the bottom of an outer tank or the bottom of the undertank insulation, shall be above the ground water table or otherwise protected from contact with ground water at all times, and the material in contact with the bottom of the container shall be selected to minimize corrosion.

9-1.7.3 If the bottom of the refrigerated LP-Gas container is in contact with the soil, a heating system shall be provided to prevent the 32°F (0°C) isotherm from extending into the soil. The heating system shall be designed to permit both functional and performance monitoring. As a minimum, the undertank temperature shall be observed and logged on a weekly basis. Where there is a discontinuity in the foundation, such as bottom piping, careful attention and separate treatment shall be given to the heating system in that zone. Heating systems shall be installed so that any heating elements or temperature sensors used for control can be replaced while the tank is in service. Provisions shall be incorporated to protect against the detrimental effects of moisture accumulation in the conduit, which could result in galvanic corrosion or other forms of deterioration within the conduit or heating element.

When in contact with the soil, the bottom of a refrigerated propane container, with its contents at about –44°F (–42°C), will transfer heat from the soil. If a heating system were not provided, the soil would freeze. This must be prevented, as freezing results in "heaving" of the soil, which will move the foundation up and probably distort it. Heating systems are designed to prevent freezing and must be maintained. It is not uncommon to install thermocouples in or beneath the foundation to monitor the effectiveness of the undertank heating system. Undertank heating systems are usually electrically powered; however, in areas where below-freezing temperatures are not common, air circulation can be used.

9-1.7.4 If the foundation of a refrigerated LP-Gas container is installed to provide adequate air circulation in lieu of a heating system, the bottom of the container shall be of materials that are suitable for the temperatures to which they will be exposed.

9-1.8 Refrigerated LP-Gas Container Instruments and Controls.

Requirements for controls are included that are not required for the smaller pressurized containers. These requirements are intended to prevent overfilling.

9-1.8.1 Each refrigerated LP-Gas container shall be equipped with at least two independent liquid level gauging devices. These devices shall be replaceable without taking the container out of service.

9-1.8.2 The refrigerated LP-Gas container shall be provided with a high-liquid-level alarm. The alarm shall be set so that the operator will have sufficient time to stop the flow without exceeding the maximum permissible filling height and shall be located so that it is audible to personnel controlling the filling. A high-liquid-level flow cutoff device shall not be considered as a substitute for the alarm.

9-1.8.3 The refrigerated LP-Gas container shall be equipped with a high-liquid-level flow cutoff device, which shall be independent from all gauges.

Exception: Refrigerated LP-Gas containers of 70,000 gal (265 m^3) or less, if attended during the filling operation, shall be permitted to be equipped with liquid trycocks in lieu of the high-liquid-level alarm, and manual flow cutoff shall be permitted.

9-1.8.4 Each refrigerated LP-Gas container shall be provided with temperature indicating devices to assist in controlling cooldown rates when placing the container in service.

9-1.9 Inspection of Refrigerated LP-Gas Containers.

Both the ASME and API codes contain requirements for test and inspection. This provision reinforces the need to comply with these requirements, details the areas that must be inspected, and reiterates that they are the responsibility of the operator. It also recognizes that the operator may not have sufficient employees to conduct all required inspections and permits designation of third parties, who may be better equipped to conduct the inspections.

9-1.9.1 During construction and prior to the initial operation or commissioning, each refrigerated LP-Gas container shall be inspected or tested in accordance with the provisions of this standard and other applicable referenced codes and standards. Such inspections or tests shall be adequate to assure compliance with the design, material specifications, fabrication methods, and quality required by this and the referenced standards.

9-1.9.2 The inspections or tests required by 9-1.9.1 shall be the responsibility of the operator who shall be permitted to delegate any part of those inspections to his or her own employees, to a third party engineering or scientific organization, or to a recognized insurance or inspection company. Each inspector shall be qualified in accordance with the code or standard that is applicable to the test or inspection being performed.

9-1.10 Marking on Refrigerated LP-Gas Containers. Each refrigerated LP-Gas container shall be identified by the attachment of a nameplate on the outer covering in an accessible and visible place marked as specified with the following:

(a) Manufacturer's name and date built.
(b) Liquid volume of the container in gal (U.S. standard) or barrels.
(c) Maximum allowable working pressure in lb per sq in.
(d) Minimum temperature in degrees Fahrenheit for which the container was designed.
(e) Maximum allowable water level to which the container shall be permitted to be filled for test purposes.
(f) Density of the product to be stored in lb per cu ft, or specific gravity, for which the container was designed.
(g) Maximum level to which the container shall be permitted to be filled with the LP-Gas for which it was designed.

9-2 Refrigerated LP-Gas Container Impoundment.

9-2.1 Each refrigerated LP-Gas container shall be located within an impoundment area that complies with this section, in order to minimize the possibility that the accidental discharge of liquid LP-Gas from the container would endanger adjoining property or lives, process equipment or structures, or reach waterways.

9-2.2 Enclosed drainage channels for LP-Gas shall be prohibited.

Exception: Container downcomers used to rapidly conduct spilled LP-Gas away from critical areas shall be permitted to be enclosed provided that an adequate drainage rate is achieved.

Enclosed drainage channels are prohibited in order to prevent accumulation of vapor in enclosed spaces, which can impede the flow of liquid or promote detonation if ignition should occur.

9-2.3 Dikes, impounding walls, and drainage for refrigerated LP-Gas and flammable refrigerant containment shall have a minimum volumetric holding capacity, including any useful holding capacity of the drainage area and with allowance made for the displacement of snow accumulation, other containers, or equipment, equal to the total liquid volume of the largest container served assuming that container is full.

It is the intent of the diking requirements to ensure that in the unlikely event of a refrigerated LP-Gas spill, the liquid will be contained and not allowed to spread on the ground. Diking provisions are not required for pressurized storage containers because the pressurized liquid will vaporize quickly after being released from its pressurized container. Refrigerated LP-Gas, however, will only partially vaporize, and a large part of the liquid will remain as liquid because there is limited heat available from the air or the ground to vaporize the liquid.

These requirements are similar to those in NFPA 30, *Flammable and Combustible Liquids Code*, for flammable liquids diking.

9-2.4 Dikes, impounding walls, and drainage for refrigerated LP-Gas and flammable refrigerant containment shall be of compacted earth, concrete, metal, or other suitable materials. It shall be permitted to locate such structures independent of the container, mounted integral to, or constructed against, the container. These structures, and any penetrations thereof, shall be designed to withstand the full hydrostatic head of impounded LP-Gas or flammable refrigerant, the effect of rapid cooling to the temperature of the liquid to be confined, any anticipated fire exposure, and natural forces such as earthquake, wind, or rain.

9-2.5 To ensure that any accidentally discharged liquid is contained within an area enclosed by a dike or an impounding wall while providing a reasonably wide margin for area configuration design, the dike or impounding wall height and distance shall be determined in accordance with Figure 9-2.5.

9-2.6 Provision shall be made to clear rain or other water from the impounding area. Automatically controlled sump pumps shall be permitted if equipped with an automatic cutoff device that will prevent their operation when exposed to LP-Gas temperatures. Piping, valves, and fittings whose failure could permit liquid to escape from the impounding area shall be suitable for continuous exposure to LP-Gas temperatures. If gravity drainage is employed for water removal, provision shall be made to prevent the escape of LP-Gas by way of the drainage system.

9-2.7 Insulation systems used for impounding surfaces shall be, in the installed condition, noncombustible and suitable for the intended service considering the anticipated thermal and mechanical stresses and loadings. If flotation is a problem, mitigation measures shall be provided. Such insulation systems shall be inspected as appropriate for their intended service.

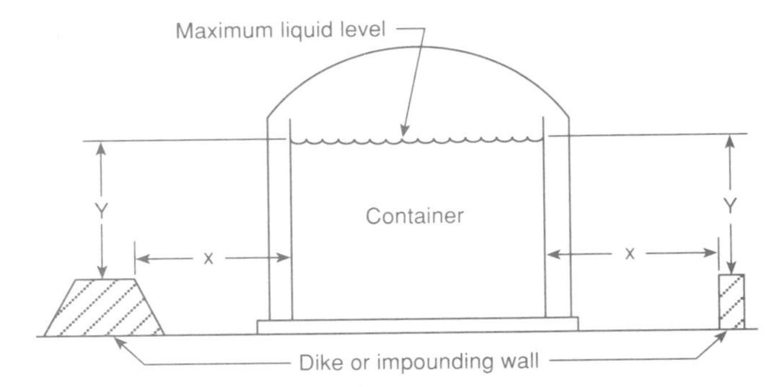

Notes:
Dimension "X" shall equal or exceed the sum of dimension "Y" plus the equivalent head in LP-Gas of the pressure in the vapor space above the liquid.
Exception: When the height of the dike or impounding wall is equal to, or greater than, the maximum liquid level, "X" may have any value.
Dimension "X" is the distance from the inner wall of the container to the closest face of the dike or impounding wall.
Dimension "Y" is the distance from the maximum liquid level in the container to the top of the dike or impounding wall.

Figure 9-2.5 *Dike or impounding wall proximity to containers.*

9-3 Locating Aboveground Refrigerated LP-Gas Containers.

9-3.1 Containers shall be located outside of buildings.

9-3.2 A refrigerated LP-Gas container having a capacity of 70,000 gal (265 m^3) or less shall be installed in accordance with Table 3-2.2.2. Refrigerated LP-Gas containers of 70,001 gal (265 m^3) or more shall be located 100 ft (30.5 m) or more from any occupied building, storage containers for flammable or combustible liquids, and from a property line that may be built upon.

9-3.3 The edge of a dike, impoundment, or drainage system intended for a refrigerated LP-Gas container shall be 100 ft (31 m) or more from a property line that can be built upon, a public way, or a navigable waterway.

9-3.4 Refrigerated LP-Gas containers shall not be located within dikes enclosing flammable liquid tanks or within dikes or impoundments enclosing nonrefrigerated pressurized LP-Gas containers.

9-3.5 Refrigerated LP-Gas containers shall not be installed one above the other.

This requirement is consistent with requirements for nonrefrigerated containers, which may not be stacked.

9-3.6 The minimum distance between aboveground refrigerated LP-Gas containers of 70,001 gal (265 m^3) or larger shall be one-half the diameter of the larger container.

9-3.7 The ground within 25 ft (7.6 m) of any aboveground refrigerated LP-Gas containers and all ground within a diked, impoundment, or drainage area shall be kept clear of readily ignitible materials such as weeds and long dry grass.

This requirement is similar to that for nonrefrigerated containers except that the distance required to be kept clear is greater than the 10 ft (3 m) required for nonrefrigerated containers. This recognizes the larger size of refrigerated containers. Most refrigerated container installation facilities are designed to prevent the growth of vegetation near containers by the use of gravel or crushed stone surfacing.

Plant operators must take steps to ensure that any brush-clearing operations do not create hazards. At least one incident has occurred where a portable gasoline-powered weed cutter cut a pipe, resulting in liquid release and ignition.

References Cited in Commentary

The following publication is available from the National Fire Protection Association, 1 Batterymarch Park, P.O. Box 9101, Quincy, MA 02269-9101.

NFPA 30, *Flammable and Combustible Liquids Code*, 1990 edition.

The following publication is available from the American Petroleum Institute, 1220 L Street, NW, Washington, DC 20005.

API 2510-89, *Design and Construction of Liquefied Petroleum Gas Installations (LPG).*

10

Marine Shipping and Receiving

This section was added in the 1989 edition in conjunction with the inclusion of marine terminals in the scope of the standard. In the 1992 edition, it was relocated to its own chapter. It was expanded in the 1995 edition. Much of the original material was drawn from NFPA 59A, *Standard for the Production, Storage, and Handling of Liquefied Natural Gas (LNG)*, and the existing practices at marine terminals as required by the U.S. Coast Guard and other authorities having jurisdiction. This section also is an outgrowth of the requirements of NFPA 30, *Flammable and Combustible Liquids Code*, and NFPA 59, *Standard for the Storage and Handling of Liquefied Petroleum Gases at Utility Gas Plants*, which reflect the experience and practices of the marine industry.

10-1 Piers.

10-1.1 Design, construction, and operation of piers, docks, and wharves shall comply with relevant regulations and the requirements of the authorities having jurisdiction.

NOTE: Federal regulations applicable to marine terminals are contained in Title 33 of the *U.S. Code of Federal Regulations*.

NFPA 30 provides requirements for locating wharves handling flammable liquids.

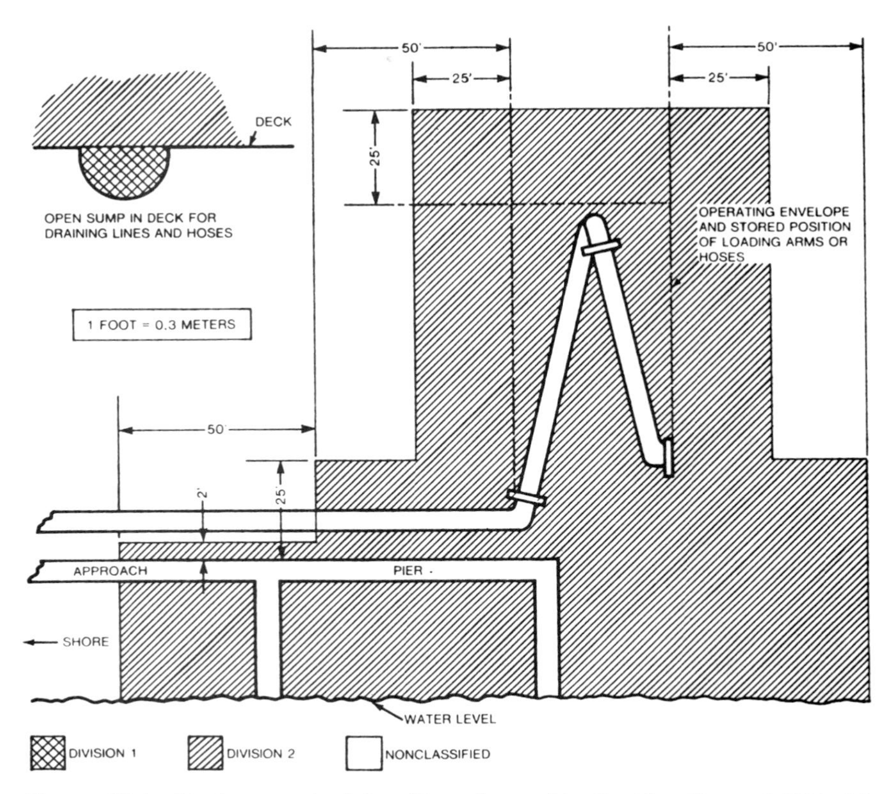

Figure 10.1 Marine terminal handling flammable liquids. (From NFPA 30, Flammable and Combustible Liquids Code.)

Notes to Figure 10.1

(1) The "source of vapor" shall be the operating envelope and stored position of the outboard flange connection of the loading arm (or hose).

(2) The berth area adjacent to tanker and barge cargo tanks is to be Division 2 to the following extent:

a. 25 ft (7.6 m) horizontally in all directions on the pier side from that portion of the hull containing cargo tanks.

b. From the water level to 25 ft (7.6 m) above the cargo tanks at their highest position.

(3) Additional locations may have to be classified as required by the presence of other sources of flammable liquids (or flammable gases) on the berth, or by Coast Guard or other regulations.

10-1.2 General cargo, flammable liquids, or compressed gases, other than ships' general stores for the LPG tank vessel, shall not be handled over a pier or dock within 100 ft (30.5 m) of the point of transfer connection while LPG or other flammable liquids are being transferred. Ship bunkering operations shall not be permitted prior to or during cargo transfer operations.

Bunkering is the process of refueling the ship. LP-Gas cargo and other flammable liquids and compressed gases cannot be handled within 100 ft of the point of transfer while bunkering operations are in process.

10-1.3 Trucks and other motorized vehicles shall be prohibited on the pier or dock within 100 ft (30.5 m) of the transfer connection while transfer operations are in progress. Authorized parking areas, if provided for in the waterfront area, shall be properly marked. Suitable warning signs or barricades shall be used to indicate when transfer operations are in progress.

Note that all vehicle traffic is prohibited, including forklift trucks, during transfer operations.

10-1.4 The shore facility shall ensure that adequate security personnel and measures are available to prevent unauthorized individuals access to the waterfront area while the LPG vessel is alongside the pier or dock. Security personnel shall maintain control of visitors, delivery trucks, and service personnel. The facility operator shall ensure that any person assigned security duty has been instructed on security procedures.

Adequate security requires a reasonable number of security personnel to control access to the berth area and to call for assistance (terminal personnel, local police, and fire department, etc.)—should the need arise.

10-1.5 The shore mooring equipment shall be designed and maintained to safely hold the vessel to the pier or dock under all anticipated weather, current, and tidal conditions.

Proper design is necessary to safely hold the vessel or barge to the berth during unexpected changes in climatic conditions.

10-1.6 All electrical equipment and wiring installed on the pier or dock shall comply with 3-7.2.1 and 3-7.2.2 and shall be suitable for the electrical area classification.

This requirement applies to all fixed electrical equipment operated while a transfer is being made. If a transfer is done during hours of darkness, the pier lighting must be checked for compliance.

10-1.7 If the terminal conducts transfers between sunset and sunrise, the pier or dock area shall have a lighting system that allows operations personnel to safely conduct operations and illuminate the transfer connection area, control valves, storage containers, other equipment, and walkways.

Lighting must be adequate during hours of darkness for personnel to conduct normal and emergency operations without the aid of a flash light or other handheld lighting equipment.

(a) Lighting shall illuminate communications, fire fighting, and other emergency equipment.

(b) All lighting shall be located or shielded so that it is not confused with any aids to navigation and does not interfere with navigation on the adjacent waterway.

(c) All lighting fixtures and wiring shall comply with 3-7.2.1 and 3-7.2.2.

10-1.8 Welding and cutting, if required, shall be conducted in accordance with NFPA 51B, *Fire Prevention in Use of Cutting and Welding Processes*. Smoking shall be permitted only in conspicuously marked, designated areas.

10-1.9 The shore facility shall ensure that medical first aid equipment and fire extinguishers are installed and available in appropriate locations, types, and quantities for any reasonably anticipated abnormal condition, while the vessel is alongside the berth, and shall ensure:

Medical first aid equipment and supplies are to handle minor injuries, prior to the arrival of professional EMTs, or for transporting an injured person to a medical facility.

(a) That extinguishers are installed and maintained in accordance with NFPA 10, *Standard for Portable Fire Extinguishers,* and are ready for use.

(b) That emergency equipment is positioned and ready to operate prior to the start of the transfer operation.

(c) That locations of all fire extinguishers are conspicuously marked and ready accessibility and lighting is maintained.

10-1.10 Prior to the start of the transfer, the shore facility shall ensure that warning signs be placed in the marine transfer area, readily visible from the shoreline and berth areas, that have the following text:

Warning
Dangerous Cargo
No Visitors
No Smoking
No Open Light

10-1.11 The shore facility shall have a flammable gas detector, capable of detecting LP-Gas, readily available for use at the berth.

The combustible gas detector should be capable of detecting propane vapor from 0–100 percent LFL.

10-1.12 The shore facility shall ensure that portable electrical equipment is not used within 100 ft (30.5 m) of the transfer connection while transfer operations are in progress. When the transfer operation is completed (secured) and the transfer piping is disconnected, the equipment used shall be in compliance with 3-7.2.1 and 3-7.2.2.

Exception: Electrical equipment listed for use in Class I, Division I locations and intrinsically safe equipment.

The use of intrinsically safe equipment is specifically permitted. Portable radios are frequently used during transfer operations.

10-1.13 The berth owner or shore facility operator shall ensure that the following life safety equipment is positioned on the berth and ready for immediate use, while personnel are working on the berth or a vessel is alongside:

Life rings and other life safety equipment should be U.S. Coast Guard approved.

1. Life rings with attendant rope of sufficient length.
2. Approved fire blanket.

10-2 Pipelines.

10-2.1 Pipelines shall be located on the dock or pier so that they are not exposed to damage from vehicular traffic or other possible cause of physical damage. Underwater pipelines shall be located or protected so that they are not exposed to damage from marine traffic, and their locations shall be posted or identified in accordance with federal regulations.

NOTE: Refer to *Code of Federal Regulations*, Title 49, Part 195.

The *Code of Federal Regulations*, Title 49, Part 195, provides safety standards for pipeline facilities used in transporting hazardous liquids. The Coast Guard recognizes the special problems presented by operations involving loading, unloading, handling, and storage of bulk cargoes of certain hazardous materials, especially if the operation occurs at a general cargo marine terminal. For this reason the Coast Guard has defined "cargoes of particular hazard" and "facilities of particular hazard" in its regulations (33 CFR 126.10 and 126.05, respectively). Included as cargoes of particular hazard are flammable liquids, flammable compressed gases, and liquefied natural gas. These cargoes may be handled only at a designated facility.

10-2.2 Isolation valving and bleed connections shall be provided at the loading or unloading manifold for both liquid and vapor return lines so that hoses and arms can be blocked off, drained or pumped out, and depressurized before disconnecting. Liquid isolation valves, regardless of size, and vapor valves 8 in. (20 mm) and larger in size shall be equipped with powered operators in addition to means for manual operation. Power-operated valves shall be capable of being closed from a remote control station located at least 50 ft (15 m) from the manifold area, as well as locally. Unless the valve will automatically fail closed on loss of power, the valve actuator and its power supply within 50 ft (15 m) of the valve shall be protected against operational failure due to fire exposure of at least 10 min. Valves shall be located at the point of hose or arm connection to the manifold. Bleeds or vents shall discharge to a safe area.

10-2.3 In addition to the isolation valves at the manifold, each vapor return and liquid transfer line shall be provided with a readily accessible isolation valve located on shore near the approach to the pier or dock. Where more than one line is involved, the valves shall be grouped in one location. Valves shall be identified as to their service. Valves 8 in. (20 mm) and larger in size shall be equipped with power operators. Means for manual operation shall be provided.

10-2.4 Pipelines used for liquid unloading only shall be provided with a check valve located at the manifold adjacent to the manifold isolation valve.

10-3 Prior to Transfer.

The *Code of Federal Regulations*, Title 33, Part 126.15, provides guidelines for the control of liquid cargo transfer systems, including actions required prior to the transfer of cargo. This section was expanded significantly in the 1995 edition to provide guidance in NFPA 58. The revised section provides an expanded body of rules that enhance the safety of product transfer in a form that is usable.

10-3.1 Prior to starting transfer operations, the officer in charge of the vessel transfer operation and the person in charge of the shore facility shall inspect their respective facilities. The inspection shall ensure that all cargo transfer equipment and hoses have been maintained, tested, and are in operating condition. Following this inspection, they shall meet to discuss the transfer procedures and, when ready, each will notify the other that each facility is ready in all respects to start transfer operations.

NOTE: For guidance refer to Title 33 of the *U.S. Code of Federal Regulations*.

10-3.2 The supervisor in charge of the shore facility and the officer in charge of vessel operations shall ensure that their respective facilities are, in all respects, ready to start transfer operations.

NOTE: For guidance refer to 46 CFR Part 35, 35-30.

10-3.3 The shore facility transfer system shall be equipped with a remotely operated emergency shutdown system.

10-3.4 The supervisor in charge of the shore facility shall ensure that a facilities Emergency Procedures Manual is readily available and contains the following information:

Readily available means that it should be located in an area that is quickly and easily accessible to terminal and emergency response personnel.

(a) LPG release response and emergency shutdown procedures.

(b) Telephone number for all emergency response organizations, U.S. Coast Guard, emergency medical facilities, and hospital(s).

(c) Description and location of the facility fire systems and emergency equipment.

10-3.5 The supervisor in charge of the shore facility shall ensure that a facilities Standard Operating Procedures Manual is readily available and contains the following information:

(a) Procedures for startup, operation, and shutdown of the transfer system and equipment. In the case of refrigerated product transfer, procedures for cooling down the transfer hose and line.

(b) Telephone numbers for all emergency response organizations, U.S. Coast Guard, emergency medical facilities, and hospital(s).

(c) Description, location, and operational guidelines for the facility fire systems and emergency equipment.

Each transfer operation shall be conducted in accordance with the Operations Manual, using established safe practices.

10-3.6 The supervisor in charge of the shore facility shall ensure that, at the completion of the transfer, and prior to disconnect of the transfer hose or arm, the transfer connection has been purged of all liquid and depressurized. The liquid and vapor pressure shall be returned either back to the vessel or to the shore facility; it shall not be vented to the atmosphere.

References Cited in Commentary

The following publications are available from the National Fire Protection Association, 1 Batterymarch Park, P.O. Box 9101, Quincy, MA 02269-9101.

NFPA 30, *Flammable and Combustible Liquids Code*, 1990 edition.

NFPA 59, *Standard for the Storage and Handling of Liquefied Petroleum Gases at Utility Gas Plants*, 1992 edition.

NFPA 59A, *Standard for the Production, Storage, and Handling of Liquefied Natural Gas (LNG)*, 1990 edition.

The following publications are available from the U.S. Government Printing Office, Washington, D.C.

Code of Federal Regulations, Title 33, Parts 126.05, 126.10, and 126.15.

Code of Federal Regulations, Title 49, Part 195.

11

Referenced Publications

11-1

The following documents or portions thereof are referenced within this standard and shall be considered part of the requirements of this document. The edition indicated for each reference is the current edition as of the date of the NFPA issuance of this document.

11-1.1 **NFPA Publications.** National Fire Protection Association, 1 Batterymarch Park, P.O. Box 9101, Quincy, MA 02269-9101.

NFPA 10, *Standard for Portable Fire Extinguishers*, 1994 edition.

NFPA 15, *Standard for Water Spray Fixed Systems for Fire Protection*, 1990 edition.

NFPA 30, *Flammable and Combustible Liquids Code*, 1993 edition.

NFPA 37, *Standard for the Installation and Use of Stationary Combustion Engines and Gas Turbines*, 1994 edition.

NFPA 50B, *Standard for Liquefied Hydrogen Systems at Consumer Sites*, 1994 edition.

NFPA 51, *Standard for the Design and Installation of Oxygen-Fuel Gas Systems for Welding, Cutting, and Allied Processes*, 1992 edition.

NFPA 51B, *Standard for Fire Prevention in Use of Cutting and Welding Processes*, 1994 edition.

NFPA 54 (ANSI Z223.1), *National Fuel Gas Code*, 1992 edition.

NFPA 59, *Standard for the Storage and Handling of Liquefied Petroleum Gases at Utility Gas Plants*, 1995 edition.

NFPA 61B, *Standard for the Prevention of Fires and Explosions in Grain Elevators and Facilities Handling Bulk Raw Agricultural Commodities*, 1989 edition.
NFPA 70, *National Electrical Code*, 1993 edition.
NFPA 82, *Standard on Incinerators and Waste and Linen Handling Systems and Equipment*, 1994 edition.
NFPA 86, *Standard for Ovens and Furnaces*, 1990 edition.
NFPA 96, *Standard for Ventilation Control and Fire Protection of Commercial Cooking Operations*, 1994 edition.
NFPA *101*, *Life Safety Code*, 1994 edition.
NFPA 251, *Standard Methods of Fire Tests of Building Construction and Materials*, 1990 edition.
NFPA 302, *Fire Protection Standard for Pleasure and Commercial Motor Craft*, 1994 edition.
NFPA 501A, *Standard for Fire Safety Criteria for Manufactured Home Installations, Sites, and Communities*, 1992 edition.
NFPA 501C, *Standard on Recreational Vehicles*, 1993 edition.
NFPA 505, *Firesafety Standard for Powered Industrial Trucks Including Type Designations, Areas of Use, Maintenance, and Operation*, 1992 edition.

11-1.2 API Publications. American Petroleum Institute, 2101 L St., NW, Washington, DC 20037.

API-ASME *Code for Unfired Pressure Vessels for Petroleum Liquids and Gases*, Pre - July 1, 1961.
API 620, *Recommended Rules for Design and Construction of Large, Welded, Low-Pressure Storage Tanks*, 1990.

11-1.3 ASCE Publication. American Society of Civil Engineers, United Engineering Center, 345 East 47th St., New York, NY 10017.

ASCE 7-1993, *Design Loads for Buildings and Other Structures.*

11-1.4 ASME Publications. American Society for Mechanical Engineers, 345 East 47th St., New York, NY 10017.

"Rules for the Construction of Unfired Pressure Vessels," Section VIII, ASME *Boiler and Pressure Vessel Code*, 1992.
ASME B31.3-1993, *Chemical Plant and Petroleum Refinery Piping.*
ASME B36.10M-1985, *Welded and Seamless Wrought Steel Pipe.*

11-1.5 ASTM Publications. American Society for Testing and Materials, 1916 Race St., Philadelphia, PA 19103.

ASTM A47-1990, *Standard Specification for Ferritic Malleable Iron Castings.*
ASTM A48-1994, *Standard Specification for Gray Iron Castings.*
ASTM A53-1993, *Standard Specification for Pipe, Steel, Black and Hot-Dipped, Zinc-Coated Welded and Seamless.*

ASTM A106-1994, *Standard Specification for Seamless Carbon Steel Pipe for High-Temperature Service.*

ASTM A395-1988, *Standard Specification for Ferritic Ductile Iron Pressure-Retaining Castings for Use at Elevated Temperatures.*

ASTM A513-1992, *Standard Specification for Electric-Resistance-Welded Carbon and Alloy Steel Mechnical Tubing.*

ASTM A536-1984, *Standard Specification for Ductile Iron Castings.*

ASTM A539-1990, *Standard Specification for Electric-Resistance-Welded Coiled Steel Tubing for Gas Fuel Oil Lines.*

ASTM B42-1993, *Standard Specification for Seamless Copper Pipe, Standard Sizes.*

ASTM B43-1991, *Standard Specification for Seamless Red Brass Pipe, Standard Sizes.*

ASTM B86-1988, *Standard Specification for Zinc-Alloy Die Casting.*

ASTM B88-1993, *Standard Specification for Seamless Copper Water Tube.*

ASTM B135-1991, *Standard Specification for Seamless Brass Tube.*

ASTM B280-1993, *Standard Specification for Seamless Copper Tube for Air Conditioning and Refrigeration Field Service.*

ASTM D2513-1994, *Standard Specification for Thermoplastic Gas Pressure Pipe, Tubing and Fittings.*

ASTM D2683-1993, *Standard Specification for Socket-Type Polyethylene (PE) Fittings for Outside Diameter Controlled Polyethylene Pipe.*

ASTM D3261-1993, *Standard Specification for Butt Heat Fusion Polyethylene (PE) Plastic Fittings for Polyethylene (PE) Plastic Pipe and Tubing.*

ASTM F1055-1993, *Standard Specification for Electrofusion Type Polyethylene Fittings for Outside Diameter Controlled Polyethylene Pipe and Tubing.*

11-1.6 **AWS Publication.** American Welding Society, 2501 NW 7th St., Miami, FL 33125.

AWS Z49.1-1988, *Safety in Welding and Cutting.*

11-1.7 **CGA Publication.** Compressed Gas Association, Inc., 1235 Jefferson Davis Highway, Arlington, VA 22202.

ANSI/CGA C-4-1990, *Method of Marking Portable Compressed Gas Containers to Identify the Material Contained.*

11-1.8 **Federal Regulations.** U.S. Government Printing Office, Washington, DC 20401.

Code of Federal Regulations, Title 49, Parts 171-192 and Parts 393 and 397. (Also available from the Association of American Railroads, American Railroads Bldg., 1920 L St. NW, Washington, DC 20036 and American Trucking Assns., Inc., 2201 Mill Rd., Alexandria, VA 22314.)

11-1.9 ICBO Publication. International Conference of Building Officials, 5360 S. Workman Mill Rd., Whittier, CA 90601.

Uniform Building Code, 1991.

11-1.10 UL Publications. Underwriters Laboratories Inc., 333 Pfingsten Rd., Northbrook, IL 60062.

UL 132-93, *Standard on Safety Relief Valves for Anhydrous Ammonia and LP-Gas.*

UL 144-94, *Standard for LP-Gas Regulators.*

UL 147A-92, *Standard for Nonrefillable (Disposal) Type Fuel Gas Cylinder Assemblies.*

UL 147B-92, *Standard for Nonrefillable (Disposal) Type Metal Container Assemblies for Butane.*

UL 567-92, *Standard Pipe Connectors for Flammable and Combustible Liquids and LP-Gas.*

Explanatory Material

This appendix is not part of the requirements of this NFPA document but is included for informational purposes only.

The material contained in Appendix A of this standard is included within the text of this handbook, and therefore is not repeated here.

Properties of LP-Gases

This appendix is not a part of the requirements of this NFPA document but is included for informational purposes only.

This appendix provides information on the physical properties of LP-Gas. This is included for the benefit of users of the standard who may require physical property information for butane and propane. Additional information not in the standard is provided in graphical form.

- Figure B.1, the pressure-temperature chart for propane-air mixtures is used to determine the dew point of mixtures. This is needed to determine the minimum temperature piping containing propane-air mixtures can be exposed to without condensing liquid propane.
- Figure B.2, dewpoints of LP-Gas–air mixtures, provides three graphs, the first duplicates B.1, and the others provide similar information for N-butane and isobutane.
- Figure B.3, pressure-density chart for propane, provides the properties of propane under all pressure and temperature conditions.
- Figure B.4, temperature-density chart for propane, is similar to B.4 but presents the same information plotted with different variables on each axis, and is also useful.

B-1 Approximate Properties of LP-Gases.

B-1.1 Source of Property Values.

B-1.1.1 The property values for the LP-Gases are based on average industry values and include values for LP-Gases coming from natural gas liquids plants as well as those coming from petroleum refineries. Thus, any particular commercial propane or butane might have properties varying slightly from the values shown. Similarly, any propane-butane mixture might have properties varying from those obtained by computation from these average values (*see B-1.2.1 for computation method used*). Since these are average values, the interrelationships between them (i.e., lb per gal, specific gravity, etc.) will not cross-check perfectly in all cases.

B-1.1.2 Such variations are not sufficient to prevent the use of these average values for most engineering and design purposes. They stem from minor variations in composition. The commercial grades are not pure (CP-Chemically Pure) propane or butane, or mixtures of the two, but might also contain small and varying percentages of ethane, ethylene, propylene, isobutane, or butylene, which can cause slight variations in property values. There are limits to the accuracy of even the most advanced testing methods used to determine the percentages of these minor components in any LP-Gas.

B-1.2 Approximate Properties of Commercial LP-Gases.

B-1.2.1 The principal properties of commercial propane and commercial butane are shown in Table B-1.2.1. Reasonably accurate property values for propane-butane mixtures can be obtained by computation, applying the percentages by weight of each in the mixture to the values for the property it is desired to obtain. Slightly more accurate results for vapor pressure are obtained by using the percentages by volume. Very accurate results can be obtained using data and methods explained in petroleum and chemical engineering data books.

B-1.3 Specifications of LP-Gases. Specifications of LP-Gases covered by this standard are listed in Gas Processors Association, Standard 2140, *Liquefied Petroleum Gas Specifications for Test Methods,* or ASTM D1835, *Specification for Liquefied Petroleum (LP) Gases.*

Table B-1.2.1 (English) Approximate Properties of LP-Gases

	Commercial Propane	Commercial Butane
Vapor Pressure in psi at:		
70°F	127	17
100°F	196	37
105°F	210	41
130°F	287	69
Specific Gravity of Liquid at 60°F	0.504	0.582
Initial Boiling Point at 14.7 psia, °F	–44	15
Weight per Gallon of Liquid at 60°F, lb	4.20	4.81
Specific Heat of Liquid, Btu/lb at 60°F	0.630	0.549
Cu ft of Vapor per Gallon at 60°F	36.38	31.26
Cu ft of Vapor per Pound at 60°F	8.66	6.51
Specific Gravity of Vapor (Air = 1) at 60°F	1.50	2.01
Ignition Temperature in Air, °F	920–1,120	900–1,000
Maximum Flame Temperature in Air, °F	3,595	3,615
Limits of Flammability in Air, Percent of Vapor in Air-Gas Mixture:		
(a) Lower	2.15	1.55
(b) Upper	9.60	8.60
Latent Heat of Vaporization at Boiling Point:		
(a) Btu per Pound	184	167
(b) Btu per Gallon	773	808
Total Heating Values After Vaporization:		
(a) Btu per Cubic Foot	2,488	3,280
(b) Btu per Pound	21,548	21,221
(c) Btu per Gallon	91,502	102,032

Table B-1.2.1 (Metric) Approximate Properties of LP-Gases

	Commercial Propane	Commercial Butane
Vapor Pressure in kPa at:		
20°C	895	103
40°C	1 482	285
45°C	1 672	345
55°C	1 980	462
Specific Gravity	0.504	0.582
Initial Boiling Point at 1.00 Atm. Pressure, °C	–42	–9
Weight per Cubic Metre of Liquid at 15.56°C, kg	504	582
Specific Heat of Liquid, Kilojoule per Kilogram, at 15.56°C	1.464	1.276
Cubic Metre of Vapor per Litre of Liquid at 15.56°C	0.271	0.235
Cubic Metre of Vapor per Kilogram of Liquid at 15.56°C	0.539	0.410
Specific Gravity of Vapor (Air = 1) at 15.56°C	1.50	2.01
Ignition Temperature in Air, °C	493–549	482–538
Maximum Flame Temperature in Air, °C	1 980	2 008
Limits of Flammability in Air, % of Vapor in Air-Gas Mixture:		
(a) Lower	2.15	1.55
(b) Upper	9.60	8.60
Latent Heat of Vaporization at Boiling Point:		
(a) Kilojoule per kilogram	428	388
(b) Kilojoule per litre	216	226
Total Heating Value After Vaporization:		
(a) Kilojoule per Cubic Metre	92 430	121 280
(b) Kilojoule per Kilogram	49 920	49 140
(c) Kilojoule per Litre	25 140	28 100

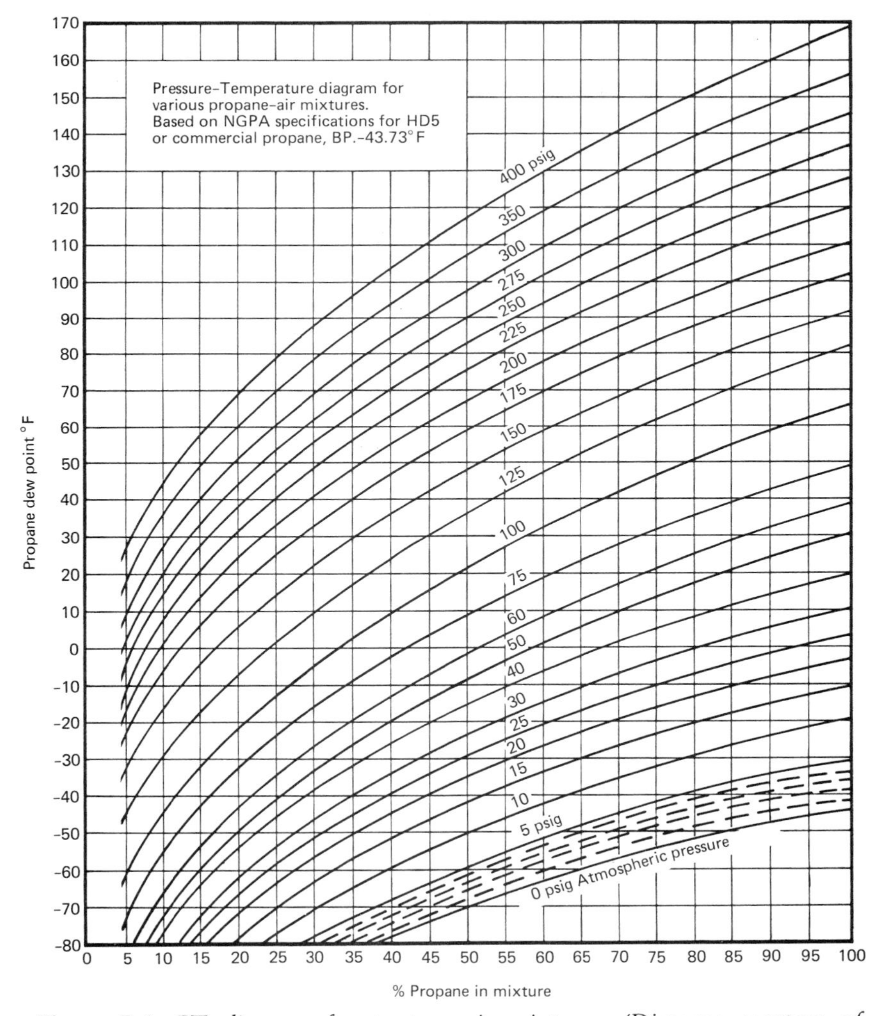

Figure B.1 *PT diagram for propane-air mixtures. (Diagram courtesy of Butane Propane Handbook.)*

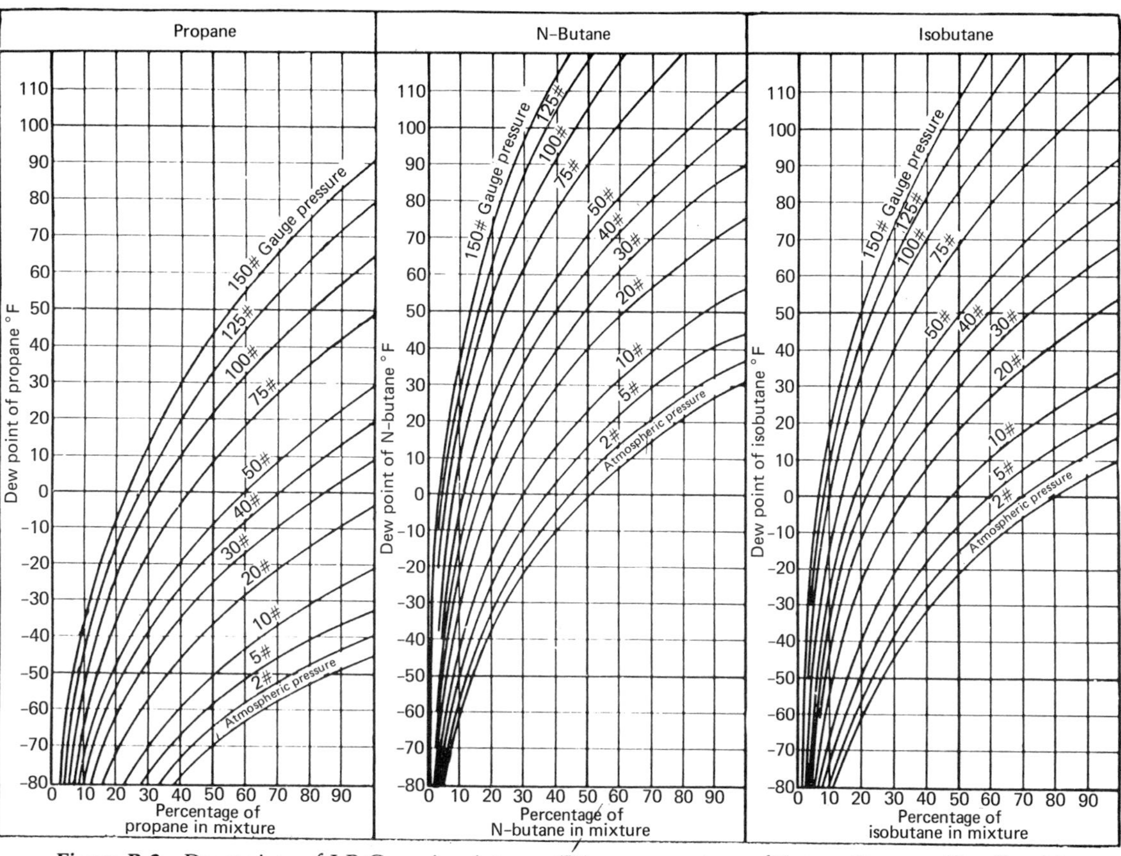

Figure B.2 Dewpoints of LP-Gas–air mixtures. (Diagram courtesy of Butane Propane Handbook.)

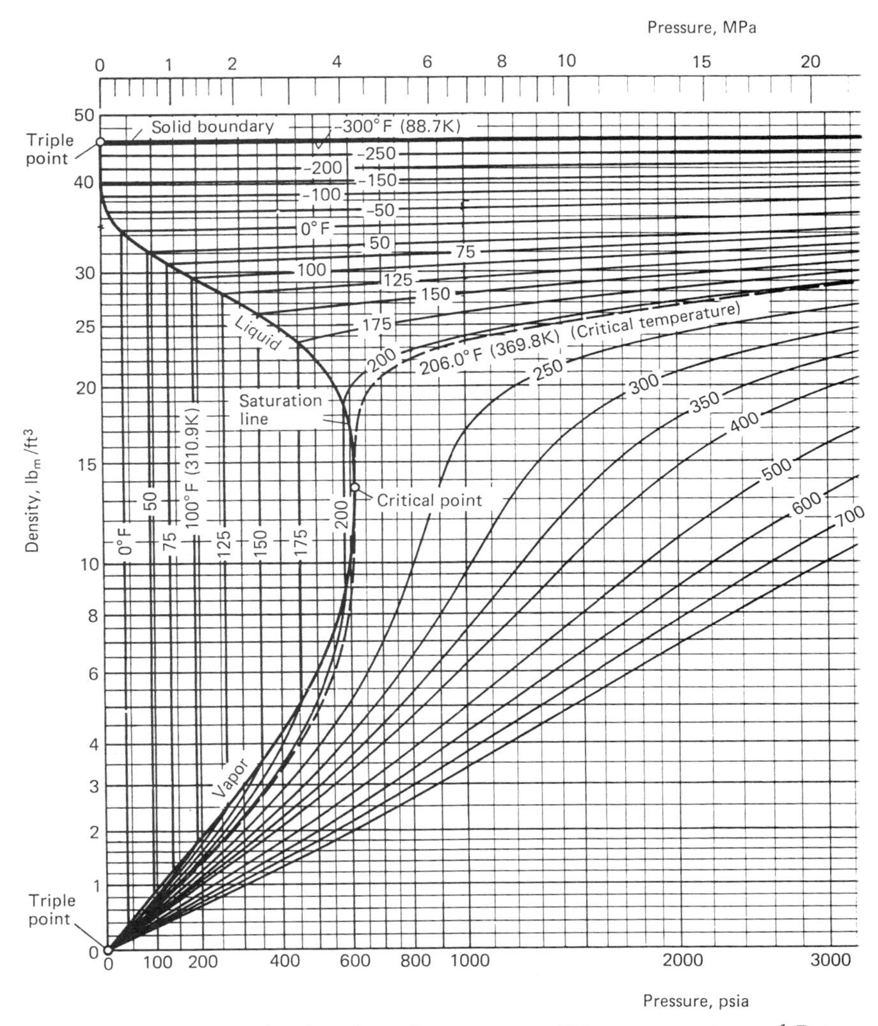

Figure B.3 Pressure-density chart for propane. (Diagram courtesy of Butane Propane Handbook.)

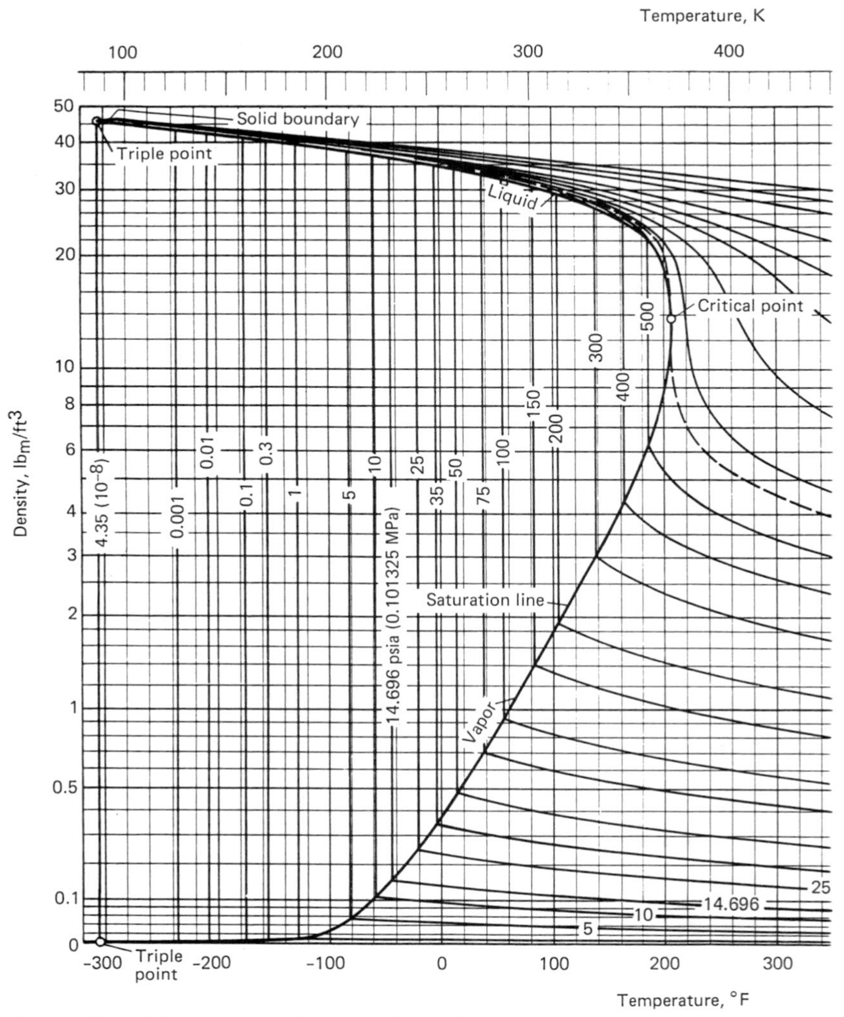

Figure B.4 *Temperature-density chart for propane (Diagram courtesy of Butane Propane Handbook).*

Design, Construction, and Requalification of DOT (ICC) Cylinder Specification Containers

This appendix is not a part of the requirements of this NFPA document but is included for informational purposes only.

C-1 Scope.

C-1.1 Application.

C-1.1.1 This appendix provides general information on DOT cylinder specification containers referred to in this standard. For complete information consult the applicable specification (*see C-2.1.1*). The water capacity of such cylinders are not permitted to be more than 1,000 lb (454 kg).

C-1.1.2 This appendix is not applicable to DOT tank car portable tank container or cargo tank specifications. Portable and cargo tanks are basically ASME containers and are covered in Appendix D.

C-1.1.3 Prior to April 1, 1967, these specifications were promulgated by the Interstate Commerce Commission (ICC). On this date, certain functions of the ICC, including the promulgation of specifications and regulations dealing with LP-Gas cylinders, were transferred to the Department of Transportation (DOT). Throughout this appendix both ICC and DOT are used; ICC applying to dates prior to April 1, 1967, and DOT to subsequent dates.

C-2 LP-Gas Cylinder Specifications.

C-2.1 Publishing of DOT Cylinder Specifications.

C-2.1.1 DOT cylinder specifications are published under Title 49, *Code of Federal Regulations*, Parts 171-190, available from U.S. Government Printing Office, Washington, D.C. The information in this publication is also issued as a Tariff at approximately three-year intervals by the Bureau of Explosives, American Railroads Building, 1920 L Street, NW, Washington, DC 20036.

C-2.2 DOT Specification Nomenclature.

C-2.2.1 The specification designation consists of a one-digit number, sometimes followed by one or more capital letters, then by a dash and a three-digit number. The one-digit number alone, or in combination with one or more capital letters, designates the specification number. The three-digit number following the dash shows the service pressure for which the container is designed. Thus, "4B-240" indicates a cylinder built to Specification 4B for a 240 psi (1,650 kPa) service pressure. (*See C-2.2.3.*)

C-2.2.2 The specification gives the details of cylinder construction, such as material used, method of fabrication, tests required, and inspection method, and prescribes the service pressure, or range of service pressures for which that specification can be used.

C-2.2.3 The term "service pressure" is analagous to, and serves the same purpose as, the ASME "design pressure." However, it is not identical, representing instead the highest pressure to which the container will normally be subjected in transit or in use but not necessarily the maximum pressure to which it might be subjected under emergency conditions in transportation. The service pressure stipulated for the LP-Gases is based on the vapor pressures exerted by the product in the container at two different temperatures, the higher pressure of the two becoming the service pressure, as follows:

(a) The pressure in the container at 70°F (21°C) must be less than the service pressure for which the container is marked, and

(b) The pressure in the container at 130°F (54.4°C) must not exceed $^5/_4$ times the pressure for which the container is marked.

EXAMPLE: Commercial propane has a vapor pressure at 70°F (21°C) of 132 psi. However, its vapor pressure at 130°F (54.4°C) is 300 psi (2070 kPa), so service pressure [$^5/_4$ times which must not exceed 300 psi (2070 kPa)] is 300 divided by $^5/_4$, or 240 psi. Thus, commercial propane requires at least 240 psi (1650 kPa) service pressure cylinder.

C-2.3 DOT Cylinder Specifications Used for LP-Gas.

C-2.3.1 A number of different specifications were approved by DOT (and its predecessor ICC) for use with LP-Gases. Some of these are no longer published or used for new construction. However, containers built under these old specifications, if properly maintained and requalified, are still acceptable for LP-Gas transportation.

C-2.3.2 DOT specifications cover primarily safety in transportation. However, in order for the product to be used, it is necessary for it to come to rest at the point of use and serve as LP-Gas storage during the period of use. Containers adequate for transportation are also deemed to be adequate for use as provided in NFPA 58. As small size ASME containers were not available at the time tank truck delivery was started, ICC (now DOT) cylinders have been equipped for tank truck deliveries and permanently installed.

C-2.3.3 The DOT cylinder specifications most widely used for the LP-Gases are shown in Table C-2.3.3. The differing materials of construction, method of fabrication, and the date of the specification reflect the progress made in knowledge of the products to be contained and improvement in metallurgy and methods of fabrication.

Table C-2.3.3

Specification No. & Marking	Material of Construction	Method of Fabrication
26–150*	Steel	Welded and Brazed
3B–300	Steel	Seamless
4–300	Steel	Welded
4B–300	Steel	2 piece Welded & Brazed
4B–240	Steel	2 piece Welded & Brazed
4BA–240	Alloy Steel	2 piece Welded & Brazed
4E–240	Aluminum	Welded and Brazed
4BW–240	Steel	3 piece Welded

*The term "service pressure" had a different connotation at the time the specification was adopted.

C-3 Requalification, Retesting, and Repair of DOT Cylinder Specification Containers.

C-3.1 Application.

C-3.1.1 This section outlines the requalification, retesting, and repair requirements for DOT cylinder specification containers but should be used only as a guide. For official information, the applicable DOT regulations should be consulted.

C-3.2 Requalification (Including Retesting) of DOT Cylinders.

C-3.2.1 DOT rules prohibit DOT cylinders from being refilled, continued in service, or transported unless they are properly qualified or requalified for LP-Gas service in accordance with DOT regulations.

C-3.2.2 DOT rules require a careful examination of every container each time it is to be filled, and it must be rejected if there is evidence of exposure to fire, bad gouges or dents, seriously corroded areas, leaks, or other conditions indicating possible weaknesses that might render it unfit for service. The following disposition is to be made of rejected cylinders:

(a) Containers subjected to fire are required to be requalified, reconditioned, or repaired in accordance with C-3.3.1, or permanently removed from service except that DOT 4E (aluminum) cylinders must be permanently removed from service.

(b) Containers showing serious physical damage, leaks, or with a reduction in the marked tare weight of 5 percent or more are required to be retested in accordance with C-3.2.4(a) or (b) and, if necessary, repaired in accordance with C-3.3.1.

C-3.2.3 All containers, including those apparently undamaged, are required to be periodically requalified for continued service. The first requalification for a new cylinder is required within 12 years after the date of manufacture. Subsequent requalifications are required within the periods specified under the requalification method used.

C-3.2.4 DOT regulations permit three alternative methods of requalification for most commonly used LP-Gas specification containers (*see DOT regulations for permissible requalification methods for specific cylinder specifications*). Two use hydrostatic testing, and the third uses a carefully made and duly recorded visual examination by a competent person. In the case of the two hydrostatic test methods, only test results are recorded, but a careful visual examination of each container is also required. DOT regulations cite in detail the data to be recorded for the hydrostatic test methods, the observations to be made during the recorded visual examination method, and the marking of containers to indicate the requalification date and the method used. The three methods are outlined as follows:

(a) The water jacket-type hydrostatic test is permitted to be used to requalify containers for 12 years before the next requalification is due. A pressure of twice the marked service pressure is applied, using a water jacket (or the equivalent) so that the total expansion of the container during the application of the test pressure can be observed and recorded for comparison with the permanent expansion of the container after depressurization. The following disposition is made of containers tested in this manner:

1. Containers that pass the retest, and the visual examination required with it (*see C-3.2.4*), are marked with the date and year of the test (Example: "6-90," indicating requalification by the water jacket test method in June 1990) and are permitted to be placed back in service.
2. Containers that leak, or for which the permanent expansion exceeds 10 percent of the total expansion (12 percent for Specification 4E aluminum cylinders), must be rejected. If rejected for leakage, containers are permitted to be repaired in accordance with C-3.3.1.

(b) The simple hydrostatic test is permitted to be used to requalify containers for seven years before the next requalification is due. A pressure of twice the marked service pressure is applied, but no provision is made for measuring total and permanent expansion during the test outlined in C-3.2.4(a) above. The container is carefully observed while under the test pressure for leaks, undue swelling, or bulging indicating weaknesses. The following disposition is made of containers tested in this matter:

1. Containers that pass the test, and the visual examination required with it (*see C-3-2.4*), are marked with the date and year of the retest followed by an "S" (Example: "8-91S," indicating requalification by the simple hydrostatic test method in August 1991) and are permitted to be placed back in service.
2. Containers developing leaks or showing undue swelling or bulging must be rejected. If rejected for leaks, containers are permitted to be repaired in accordance with C-3.3.1.

(c) The recorded visual examination is permitted to be used to requalify containers for five years before the next qualification is due provided the container has been used exclusively for LP-Gas commercially free of corroding components. Inspection is to be made by a competent person, using as a guide the Compressed Gas Association *Standard for Visual Inspection of Steel Compressed Gas Cylinders* (CGA Pamphlet C-6), and recording the inspection results as required by DOT regulations. [Note: Reference to NPGA Safety Bulletin, *Recommended Procedures for Visual Inspection and Requalification of DOT (ICC) Cylinders in LP-Gas Service*, is also recommended.] The following disposition is to be made of containers inspected in this manner:

FIVE YEAR VISUAL INSPECTION REPORT

________________ Co. Date ________________

Cylinder Identification					Protective Coating		Cylinders Checked For							Disposition		
Serial Number	Date Mf'd.	ICC Spec. Number	Symbol	Name of Manufacturer	Type	Condition	Fire Damage	Dents or Digs	Footring	Leaks	Corrosion	Collar and/or Opening	Bulges	(see) (code)	Date Inspect.	Inspector's Initials
90165	6-58	4BA240	GAS INC.	ABCyL	PAINT	GOOD	✓	✓	✓	✓	✓	✓	✓	OK	1-6-72	JHD
220196	10-70	4BW240	ABCCLy	ABCyL	GALVANIZED	FAIR	✓	SC	✓	✓	✓	✓	✓	SC	1-6-72	JHD
109640	5-60	4BA240	XYZCoy	ABCyL	PAINT	GOOD	✓	✓	✓	✓	✓	R	✓	R	1-6-72	JHD

Disposition Code
OK—Return To Service
SC—Scrap
RM—Return To Manufacturer For Repair

Figure C.1 *Visual inspection report form. (Form courtesy of National Propane Gas Association.)*

The Compressed Gas Association's *Standard for Visual Inspection of Steel Compressed Gas Cylinders*, Pamphlet C-6, is a guide that can be used to requalify cylinders as required by DOT regulations. It contains examples of various types of pitting, corrosion, and failed cylinders and is very useful for persons not completely familiar with the subject.

The National Propane Gas Association publishes two safety pamphlets that provide useful information on requalifying cylinders: NPGA 117, *Visual Cylinder Inspection*, which is a guide for visual inspection of cylinders prior to refilling, and NPGA 118, *Recommended Procedures for Visual Inspection and Requalification of DOT (ICC) Cylinders in LP-Gas Service*, which provides the complete set of DOT requirements along with illustrations demonstrating the use of the gauges needed to complete the inspection and a "visual inspection report form." (*See Figure C.1.*)

1. Containers that pass the visual examination are marked with the date and year of the examination followed by an "E" (Example: "7-90E," indicating requalification by the recorded visual examination method in July 1990) and are permitted to be placed back in service.

2. Containers that leak, show serious denting or gouging, or excessive corrosion must either be scrapped or repaired in accordance with C-3.3.1.

C-3.3 Repair of DOT Cylinder Specification Containers.

C-3.3.1 Repair of DOT cylinders is required to be performed by a manufacturer of the type of cylinder to be repaired or by a repair facility authorized by DOT.

Repairs normally made are for fire damage, leaks, denting, gouges, and for broken or detached valve protecting collars or foot rings.

Design of ASME and API-ASME Containers

This appendix is not a part of the requirements of this NFPA document but is included for informational purposes only.

D-1 General.

D-1.1 Application.

D-1.1.1 This appendix provides general information on containers designed and constructed in accordance with ASME or API-ASME codes, usually referred to as ASME containers. For complete information on either ASME or API-ASME containers, the applicable code should be consulted. Construction of containers to the API-ASME Code has not been authorized since July 1, 1961.

D-1.1.2 DOT (ICC) specification portable tank containers and cargo tanks are either ASME or API-ASME containers. In writing these specifications, which should be consulted for complete information, additions were made to these pressure vessel codes to cover the following:

(a) Protection of container valves and appurtenances against physical damage in transportation.

(b) Holddown devices for securing cargo containers to conventional vehicles.

(c) Attachments to relatively large [6,000 gal (22.7 m^3) or more water capacity] cargo containers in which the container serves as a stress member in lieu of a frame.

D-1.2 Development of ASME and API-ASME Codes.

D-1.2.1 ASME-type containers of approximately 12,000 gal (45.4 m^3) water capacity or more were initially used for bulk storage in processing, distribution, and industrial plants. As the industry expanded and residential and commercial usage increased, the need for small ASME containers with capacities greater than the upper limit for DOT cylinders grew. This ultimately resulted in the development of cargo containers for tank trucks and the wide use of ASME containers ranging in size from less than 25 gal (0.1 m^3) to 120,000 gal (454 m^3) water capacity.

D-1.2.2 The American Society of Mechanical Engineers (ASME) in 1911 set up the Boiler and Pressure Vessel Committee to formulate "standard rules for the construction of steam boilers and other pressure vessels." The ASME *Boiler and Pressure Vessel Code*, first published in 1925, has been revised regularly since that time. During this period there have been changes in the Code as materials of construction improved and more was known about them, and as fabrication methods changed and inspection procedures were refined.

D-1.2.3 One major change involved the so-called "factor of safety" (the ratio of the ultimate strength of the metal to the design stress used). Prior to 1946, a 5:1 safety factor was used. Fabrication changed from the riveting widely used when the Code was first written (some forge welding was used) to fusion welding. This latter method was incorporated into the Code as welding techniques were perfected and now predominates.

D-1.2.4 The safety factor change in the ASME Code was based on the technical progress made since 1925 and on experience with the use of the API-ASME Code. This offshoot of the ASME Code, initiated in 1931, was formulated and published by the American Petroleum Institute (API) in cooperation with the ASME. It justified the 4:1 safety factor on the basis of certain quality and inspection controls not at that time incorporated in the ASME Code editions.

D-1.2.5 *ASME Code Case Interpretations and Addenda* are published between Code editions and normally become part of the Code in the new edition. Adherence to these is considered compliance with the Code. [*See 2-2.1.3(a).*]

D-2 Design of Containers for LP-Gas.

D-2.1 ASME Container Design.

D-2.1.1 When ASME containers were first used to store LP-Gas, the properties of the CP grades of the principal constituents were available, but the average properties for the commercial grades of propane and butane were not. Also, there was no experience as to what temperatures and pressures to expect for product stored in areas with high atmospheric temperatures. A 200-psi (1.40-MPa) design pressure was deemed appropriate for propane [the CP grade of which has a vapor pressure of 176 psi (1.2 MPa) at 100°F (37.8°C)] and 80 psi (0.55 MPa) for butane [CP grade has vapor pressure of 37 psi (0.26 MPa) at 100°F (37.8°C)]. These containers were built with a 5:1 safety factor (*see D-1.2.3*).

D-2.1.2 Pressure vessel codes, following boiler pressure relief valve practice, require that the pressure relief valve start-to-leak setting be the design pressure of the container. In specifying pressure relief valve capacity, however, they stipulate that this relieving capacity be adequate to prevent the internal pressure from rising above 120 percent of the design pressure under fire exposure conditions.

D-2.1.3 Containers built in accordance with D-2.1.1 were entirely adequate for the commercial grades of the LP-Gases [the vapor pressure of propane at 100°F (38°C) is 205 psi (1.41 MPa); the vapor pressure of butane at 100°F (38°C) is 37 psi (0.26 MPa)]. However, as they were equipped with pressure relief valves set to start-to-leak at the design pressure of the container, these relief valves occasionally opened on an unusually warm day. Since any unnecessary release of a flammable gas is potentially dangerous, and considering recommendations of fire prevention and insurance groups as well as to the favorable experience with API-ASME containers (*see D-2.2.1*), relief valve settings above the design pressure [up to 250 psi (1.7 MPa) for propane and 100 psi (0.69 MPa) for butane] were widely used.

D-2.1.4 In determining safe filling limits for compressed liquefied gases, DOT (ICC) uses the criterion that the container not become liquid full at the highest temperature the liquid may be expected to reach due to the normal atmospheric conditions to which the container may be exposed. For containers of more than 1,200 gal (4.5 m^3) water capacity, the liquid temperature selected is 115°F (46°C). The vapor pressure of the gas to be contained at 115°F (46°C) is specified by DOT as the minimum design pressure for the container. The vapor pressure of CP propane at 115°F (46.1°C) is 211 psi (1450 kPa), and of commercial propane, 243 psi (1670 kPa). The vapor pressure of both normal butane and commercial butane at 115°F (46.1°C) is 51 psi (350 kPa).

D-2.1.5 The ASME *Boiler and Pressure Vessel Code* editions generally applicable to LP-Gas containers, and the design pressures, safety factors, and exceptions to these editions for LP-Gas use, are shown in Table D-2.1.5. These reflect the use of the information in D-2.1.1 through D-2.1.4.

Table D-2.1.5

Year ASME Code Edition Published	Design Pressure, psi (MPa) Butane	Propane	Safety Factor
1931 through 46[2]	100[1] (0.7)	200 (1.4)	5:1
1949 Par. U-68 & U-69[2]	100 (0.7)	200 (1.4)	5:1
1949 Par. U-200 & U-201[2]	125 (0.9)	250 (1.7)	4:1
1952 through 80	125 (0.9)	250[3] (1.7)	4:1

[1]Until December 31, 1947, containers designed for 80 psi (0.6 MPa) under prior (5:1 safety factor) codes were authorized for butane. Since that time, either 100 psi (0.7 MPa) (under prior codes) or 125 psi (0.9 MPa) (under present codes) is required.

[2]Containers constructed in accordance with 1949 and prior editions of the ASME Code were not required to be in compliance with paragraphs U-2 to U-10 inclusive, or with paragraph U-19. Construction in accordance with paragraph U-70 of these editions was not authorized.

[3]Higher design pressure [312.5 psi (2.2 MPa)] is required for small ASME containers used for vehicular installations (such as forklift trucks used in buildings or those installed in enclosed spaces) because they may be exposed to higher temperatures and consequently develop higher internal pressure.

D-2.2 API-ASME Container Design.

D-2.2.1 The API-ASME Code was first published in 1931. Based on petroleum industry experience using certain material quality and inspection controls not at that time incorporated in the ASME Code, the 4:1 safety factor was first used. Many LP-Gas containers were built under this Code with design pressures of 125 psi (0.86 MPa) [100 psi (0.69 MPa) until December 31, 1947] for butane and 250 psi (1.7 MPa) for propane. Containers constructed in accordance with the API-ASME Code were not required to comply with Section 1, or the appendix to Section 1. Paragraphs W-601 through W-606 of the 1943 and earlier editions were not applicable to LP-Gas containers.

D-2.2.2 The ASME Code, by changing from the 5:1 to the 4:1 safety factor through consideration of the factors described in D-2.1.1 through D-2.1.4, became nearly identical in effect to the API-ASME Code by the 1950s. Thus, the API-ASME Code was phased out, and construction was not authorized after July 1, 1961.

D-2.3 Design Criteria for LP-Gas Containers.

D-2.3.1 To prevent confusion in earlier editions of this standard, the nomenclature "container type" was used to designate the design pressure of the container to be used for various types of LP-Gases. With the adoption of the 4:1 safety factor in the ASME Code and the phasing out of the API-ASME Code, the need for "container type" ceased to exist.

D-2.4 DOT (ICC) Specifications Utilizing ASME or API-ASME Containers.

D-2.4.1 DOT (ICC) Specifications for portable tank containers and cargo tanks require ASME or API-ASME construction for the container proper (*see D-1.1.2*). Several such specifications were written by the ICC prior to 1967, and DOT has continued this practice.

D-2.4.2 ICC Specifications written prior to 1946, and to some extent through 1952, used ASME containers with a 200 psi (1380 kPa) design pressure for propane and 80 psi (550 kPa) for butane [100 psi (690 kPa) after 1947] with a 5:1 safety factor. During this period and until 1961, ICC Specifications also permitted API-ASME containers with a 250 psi (1720 kPa) design pressure for propane and 100 psi (690 kPa) for butane [125 psi (862 kPa) after 1947].

D-2.4.3 To prevent any unnecessary release of flammable vapor during transportation (*see D-2.1.3*), the use of safety relief valve settings 25 percent above the design pressure was common for ASME 5:1 safety factor containers. To eliminate confusion, and in line with the good experience with API-ASME containers, the ICC permitted the rerating of these particular ASME containers used under its specifications to 125 percent of the originally marked design pressure.

D-2.4.4 DOT (ICC) Specifications applicable to portable tank containers and cargo tanks currently in use are listed in Table D-2.4.4. New construction is not permitted under the older specifications. However, these older containers are permitted to continue to be used provided they have been maintained in accordance with DOT (ICC) regulations.

Table D-2.4.4

Spec. Number	ASME Construction			API-ASME Construction		
	Design Pressure, psi		Safety Factor	Design Pressure, psi		Safety Factor
	Propane	Butane		Propane	Butane	
ICC-50[1]	200[3]	100[3]	5:1	250	125	4:1
ICC-51[1]	250	125	4:1	250	125	4:1
MC-320[2,4]	200[3]	100[3]	5:1	250	125	4:1
MC-330[2]	250	125	4:1	250	125	4:1
MC-331[2]	250	125	4:1	250	125	4:1

[1]Portable tank container.

[2]Cargo tank.

[3]Permitted to be rerated to 125 percent of original ASME design pressure.

[4]Require DOT exemption.

For SI Units

100 psi = 0.69 MPa; 125 psi = 0.86 MPa; 200 psi = 1.40 MPa; 250 psi = 1.72 MPa

D-3 Underground ASME or API-ASME Containers.

D-3.1 Use of Containers Underground.

D-3.1.1 ASME or API-ASME containers are used for underground or partially underground installation in accordance with 3-2.4.8 or 3-2.4.9. The temperature of the soil is normally low so that the average liquid temperature and vapor pressure of product stored in underground containers will be lower than in aboveground containers.

D-3.1.2 Containers listed to be used interchangeably for installation either aboveground or underground must comply as to pressure relief valve rated relieving capacity and filling density with aboveground provisions when installed aboveground [*see 2-3.2.4(a)*]. When installed underground the pressure relief valve rated relieving capacity and filling density can be in accordance with underground provisions (*see E-2.3.1*), provided all other underground installation provisions are met. Partially underground containers are considered as aboveground insofar as loading density and pressure relief valve rated relieving capacity are concerned.

Pressure Relief Devices

This appendix is not a part of the requirements of this NFPA document but is included for informational purposes only.

(This appendix contains non-NFPA mandated provisions.)

E-1 Pressure Relief Devices for DOT (ICC) Cylinders.

E-1.1 Source of Provisions for Relief Devices.

E-1.1.1 The requirements for relief devices on DOT cylinders are established by the DOT. Complete technical information regarding these requirements will be found in the Compressed Gas Association (CGA) Publication S-1.1, *Pressure-Relief Device Standards*, Part 1—Cylinders for Compressed Gases.

E-1.2 Essential Requirements of LP-Gas Cylinder Relief Devices.

E-1.2.1 CGA Publication S-1.1 provides that LP-Gas cylinders be equipped with fusible plugs, spring-loaded pressure relief valves, or a combination of the two. Fusible plugs are not permitted on cylinders used in certain vehicular installations [*see 8-2.3(a)4*]. The provisions of E-1.2.2 and E-1.2.3 outline the generally accepted industry practice in the use of fusible plugs and safety pressure devices on LP-Gas cylinders.

E-1.2.2 If fusible plugs constitute the only relief devices, the plugs used must comply with the flow capacity requirements of CGA S-1.1 with a nominal melting or yield point of 165°F (74°C) [not less than 157°F (69°C) nor more than 170°F (77°C)]. For cylinders over 30 in. (0.76 m) long (exclusive of neck), a plug is required in each end of the cylinder.

E-1.2.3 If a spring-loaded pressure relief valve(s) constitutes the only relief device, the valves used must comply with the flow capacity requirements of CGA S-1.1, with the set pressure not less than 75 percent nor more than 100 percent of the minimum required test pressure of the cylinder. For example, the test pressure for a 240-psi (1.65-MPa) service pressure is 480 psi (3.31 MPa); 75 percent of this is 360 psi (2.48 MPa). In practice, such valves are set at 375 psi (2.59 MPa).

E-2 Pressure Relief Devices for ASME Containers.

E-2.1 Source of Provisions for Relief Devices.

E-2.1.1 Capacity requirements for relief devices are in accordance with the applicable provisions of Compressed Gas Association (CGA) Publication S-1.2, *Pressure-Relief Device Standards*, Part 2—Cargo and Portable Tanks for Compressed Gases; or with CGA Publication S-1.3, *Safety Relief-Device Standards*, Part 3—Compressed Gas Storage Containers.

E-2.2 Spring-Loaded Pressure Relief Valves for Aboveground and Cargo Containers.

E-2.2.1 The minimum rate of discharge for spring-loaded pressure relief valves is based on the outside surface of the containers on which the valves are installed. Paragraph 2-2.6.5(h) provides that new containers be marked with the surface area in sq ft. The surface area of containers not so marked (or not legibly marked) can be computed by use of the applicable formula:

(a) Cylindrical container with hemispherical heads:

Surface area = overall length × outside diameter × 3.1416.

(b) Cylindrical container with other than hemispherical heads:

Surface area = (overall length + 0.3 outside diameter) × outside diameter × 3.1416.

NOTE: This formula is not precise, but will give results with limits of practical accuracy in sizing relief valves.

(c) Spherical containers:

Surface area = outside diameter squared × 3.1416.

E-2.2.2 The minimum required relieving capacity in cu ft per min of air at 120 percent of the maximum permitted start-to-leak pressure (or Flow Rate CFM Air), under standard conditions of 60°F (16°C) and atmospheric pressure [14.7 psia (an absolute pressure of 0.1 MPa)], must be as shown in Table E-2.2.2 for the surface area in sq ft of the container on which the pressure relief valve is to be installed. The flow rate can be interpolated for intermediate values of surface area. For containers with a total outside surface area exceeding 2,000 sq ft (185.8 m^2), the required flow rate should be calculated using the formula:

Flow Rate CFM Air $= 53.632 \times A^{0.82}$ where A = total outside surface area of container in sq ft.

Table E-2.2.2

Surface Area Sq Ft	Flow Rate CFM Air	Surface Area Sq Ft	Flow Rate CFM Air	Surface Area Sq Ft	Flow Rate CFM Air
20 or less	626	170	3620	600	10170
25	751	175	3700	650	10860
30	872	180	3790	700	11550
35	990	185	3880	750	12220
40	1100	190	3960	800	12880
45	1220	195	4050	850	13540
50	1330	200	4130	900	14190
55	1430	210	4300	950	14830
60	1540	220	4470	1000	15470
65	1640	230	4630	1050	16100
70	1750	240	4800	1100	16720
75	1850	250	4960	1150	17350
80	1950	260	5130	1200	17960
85	2050	270	5290	1250	18570
90	2150	280	5450	1300	19180
95	2240	290	5610	1350	19780
100	2340	300	5760	1400	20380
105	2440	310	5920	1450	20980
110	2530	320	6080	1500	21570
115	2630	330	6230	1550	22160
120	2720	340	6390	1600	22740
125	2810	350	6540	1650	23320
130	2900	360	6690	1700	23900
135	2990	370	6840	1750	24470
140	3080	380	7000	1800	25050
145	3170	390	7150	1850	25620
150	3260	400	7300	1900	26180
155	3350	450	8040	1950	26750
160	3440	500	8760	2000	27310
165	3530	550	9470		

E-2.3 Spring-Loaded Pressure Relief Valves for Underground or Mounded Containers.

E-2.3.1 In the case of containers installed underground or mounded, the pressure relief valve relieving capacities are permitted to be as small as 30 percent of those specified in Table E-2.2.2 provided the container is empty of liquid when installed, that no liquid is placed in it until it is completely covered with earth, and that it is not uncovered for removal until all liquid has been removed.

E-2.3.2 Containers partially underground must have pressure relief valve relieving capacities in accordance with 2-3.2.4.

E-2.4 Provisions for Fusible Plugs.

E-2.4.1 Fusible plugs, supplementing spring-loaded pressure relief valves and complying with 2-3.2.4(e), are permitted only with aboveground stationary containers of 1,200 gal (4.5 m^3) or less water capacity. They should not be used on larger containers or on portable or cargo containers of ASME construction. The total fusible plug discharge area is limited to 0.25 sq in. (1.6 cm^2) per container.

E-2.5 Pressure Relief Valve Testing.

E-2.5.1 Frequent testing of pressure relief valves on LP-Gas containers is not considered necessary for the following reasons:

(a) The LP-Gases are so-called "sweet gases" having no corrosive or other deleterious effect on the metal of the containers or relief valves.

(b) The relief valves are constructed of corrosion-resistant materials and are installed so as to be protected against the weather. The variations of temperature and pressure due to atmospheric conditions are not sufficient to cause any permanent set in the valve springs.

(c) The required odorization of the LP-Gases makes escape almost instantly evident.

(d) Experience over the years with the storage of LP-Gases has shown a good safety record on the functioning of pressure relief valves.

E-2.5.2 Since no mechanical device can be expected to remain in operative condition indefinitely, it is suggested that the pressure relief valves on containers of more than 2,000 gal (7.6 m^3) water capacity be tested at approximately 10-year intervals. Some types of valves can be tested by the use of an external lifting device having an indicator to show the pressure equivalent at which the valve may be expected to open. Others must be removed from the container for testing, requiring that the container first be emptied.

Liquid Volume Tables, Computations, and Graphs

This appendix is not a part of the requirements of this NFPA document but is included for informational purposes only.

F-1 Scope.

F-1.1 Application. This appendix explains the basis for Table 4-4.2.1, includes the LP-Gas liquid volume temperature correction table, Table F-3.1.3, and describes its use. It also explains the methods of making liquid volume computations to determine the maximum permissible LP-Gas content of containers in accordance with Tables 4-4.2.2(a), (b), and (c).

F-2 Basis for Determination of LP-Gas Container Capacity.

F-2.1 The basis for determination of the maximum permitted filling limits shown in Table 4-4.2.1 is the maximum safe quantity, which will assure that the container will not become liquid full when the liquid is at the highest anticipated temperature.

(a) For portable containers built to DOT specifications and other aboveground containers with water capacities of 1,200 (4.5 m^3) gal or less, this temperature is assumed to be 130°F (54°C).

(b) For other aboveground uninsulated containers with water capacities in excess of 1,200 gal (4.5 m^3), including those built to DOT portable or cargo tank specifications, this temperature is assumed to be 115°F (46°C).

(c) For all containers installed underground, this temperature is assumed to be 105°F (41°C).

F-3 Liquid Volume Correction Table.

Correction of Observed Volume to Standard Temperature Condition (60°F and Equilibrium Pressure).

F-3.1.1 The volume of a given quantity of LP-Gas liquid in a container is directly related to its temperature, expanding as temperature increases and contracting as temperature decreases. Standard conditions, often used for weights and measures purposes and, in some cases, to comply with safety regulations, specify correction of the observed volume to what it would be at 60°F (16°C).

F-3.1.2 To correct the observed volume to 60°F (16°C), the specific gravity of LP-Gas at 60°F (16°C) in relation to water at 60°F (16°C) (usually referred to as "60°/60°F") and its average temperature must be known. The specific gravity normally appears on the shipping papers. The average liquid temperature can be obtained as follows:

(a) Insert a thermometer in a thermometer well in the container into which the liquid has been transferred, and read the temperature after the completion of the transfer [*see F-3.1.2(c) for proper use of a thermometer*].

(b) If the container is not equipped with a well, but is essentially empty of liquid prior to loading, the temperature of the liquid in the container from which liquid is being withdrawn can be used. Otherwise, a thermometer can be inserted in a thermometer well or other temperature sensing device installed in the loading line at a point close to the container being loaded, reading temperatures at intervals during transfer and averaging. [*See F-3.1.2(c).*]

(c) A suitable liquid should be used in thermometer wells to obtain an efficient heat transfer from the LP-Gas liquid in the container to the thermometer bulb. The liquid used should be noncorrosive and should not freeze at the temperatures to which it will be subjected. Water should not be used.

F-3.1.3 The volume observed or measured is corrected to 60°F (16°C) by use of Table F-3.1.3. The column headings, across the top of the tabulation, list the range of specific gravities for the LP-Gases. Specific gravities are shown from 0.500 to 0.590 by 0.010 increments, except that special columns are inserted for chemically

Table F-3.1.3 Liquid Volume Correction Factors

Observed Temperature Degrees Fahrenheit	Specific Gravities at 60°F/60°F												
	0.500	Propane 0.5079	0.510	0.520	0.530	0.540	0.550	0.560	iso-Butane 0.5631	0.570	0.580	n-Butane 0.5844	0.590
		Volume Correction Factors											
–50	1.160	1.155	1.153	1.146	1.140	1.133	1.127	1.122	1.120	1.116	1.111	1.108	1.106
–45	1.153	1.148	1.146	1.140	1.134	1.128	1.122	1.117	1.115	1.111	1.106	1.103	1.101
–40	1.147	1.142	1.140	1.134	1.128	1.122	1.117	1.111	1.110	1.106	1.101	1.099	1.097
–35	1.140	1.135	1.134	1.128	1.122	1.116	1.112	1.106	1.105	1.101	1.096	1.094	1.092
–30	1.134	1.129	1.128	1.122	1.116	1.111	1.106	1.101	1.100	1.096	1.092	1.090	1.088
–25	1.127	1.122	1.121	1.115	1.110	1.105	1.100	1.095	1.094	1.091	1.087	1.085	1.083
–20	1.120	1.115	1.114	1.109	1.104	1.099	1.095	1.090	1.089	1.086	1.082	1.080	1.079
–15	1.112	1.109	1.107	1.102	1.097	1.093	1.089	1.084	1.083	1.080	1.077	1.075	1.074
–10	1.105	1.102	1.100	1.095	1.091	1.087	1.083	1.079	1.078	1.075	1.072	1.071	1.069
–5	1.098	1.094	1.094	1.089	1.085	1.081	1.077	1.074	1.073	1.070	1.067	1.066	1.065
0	1.092	1.088	1.088	1.084	1.080	1.076	1.073	1.069	1.068	1.066	1.063	1.062	1.061
2	1.089	1.086	1.085	1.081	1.077	1.074	1.070	1.067	1.066	1.064	1.061	1.060	1.059
4	1.086	1.083	1.082	1.079	1.075	1.071	1.068	1.065	1.064	1.062	1.059	1.058	1.057
6	1.084	1.080	1.080	1.076	1.072	1.069	1.065	1.062	1.061	1.059	1.057	1.055	1.054
8	1.081	1.078	1.077	1.074	1.070	1.066	1.063	1.060	1.059	1.057	1.055	1.053	1.052
10	1.078	1.075	1.074	1.071	1.067	1.064	1.061	1.058	1.057	1.055	1.053	1.051	1.050
12	1.075	1.072	1.071	1.068	1.064	1.061	1.059	1.056	1.055	1.053	1.051	1.049	1.048
14	1.072	1.070	1.069	1.066	1.062	1.059	1.056	1.053	1.053	1.051	1.049	1.047	1.046
16	1.070	1.067	1.066	1.063	1.060	1.056	1.054	1.051	1.050	1.048	1.046	1.045	1.044
18	1.067	1.065	1.064	1.061	1.057	1.054	1.051	1.049	1.048	1.046	1.044	1.043	1.042
20	1.064	1.062	1.061	1.058	1.054	1.051	1.049	1.046	1.046	1.044	1.042	1.041	1.040
22	1.061	1.059	1.058	1.055	1.052	1.049	1.046	1.044	1.044	1.042	1.040	1.039	1.038
24	1.058	1.056	1.055	1.052	1.049	1.046	1.044	1.042	1.042	1.040	1.038	1.037	1.036
26	1.055	1.053	1.052	1.049	1.047	1.044	1.042	1.039	1.039	1.037	1.036	1.036	1.034
28	1.052	1.050	1.049	1.047	1.044	1.041	1.039	1.037	1.037	1.035	1.034	1.034	1.032
30	1.049	1.047	1.046	1.044	1.041	1.039	1.037	1.035	1.035	1.033	1.032	1.032	1.030
32	1.046	1.044	1.043	1.041	1.038	1.036	1.035	1.033	1.033	1.031	1.030	1.030	1.028
34	1.043	1.041	1.040	1.038	1.036	1.034	1.032	1.031	1.030	1.029	1.028	1.028	1.026
36	1.039	1.038	1.037	1.035	1.033	1.031	1.030	1.028	1.028	1.027	1.025	1.025	1.024
38	1.036	1.035	1.034	1.032	1.031	1.029	1.027	1.026	1.025	1.025	1.023	1.023	1.022
40	1.033	1.032	1.031	1.029	1.028	1.026	1.025	1.024	1.023	1.023	1.210	1.021	1.020
42	1.030	1.029	1.028	1.027	1.025	1.024	1.023	1.022	1.021	1.021	1.019	1.019	1.018
44	1.027	1.026	1.025	1.023	1.022	1.021	1.020	1.019	1.019	1.018	1.017	1.017	1.016
46	1.023	1.022	1.022	1.021	1.020	1.018	1.018	1.017	1.016	1.016	1.015	1.015	1.014
48	1.020	1.019	1.019	1.018	1.017	1.016	1.015	1.014	1.014	1.013	1.013	1.013	1.012
50	1.017	1.016	1.016	1.015	1.014	1.013	1.013	1.012	1.012	1.011	1.011	1.011	1.010
52	1.014	1.013	1.012	1.012	1.011	1.010	1.010	1.009	1.009	1.009	1.009	1.009	1.008
54	1.010	1.010	1.009	1.009	1.008	1.008	1.007	1.007	1.007	1.007	1.006	1.006	1.006
56	1.007	1.007	1.006	1.006	1.005	1.005	1.005	1.005	1.005	1.005	1.004	1.004	1.004
58	1.003	1.003	1.003	1.003	1.003	1.003	1.002	1.002	1.002	1.002	1.002	1.002	1.002
60	1.000	1.000	1.000	1.000	1.000	1.000	1.000	1.000	1.000	1.000	1.000	1.000	1.000
62	0.997	0.997	0.997	0.997	0.997	0.997	0.997	0.998	0.998	0.998	0.998	0.998	0.998
64	0.993	0.993	0.994	0.994	0.994	0.994	0.995	0.995	0.995	0.995	0.996	0.996	0.996

Table F-3.1.3 (continued)

Observed Temperature Degrees Fahrenheit	0.500	Propane 0.5079	0.510	0.520	0.530	0.540	0.550	0.560	iso-Butane 0.5631	0.570	0.580	n-Butane 0.5844	0.590
	Specific Gravities at 60°F/60°F — Volume Correction Factors												
66	0.990	0.990	0.990	0.990	0.991	0.992	0.992	0.993	0.993	0.993	0.993	0.993	0.993
68	0.986	0.986	0.987	0.987	0.988	0.989	0.990	0.990	0.990	0.990	0.991	0.991	0.991
70	0.983	0.983	0.984	0.984	0.985	0.986	0.987	0.988	0.988	0.988	0.989	0.989	0.989
72	0.979	0.980	0.981	0.981	0.982	0.983	0.984	0.985	0.986	0.986	0.987	0.987	0.987
74	0.976	0.976	0.977	0.978	0.980	0.980	0.982	0.983	0.983	0.984	0.985	0.985	0.985
76	0.972	0.973	0.974	0.975	0.977	0.978	0.979	0.980	0.981	0.981	0.982	0.982	0.983
78	0.969	0.970	0.970	0.972	0.974	0.975	0.977	0.978	0.978	0.979	0.980	0.980	0.981
80	0.965	0.967	0.967	0.969	0.971	0.972	0.974	0.975	0.976	0.977	0.978	0.978	0.979
82	0.961	0.963	0.963	0.966	0.968	0.969	0.971	0.972	0.973	0.974	0.976	0.976	0.977
84	0.957	0.959	0.960	0.962	0.965	0.966	0.968	0.970	0.971	0.972	0.974	0.974	0.975
86	0.954	0.956	0.956	0.959	0.961	0.964	0.966	0.967	0.968	0.969	0.971	0.971	0.972
88	0.950	0.952	0.953	0.955	0.958	0.961	0.963	0.965	0.966	0.967	0.969	0.969	0.970
90	0.946	0.949	0.949	0.952	0.955	0.958	0.960	0.962	0.963	0.964	0.967	0.967	0.968
92	0.942	0.945	0.946	0.949	0.952	0.955	0.957	0.959	0.960	0.962	0.964	0.965	0.966
94	0.938	0.941	0.942	0.946	0.949	0.952	0.954	0.957	0.958	0.959	0.962	0.962	0.964
96	0.935	0.938	0.939	0.942	0.946	0.949	0.952	0.954	0.955	0.957	0.959	0.960	0.961
98	0.931	0.934	0.935	0.939	0.943	0.946	0.949	0.952	0.953	0.954	0.957	0.957	0.959
100	0.927	0.930	0.932	0.936	0.940	0.943	0.946	0.949	0.950	0.952	0.954	0.955	0.957
105	0.917	0.920	0.923	0.927	0.931	0.935	0.939	0.943	0.943	0.946	0.949	0.949	0.951
110	0.907	0.911	0.913	0.918	0.923	0.927	0.932	0.936	0.937	0.939	0.943	0.944	0.946
115	0.897	0.902	0.904	0.909	0.915	0.920	0.925	0.930	0.930	0.933	0.937	0.938	0.940
120	0.887	0.892	0.894	0.900	0.907	0.912	0.918	0.923	0.924	0.927	0.931	0.932	0.934
125	0.876	0.881	0.884	0.890	0.898	0.903	0.909	0.916	0.916	0.920	0.925	0.927	0.928
130	0.865	0.871	0.873	0.880	0.888	0.895	0.901	0.908	0.909	0.913	0.918	0.921	0.923
135	0.854	0.861	0.863	0.871	0.879	0.887	0.894	0.901	0.902	0.907	0.912	0.914	0.916
140	0.842	0.850	0.852	0.861	0.870	0.879	0.886	0.893	0.895	0.900	0.905	0.907	0.910

pure propane, isobutane, and normal butane. To obtain a correction factor, follow down the column for the specific gravity of the particular LP-Gas to the factor corresponding with the liquid temperature. Interpolation between the specific gravities and temperatures shown can be used if necessary.

F-3.2 Use of Liquid Volume Correction Factors, Table F-3.1.3.

F-3.2.1 To correct the observed volume in gal for any LP-Gas (the specific gravity and temperature of which is known) to gal at 60°F (16°C), Table F-3.1.3 is used as follows:

(a) Obtain the correction factor for the specific gravity and temperature as described in F-3.1.3.

(b) Multiply the gal observed by this correction factor to obtain the gal at 60°F (16°C).

EXAMPLE: A container has in it 4,055 gal of LP-Gas with a specific gravity of 0.560 at a liquid temperature of 75°F. The correction factors in the 0.560 column are 0.980 at 76°F and 0.983 at 74°F, or, interpolating, 0.9815 for 75°F. The volume of liquid at 60°F is 4,055 × 0.9815, or 3,980 gal.

F-3.2.2 To determine the volume in gal of a particular LP-Gas at temperature "t" to correspond with a given number of gal at 60°F (16°C), Table F-3.1.3 is used as follows:

(a) Obtain the correction factor for the LP-Gas, using the column for its specific gravity and reading the factor for temperature "t."

(b) Divide the number of gal at 60°F (16°C) by this correction factor to obtain the volume at temperature "t."

EXAMPLE: It is desired to pump 800 gal (3.03 m^3) at 60°F (15.5°C) into a container. The LP-Gas has a specific gravity of 0.510 and the liquid temperature is 44°F. The correction factor in the 0.510 column for 44°F is 1.025. Volume to be pumped at 44°F is 800 / 1.025 = 780 gal (2.95 m^3).

F-4 Maximum Liquid Volume Computations.

F-4.1 Maximum Liquid LP-Gas Content of a Container at Any Given Temperature.

F-4.1.1 The maximum liquid LP-Gas content of any container depends upon the size of the container, whether it is installed aboveground or underground, the maximum permitted filling limit, and the temperature of the liquid [*see Tables 4-4.2.2(a), (b), and (c).*]

F-4.1.2 The maximum volume "V_t" (in percent of container capacity) of an LP-Gas at temperature "t," having a specific gravity "G" and a filling limit of "L," is computed by use of the formula:

$$V_t = \frac{L}{G} \div F, \text{ or } V_t = \frac{L}{G \times F} \quad \text{where:}$$

V_t = percent of container capacity that can be filled with liquid
L = filling limit
G = specific gravity of particular LP-Gas
F = correction factor to correct volume at temperature "t" to 60°F (16°C).

EXAMPLE 1: The maximum liquid content, in percent of container capacity, for an aboveground 500-gal (1.89-m^3) water capacity container of an LP-Gas having a specific gravity of 0.550 and at a liquid temperature of 45°F (7.2°C) is computed as follows:

From Table 4-4.2.1, L = 0.47, and from Table F-3.1.3, °F = 1.019.

$$\text{Thus } V_{45} = \frac{0.47}{0.550 \times 1.019} = 0.838 \text{ (83\%), or 415 gal. (1.57m}^3\text{)}$$

EXAMPLE 2: The maximum liquid content, in percent of container capacity, for an aboveground 30,000-gal (114-m^3) water capacity container of LP-Gas having a specific gravity of 0.508 and at a liquid temperature of 80°F (27°C) is computed as follows:

From Table 4-4.2.1, L = 0.45, and from Table F-3.1.3, °F = 0.967.

$$\text{Thus } V_{80} = \frac{0.45}{0.508 \times 0.967} = 0.915 \text{ (91\%), or 27,300 gal. (103 m}^3\text{)}$$

F-4.2 Alternate Method of Filling Containers.

F-4.2.1 Containers equipped only with fixed maximum level gauges or only with variable liquid level gauges, when temperature determinations are not practical, can be filled with either gauge provided the fixed maximum liquid level gauge is installed, or the variable gauge is set, to indicate the volume equal to the maximum permitted filling limit as provided in 4-4.2.2(a). This level is computed on the basis that the liquid temperature will be 40°F (4.4°C) for aboveground containers or 50°F (10°C) for underground containers.

F-4.2.2 The percentage of container capacity that can be filled with liquid is computed by use of the formula shown in F-4.1.2, substituting the appropriate values as follows:

$$V_t = \frac{L}{G \times F}, \text{ where:}$$

t = the liquid temperature. Assumed to be 40°F (4.4°C) for aboveground containers or 50°F (10°C) for underground containers.

L = the loading limit obtained from Table 4-4.2.1 for:

(a) the specific gravity of the LP-Gas to be contained.

(b) the method of installation, aboveground or underground, and if aboveground, then:

1. for containers of 1,200 gal (4.5 m^3) water capacity or less.
2. for containers of more than 1,200 gal (4.5 m^3) water capacity.

G = the specific gravity of the LP-Gas to be contained.

F = the correction factor. Obtained from Table F-3.1.3, using G and 40°F (4°C) for aboveground containers or 50°F (10°C) for underground containers.

EXAMPLE: The maximum volume of LP-Gas with a specific gravity of 0.550 that can be in a 1,000-gal (3.8-m^3) water capacity aboveground container that is filled by use of a fixed maximum liquid level gauge is computed as follows:

t is 40°F for an aboveground container.

L for 0.550 specific gravity, and an aboveground container of less than 1,200 gal (4.5 m^3) water capacity, from Table 4-4.2.1, is 47 percent.

G is 0.550.

F for 0.550 specific gravity at 40°F (4.4°C) from Table F-3.1.3 is 1.025.

$$\text{Thus } V_{40} = \frac{0.47}{0.550 \times 1.025} = 0.834\ (83\%),\ \text{or } 830 \text{ gal. } (3.1\ m^3)$$

F-4.2.3 Percentage values, such as in the example in F-4.2.2, are rounded off to the next lower full percentage point, or to 83 percent in this example.

F-4.3 Location of Fixed Maximum Liquid Level Gauges in Containers.

F-4.3.1 Due to the diversity of fixed liquid gauges, and the many sizes [from DOT cylinders to 120,000 gal (454 m^3) ASME vessels] and types (vertical, horizontal, cylindrical, and spherical) of containers in which gauges are installed, it is not possible to tabulate the liquid levels such gauges should indicate for the maximum permitted filling limits [*see Tables 4-4.2.1 and 4-4.2.2(a)*].

F-4.3.2 The percentage of container capacity that these gauges should indicate is computed by use of the formula in F-4.1.2. The liquid level this gauge should indicate is obtained by applying this percentage to the water capacity of the container in gal [water at 60°F (16°C)] and then using the strapping table for the container (obtained from its manufacturer) to determine the liquid level for this gallonage. If such a table is not available, this liquid level is computed from the internal dimensions of the container, using data from engineering handbooks.

F-4.3.3 The formula of F-4.1.2 is used to determine the maximum LP-Gas liquid content of a container to comply with Table 4-4.2.1 and 4-4.2.2(a), as follows:

$$\text{Volumetric percentage, or } V_t = \frac{L}{G \times F}, \text{ and}$$

Volume in Gallons = V_t × Container Gallons Water Capacity, or
Vol. in Gal. at t =

$$\frac{\text{L (Table 4-4.2.1)} \times \text{Container Gallons Water Capacity}}{\text{G (Spec. Grav.)} \times \text{F (For G and at temperature t)}}$$

EXAMPLE 1: Assume a 100-gal (379-L) water capacity container for underground storage of propane with a specific gravity of 0.510. From Table 4-4.2.1, L = 46 percent; from 4-4.2.2(a), t = 50°F; and from Table F-3.1.3, °F for 0.510 specific gravity and a temperature of 50°F (10°C) is 1.016; or

$$\text{Vol. in Gal. at } 50°\text{F} = \frac{0.46 \times 100}{0.510 \times 1.016} = 88.7 \text{ gal. } (335 \text{ l})$$

EXAMPLE 2: Assume an 18,000-gal (68.1-m^3) water capacity container for aboveground storage of a mixture with a specific gravity of 0.550. From Table 4-4.2.1, L = 50 percent; from 4-4.2.2(a), t = 40°F; and from Table F-3.1.3, °F for 0.550 specific gravity and 40°F (4.4°C) temperature is 1.025; or

$$\text{Vol. in Gal. at } 40°\text{F} = \frac{0.50 \times 18{,}000}{0.550 \times 1.025} = 15{,}950 \text{ gal. } (60.3 \text{ m}^3)$$

(c) Before exposure to the torch fire, none of the thermocouples on the thermal insulation system steel plate configuration shall indicate a plate temperature in excess of 100°F (38°C) or less than 32°F (0°C).

(d) The entire outside surface of the thermal insulation system shall be exposed to the torch fire environment.

(e) A torch fire test shall be run for a minimum of 50 min. The thermal insulation system shall retard the heat flow to the steel plates so that none of the thermocouples on the uninsulated side of the steel plate indicates a plate temperature in excess of 800°F (427°C).

H-4 Hose Stream Resistance Test.

H-4.1 After 20 min exposure to the torch test, the test sample shall be hit with a hose stream concurrently with the torch for a period of 10 min. The hose stream test shall be conducted in the following manner:

(a) The stream shall be directed first at the middle and then at all parts of the exposed surface, making changes in direction slowly.

(b) The hose stream shall be delivered through a 2½-in. (64-mm) hose discharging through a National Standard Playpipe of corresponding size equipped with 1⅛-in. (29-mm) discharge tip of the standard-taper smooth-bore pattern without shoulder at the orifice. The water pressure at the base of the nozzle and for the duration of the test shall be 30 psi (207 kPa). [Estimated delivery rate is 205 gpm (776 L/min).]

(c) The tip of the nozzle shall be located 20 ft (6 m) from and on a line normal to the center of the test specimen. If impossible to be so located, the nozzle may be on a line deviating not to exceed 30 degrees from the line normal to the center of the test specimen. When so located, the distance from the center shall be less than 20 ft (6 m) by an amount equal to 1 ft (0.3 m) for each 10 degrees of deviation from the normal.

(d) Subsequent to the application of the hose stream, the torching shall continue until any thermocouple on the uninsulated side of the steel plate indicates a plate temperature in excess of 800°F (427°C).

(e) The thermal insulation system shall be judged to be resistant to the action of the hose stream if the time from initiation of torching for any thermocouple on the uninsulated side of the steel plate to reach in excess of 800°F (427°C) is 50 min or greater.

(f) One (1) successful combination torch fire and hose stream test shall be required for certification.

I

Container Spacing

This appendix is not a part of the requirements of this NFPA document but is included for informational purposes only.

This appendix contains three drawings that illustrate the separation distance required for the installation of LP-Gas containers up to 2,000 gal. They incorporate the distances required in Table 3-2.2.2, and its notes. As this table is the single most used item in NFPA 58, the need for clarity and unambiguous implementation of the table is of great importance, and these drawings make it much easier for all users to apply the table, and the important notes properly.

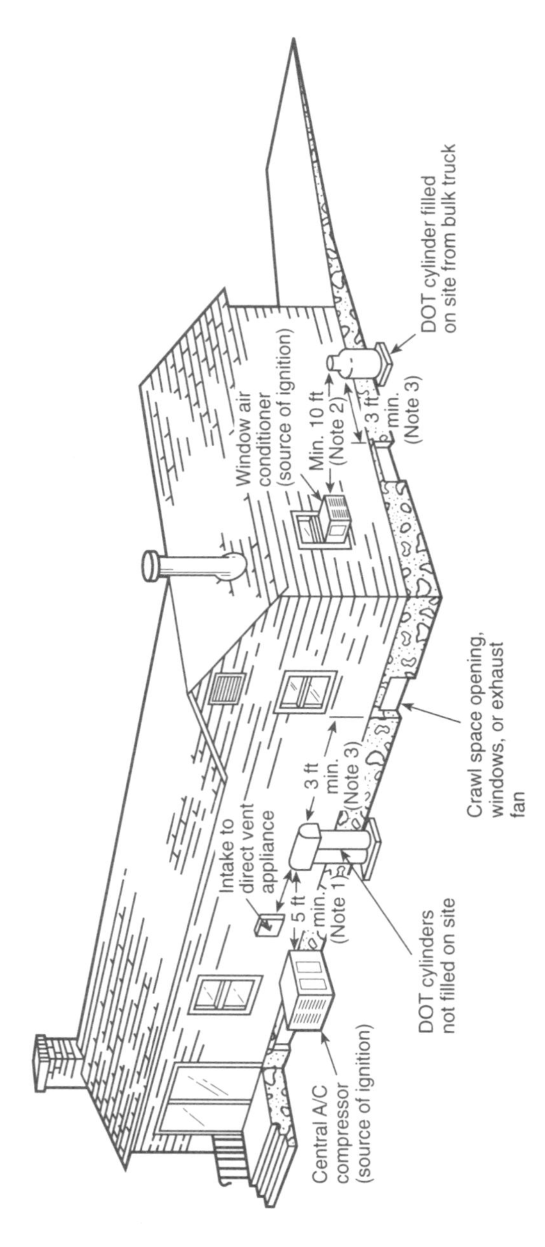

Note 1: 5-ft minimum from relief valve in any direction away from any exterior source of ignition, openings into direct vent appliances, or mechanical ventilation air intakes. Refer to Note (b) (1) under Table 3-2.2.2.

(For SI units: 1 ft = 0.3048 m)

Note 2: If the DOT cylinder is filled on site from a bulk truck, the filling connection and vent valve must be at least 10 ft from any exterior source of ignition, openings into direct-vent appliances, or mechanical ventilation air intakes. Refer to Note (b) (3) under Table 3-2.2.2.

Note 3: Refer to Note (b) (1) under Table 3-2.2.2.

***Figure I-1** DOT cylinders. (This figure for illustrative purposes only; text shall govern.)*

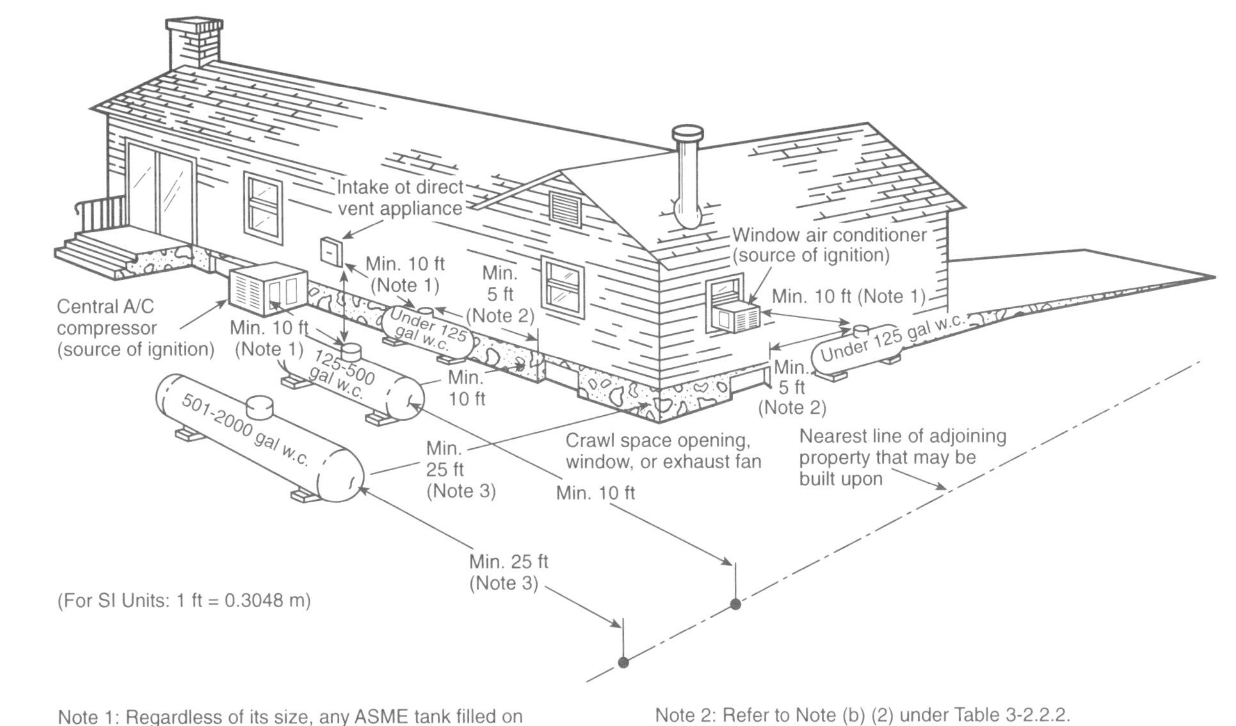

Note 1: Regardless of its size, any ASME tank filled on site must be located so that the filling connection and fixed liquid level gauge are at least 10 ft from any external source of ignition (i.e., open flame, window A/C, compressor, etc.). Intake to direct vented gas appliance or intake to a mechanical ventilation system. Refer to Note (b) (3) under Table 3-2.2.2.

Note 2: Refer to Note (b) (2) under Table 3-2.2.2.

Note 3: This distance may be reduced to no less than 10 ft (3 m) for a single container of 1,200-gal (4.5-m^3) water capacity or less provided such container is at least 25 ft (7.6 m) from any other LP-Gas container of more than 125-gal (0.5-m^3) water capacity. Refer to Note (c) under Table 3-2.2.2.

Figure I-2 *Aboveground ASME containers. (This figure for illustrative purposes only; text shall govern.)*

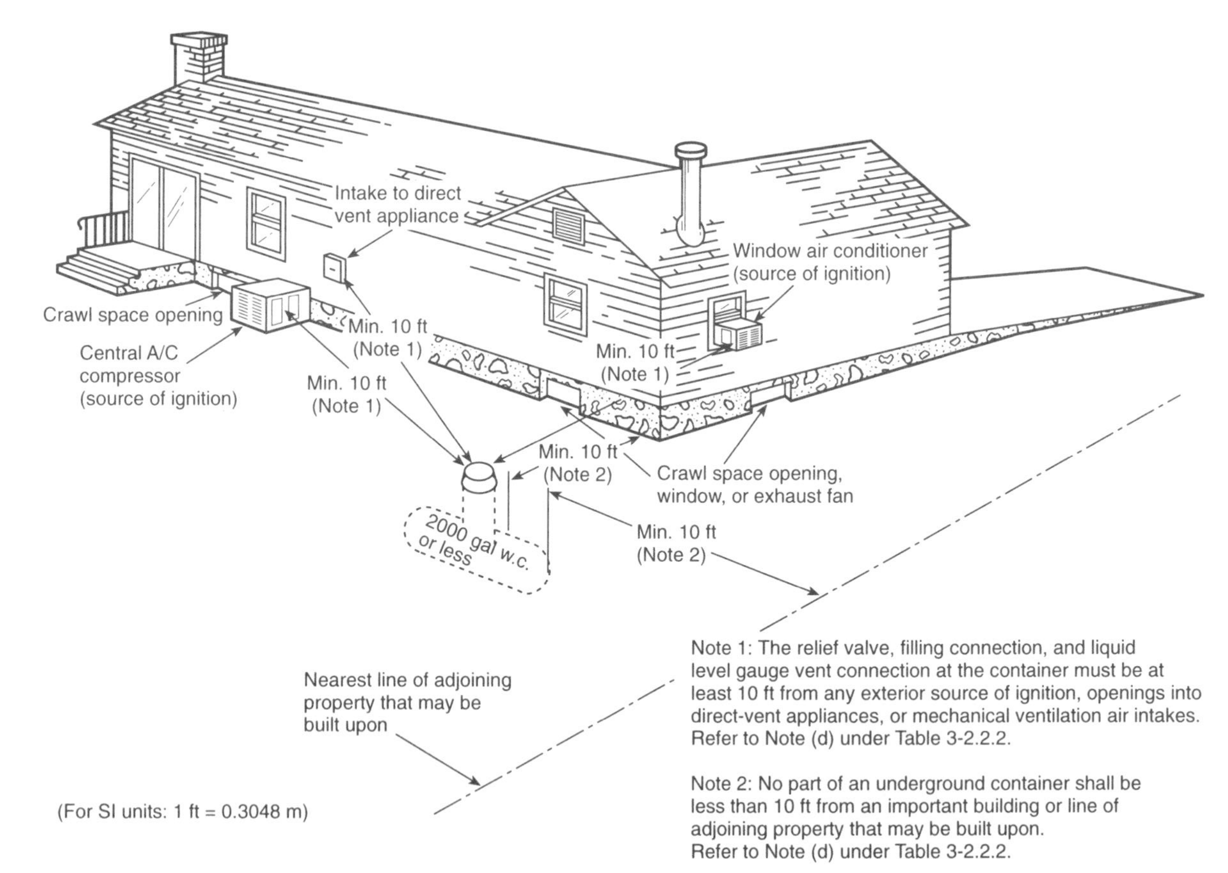

Figure I-3 *Underground ASME containers. (This figure for illustrative purposes only; text shall govern.)*

J

Referenced Publications

J-1

The following documents or portions thereof are referenced within this standard for informational purposes only and thus are not considered part of the requirements of this document. The edition indicated for each reference is the current edition as of the date of the NFPA issuance of this document.

J-1.1 NFPA Publications. National Fire Protection Association, 1 Batterymarch Park, P.O. Box 9101, Quincy, MA 02269-9101.

NFPA 10, *Standard for Portable Fire Extinguishers*, 1994 edition.

NFPA 37, *Standard for the Installation and Use of Stationary Combustion Engines and Gas Turbines*, 1994 edition.

NFPA 50, *Standard for Bulk Oxygen Systems at Consumer Sites*, 1990 edition.

NFPA 50A, *Standard for Gaseous Hydrogen Systems at Consumer Sites*, 1994 edition.

NFPA 51, *Standard for the Design and Installation of Oxygen-Fuel Gas Systems for Welding, Cutting, and Allied Processes*, 1992 edition.

NFPA 61B, *Standard for the Prevention of Fires and Explosions in Grain Elevators and Facilities Handling Bulk Raw Agricultural Commodities*, 1989 edition.

NFPA 68, *Guide for Venting of Deflagrations*, 1994 edition.

NFPA 77, *Recommended Practice on Static Electricity*, 1993 edition.

NFPA 80, *Standard for Fire Doors and Fire Windows*, 1992 edition.

NFPA 220, *Standard on Types of Building Construction*, 1992 edition.

NFPA 251, *Standard Methods of Fire Tests of Building Construction and Materials*, 1990 edition.

NFPA 252, *Standard Methods of Fire Tests of Door Assemblies*, 1995 edition.

NFPA 321, *Standard on Basic Classification of Flammable and Combustible Liquids*, 1991 edition.

NFPA 780, *Lightning Protection Code*, 1992 edition.

J-1.2 API Publications. American Petroleum Institute, 2101 L St., NW, Washington, DC 20037.

API 620, *Design and Construction of Large, Welded, Low-Pressure Storage Tanks*, 1990.

API 1632, *Cathodic Protection of Underground Petroleum Storage Tanks and Piping Systems*, 1983.

API 2510, *Design and Construction of LP-Gas Installations*, 1989.

API-ASME *Code for Unfired Pressure Vessels for Petroleum Liquids and Gases.*

J-1.3 ASCE Publication. American Society of Civil Engineers, United Engineering Center, 345 East 47th St., New York, NY 10017.

ASCE 56, *Sub-Surface Investigation for Design and Construction of Foundation for Buildings.*

J-1.4 ASME Publication. American Society for Mechanical Engineers, 345 East 47th St., New York, NY 10017.

ASME *Boiler and Pressure Vessel Code*, 1992.

J-1.5 ASTM Publication. American Society for Testing and Materials, 1916 Race St., Philadelphia, PA 19103.

ASTM D1835-1991, *Specification for Liquefied Petroleum (LP) Gases.*

J-1.6 CGA Publications. Compressed Gas Association, Inc., 1235 Jefferson Davis Highway, Arlington, VA 22202.

CGA Pamphlet C-6-1984, *Standards for Visual Inspection of Steel Compressed Gas Cylinders.*

Pressure-Relief Device Standards:

S-1.1-1994, *Cylinders for Compressed Gases* (Errata, 1982).

S-1.2-1980, *Cargo and Portable Tanks for Compressed Gases.*

S-1.3-1980, *Compressed Gas Storage Containers.*

J-1.7 Federal Publications. National Technical Information Service, U.S. Dept. of Commerce, Springfield, VA 22161.

BERC/RI-77-1, September 1977, *New Look at Odorization Levels for Propane Gas*, United States Energy Research and Development Administration, Technical Information Center.

Code of Federal Regulations, Title 46, Part 33.

Code of Federal Regulations, Title 49.

Code of Federal Regulations, Title 49, Part 179.105-4, "Thermal Protection."

Code of Federal Regulations, Title 49, Part 195, "Transportation of Hazardous Liquids by Pipeline."

J-1.8 GPA Publications. Gas Processors Association, 1812 First National Bank Bldg., Tulsa, OK 74103.

Standard 2140-1992, *Liquefied Petroleum Gas Specifications for Test Methods.*

Standard 2188-1989, *Tentative Method for the Determination of Ethyl Mercaptan in LP-Gas Using Length of Stain Tubes.*

J-1.9 NACE Publications. National Association of Corrosion Engineers, 1440 South Creek Drive, Houston, TX 77084.

RP0169-1992, *Control of External Corrosion on Underground or Submerged Metallic Piping Systems* (Rev. 1983).

RP0285-1985, *Control of External Corrosion on Metallic Buried, Partially Buried or Submerged Liquid Storage Systems.*

J-1.10 NPGA Publications. National Propane Gas Association, 1600 Eisenhower Lane, Lisle, IL 60532.

NPGA Safety Bulletin 122-1992, *Recommendations for Prevention of Ammonia Contamination of LP-Gas.*

NPGA Safety Bulletin 118-1979, *Recommended Procedures for Visual Inspection and Requalification of DOT (ICC) Cylinders in LP-Gas Service.*

J-1.11 UL Publication. Underwriters Laboratories Inc., 333 Pfingsten Rd., Northbrook, IL 60062.

UL 1746-93, *External Corrosion Protection Systems for Steel Underground Storage Tanks.*

J-1.12 ULC Publication. Underwriters Laboratories of Canada, 7 Crouse Road, Scarborough, Ontario MIR 3A9.

ULC-S603.1-M-1982, *Standard for Galvanic Corrosion Protection Systems for Steel Underground Tanks for Flammable and Combustible Liquids.*

J-1.13 CAN/CSGB Publication. Canadian General Standards Board, 222 Queen St., Suite 1402, Ottawa, Ontario K1A 1G6.

CAN/CGSB-3.0 No. 18.5-M89, November 1989, *Test for Ethyl Mercaptan Odorant in Propane, Field Method.*

Supplement

The following supplement is included in this handbook to provide additional information for LP-Gas users. Since it is not part of the standard, it is printed in black.

Guidelines for Conducting a Firesafety Analysis

This supplement is not a part of the code text of this handbook, but is included for information purposes only.

The following material was prepared as a Safety Bulletin by the Safety Committee of the National Propane Gas Association to assist in conducting the firesafety analysis that is required in the section on fire protection (3-10) of NFPA 58. The section on fire protection was first introduced in the 1976 edition of NFPA 58. There was confusion in the minds of both enforcement officials and the LP-Gas industry on how to conduct the safety analysis. The bulletin provides the information needed to conduct the analysis in a clear, usable form.

Special thanks are extended to the National Propane Gas Association for granting permission to reprint this material.

Introduction

NFPA 58 includes provisions in paragraph 3-10.2.3 for fire protection to augment the leak control and ignition source control provisions of the standard. A firesafety analysis may be required for installations having storage containers with an aggregate water capacity of more than 4000 gallons subject to exposure from a single fire. These guidelines are presented to aid you in conducting such a firesafety analysis.

In spite of efforts to eliminate hazards within an LP-Gas facility, prudent management must consider that accidents may occur and take reasonable steps to control them. The safety of employees and customers within the facility, its neighbors, and emergency personnel are of paramount concern. Conservation of the physical plant is also important—especially as it relates to the safety of neighbors and protection of their property.

Hazard Characterization

LP-Gas is stored, transported and handled in closed containers and piping until ultimately used in consuming equipment—usually by burning. An emergency is created whenever LP-Gas is released in other than the intended manner.

LP-Gas is present in containers and equipment in both liquid and gaseous (vapor) form. It can escape in either form. If it escapes in liquid form, it will vaporize rapidly to the gaseous form.

When escaping gas mixes with air, a flammable mixture will form. If ignited, it will result in a fire. It could also result in an explosion—depending upon how large the flammable mixture is or whether the mixture is confined inside a structure.

If fire is impinging on a container, the possibility of an explosion (a Boiling Liquid-Expanding Vapor Explosion, or BLEVE) exists. A BLEVE will produce a fireball and shock wave, the degree depending upon how much liquid LP-Gas is inside the container when it ruptures. It may also produce flying fragments in the form of container pieces.

Because ignition sources are controlled to a substantial degree in LP-Gas facilities, there is often a time interval between LP-Gas escape and ignition. There is also a time lag between flame contact with the unwetted portion of a container and container failure. This time can be used to advantage in controlling the emergency given adequate resources, planning, and training.

Emergency Control Objectives

The emergency control objectives consist of three operational phases:

1. Stopping or slowing down the rate of LP-Gas release.
2. Dissipating LP-Gas vapors and/or presenting flammable gas-air mixtures from reaching ignition sources and entering structures.
3. Keeping fire exposed containers and equipment cool.

Attaining these objectives requires the application of both equipment arrangement and human performance.

The Firesafety Analysis

An adequate degree of emergency control is subject to widely varying circumstances specific to each facility. Because of the large number of variables each cannot be specifically addressed in NFPA 58. Furthermore, while it is possible that a facility could be made entirely self-sufficient in fire protection, it is seldom necessary or cost-efficient to do so. These factors mean that judgement must be brought to bear and agreements reached with local authorities. While parties involved share a common goal, there may be differences in views as to how best to achieve the goal and these views must be resolved. These guidelines are offered to aid in such decision-making.

A. No Fire Protection Needed

There may be situations where no fire protection is needed beyond those provided in the facility in compliance with specific provisions in NFPA 58. A candidate for such a decision would be a well isolated facility.

If an adequate water supply is not available, agreement should be reached with public safety authorities to limit their emergency forces to control of onlookers.

B. Fire Protection Needed

Unless the circumstances described in item A exist, it must be recognized that a degree of exposure exists to the facility, its employees, the persons and property of neighbors, and that fire protection is needed beyond that specifically required by NFPA 58. This fire protection could be provided by a plant fire brigade and/or fire department.

Such a facility will undoubtedly be within an area served by one or more emergency services—e.g., public fire and police departments and ambulance services. Facility management has a right to expect assistance from these sources. Management has a duty to obtain this assistance without requiring the emergency personnel to accept undue risks. The resolution of this risk factor is a key element in the firesafety analysis process.

If we study the previously cited Emergency Control Objectives, it will be evident that a fire department is suited to accomplish Objectives 2 and 3. Water in the proper form can disperse gas clouds and cool containers and equipment. Aside from its rescue function, a fire department is essentially a man-machine system designed to apply water.

These are the reasons why NFPA 58 states that the first consideration in the analysis should be directed at the fire department's capabilities.

C. Fire Department Capability

The most important step in evaluating the fire department's adequacy is to recognize that the facility management and fire department officers must communicate with each other at all stages of the analysis. Neither should make any assumptions about the other's capability. Often the initial tendency of the facility management is to overestimate the fire department's capability. Conversely, often the initial tendency of the fire department is to underestimate its own capability.

All too often, the basic reason why a fire department will underestimate its capability is due to lack of knowledge. Most are accustomed to fighting fires in buildings and Class A and B combustibles in which the tactic consists of "attack-and-extinguish." The "control-and-not-extinguish" tactic for LP-Gas (or any other flammable gas) incidents are alien to their basic training. Furthermore, LP-Gas incidents are infrequent and fire departments don't get much on-the-job training. When the need for fast action and the potential severity of an incident are added to these factors their position is understandable (accounts of severe incidents are readily available—accounts of those readily handled are not).

The answer to this problem is adequate education and training. It will be necessary during this training to consider the particular character of the facility and the general knowledge of the fire department.

In those instances where the general knowledge is not present, the facility management should do all that it can to encourage the department to obtain it. If feasible, key fire department personnel should be directed to attend a school that offers "hands-on" field exercises in its curriculums. This may require financial assistance. Facility management should also be prepared to provide instructional material which includes visual aids. Good detailed accounts of both successfully handled, and not so successfully handled, incidents involving facilities similar to the one being analyzed have proven to be useful tools.

In any event, it must be recognized that it will be difficult, if not impossible, to reach a sound agreement on the fire department's capability to handle an incident at a specific facility until this education and training has been achieved.

D. Analysis Factors

Once a basic appreciation for the problems has been obtained, the analysis can proceed on a sound basis. As noted in NFPA 58, the analysis should include the following factors:

1. Local conditions of hazards within the container site.
2. Exposure to and from other properties.
3. The available water supply.

4. The probable effectiveness of plant fire brigades.
5. The time of response and probable effectiveness of the fire department(s).

Aspects relating to fire department capability include the following (remember, we are considering a fire department as a man-machine water distribution system in the analysis):

1. Local Conditions of Hazards Within the Facility

Experience has shown that by far the most frequent cause of accidents leading to leaks and fires in an LP-Gas facility are associated with liquid transfer operations. The character and arrangement of these are important factors. The number of simultaneous operations involving small containers, transports, delivery vehicles, and tank cars should be identified and located with respect to each other and to the storage containers and facility structures. The fewer such operations and the greater the distance between them, the greater the fire department's chances for successful containment.

Other features pertinent to access by fire apparatus and personnel—e.g., fence gates, roadways—should be noted. The position of the long axis of storage containers and cargo vehicles is important in this context because apparatus and personnel should be positioned to the sides of such containers as much as possible.

The presence of any overhead power lines should be noted.

If tank cars are involved, it is important to note that they are all insulated and do not present immediate BLEVE hazards.

Features for leak shutdown should be noted. Consider the (automatic) leak control provisions of the recently available Emergency Shut-Off Valves (ESVs).

2. Exposure to and from Other Properties

If the facility is located in what NFPA 58 refers to as a "heavily populated or congested area," the fire department will tend to place less reliance on their capability. This is so, even though their tactics would remain the same. The consequences of an unsuccessful operation could be more serious.

Facility management should recognize that vulnerable locations for property damage and/or loss of life are prime candidates for Special Protection (*see definition in the standard*). It should be recognized that pieces of large storage and cargo containers that BLEVE, may travel 1000-2000 feet and occasionally 2500-3000 feet.

3. Available Water Supply

One of the major points of consideration in this analysis will undoubtedly be the water supply. The time required to make it available, quantity, and reliability are the essential factors.

The first arriving fire company may find either unignited escaping gas or burning LP-Gas. If ignition has not occurred, its job is to try to prevent ignition and the migration of gas into an enclosed structure. A water fog is generally used to dissipate LP-Gas vapors. While the quantity of water needed will vary greatly with the size of the leak and its location, a leak that can be controlled with manpower available at this stage cannot be very large. If a large leak has occurred, or is in progress, the gas may have spread to outside the facility by the time the fire department arrives and could be beyond currently available control measures. Water would have to be supplied from a source not in the vicinity of the facility. Fortunately, experience shows that the likelihood of a leak large enough to escape beyond the facility boundaries is of a lesser degree at facilities complying with the leak control provisions in NFPA 58. Ignited escaping gas that is feasible to control does not pose a severe hazard to the facility, its neighbors, or the emergency personnel, unless the fire impinges on containers which do not have Special Protection.

Obviously, LP-Gas facilities may have conventional fires in offices, shops, garages, and warehouses. Because these present lesser control problems, a water supply designed for the hazards in this analysis will more than suffice for the ordinary hazards.

If a fire is impinging on containers not provided with Special Protection, a most severe hazard is presented. The larger the container, the more severe the BLEVE hazard. Therefore, the water supply needed is determined by the magnitude of the BLEVE hazard.

The quantity of water needed for adequate container cooling can be determined by the container area directly contacted by flames. Considerable test work indicates that around 0.20 to 0.25 gpm per sq ft of container area directly contacted by flames is needed. Based upon reports of fire exposing 30,000 gallon containers, exposure to about ⅓ of the area (700 sq ft) is not unusual. On this basis, the water needed would be about 140 to 175 gpm per each container exposed to a single fire.

A more important consideration exists in the initial stages of fire department control activity. Experience reveals rather conclusively that a container is in danger of a BLEVE after about 10 minutes of intense flame impingement on the unwetted portion of the shell. This time span will often coincide rather closely with the time it takes the fire department to get streams into operation. The instant the initial stream of water contacts the container is usually the most dangerous moment in the entire operation. Therefore, those manning the nozzle must do so from a reasonably safe position. Unless protective structures (i.e., buildings, unexposed storage tanks, tank cars, cargo vehicles, strongwalls, etc.) are available, then maintaining suitable distance becomes most important. Firefighters should be able to position a nozzle at least 50 ft away, preferably more.

The hydraulics of fire streams are such that a discharge of 250-500 gpm is needed to carry an effective stream 50-100 ft. Therefore, strictly from a cooling standpoint, this consideration will result in roughly twice the quantity needed.

More than 1000 gallons cannot be carried to the facility by the first arriving company. This 1000 gallons of water is only 2-4 minutes of operation and usually there is little prospect of more arriving before the first unit runs out. Therefore, it is necessary for a water supply to be located at or very near the facility. This can be obtained from a body of water or a reliable public or private water system and used during the time necessary to permit the transportation of water to the site.

Fire departments have demonstrated their ability to transport very large quantities of water to LP-Gas facilities. However, it takes time to set up such a system. As a minimum, a one-half to one hour supply should be available at or very near the facility.

In the most severe exposure locations, of course, it is likely that a public water system is in place. This is also the case in the larger industrial locations—often from a private water system. In such circumstances, the problem is one of being able to provide a suitable rate of flow.

Note any features present that relate to water supply in your analysis. The fire department and water department are in a much better position than facility management to evaluate this factor.

4. Plant Fire Brigade

This term is used in NFPA 58 in the context of what any plant employee can do to accomplish the Emergency Control Objectives. In any facility in which the storage and distribution of LP-Gas is the primary purpose, the employee's role reflects primarily Objective 1 and 2. Usually the number of employees in these facilities is small and totally inadequate to accomplish Objective 3. This must be accomplished through use of fire department equipment and techniques. (In this respect, the facility management should avoid providing connected hoses, nozzles, etc. at the facility. The fire department may not trust their condition and probably will not use them.)

With respect to Objective 1 (stopping or slowing down the rate of LP-Gas release), facility employees are in a far better position to accomplish this than the fire department or anyone else. This should be their primary activity.

With respect to Objective 2 (preventing flammable gas-air mixtures from reaching ignition sources and entering structures), employees can shut down equipment which constitute ignition sources, close doors and windows, etc.

Every facility should have a written emergency plan which includes activities to accomplish Objectives 1 and 2. An evaluation of this emergency plan should be part of the fire analysis. (*See NPGA Bulletin "Guidelines for Developing Plant Emergency Procedures."*)

If Special Protection of a water system type is present, facility employees could be involved with Objective 3 (keeping fire exposed containers and equipment cool).

Some large industrial facilities do maintain highly organized fire brigades large enough to address Objective 3. They are essentially fire departments.

5. Time of Response and Probable Effectiveness of Fire Department

Several aspects of fire department effectiveness have been addressed in the preceding four factors. There are others that must also be considered in the analysis.

Response time is very important and the shorter the response, the better. The importance of 10 minutes has been discussed. If this is exceeded, this alone can result in a decision that the fire department cannot effectively handle the emergency.

Response time from the receipt of an alarm to the first control activity can be predetermined by the fire department. A key component, which cannot be determined by the fire department alone, is the time between start of the emergency and receipt of the alarm. Delayed alarms are a major problem and a common occurrence in the experience of all fire departments. They will carefully evaluate this with respect to the facility.

LP-Gas facility management should emphasize to their employees the importance of prompt notification of the fire department. The written emergency plan should require prompt notification of any fire except the most minor unignited leak situations. Late notification could result in the fire department not effectively handling the emergency.

It must also be recognized that delayed alarms can occur in spite of the best efforts of facility employees. As has been noted earlier, distribution facilities do not have many employees. A leak may occur where no one is present or an employee who is in the vicinity may be immobilized as a consequence of the leak. Also, facilities do not usually operate 24 hours a day; however, accidents rarely occur during non-operating periods.

The telephone may not be a reliable alarm communicator. Telephone lines may not operate because of overload, physical damage, or they may be slow. In some cases, radios or special alarm circuits may be needed. The exact alarm procedure should be written in the emergency plan.

The nature of the fire department equipment and the manning of it is very important, especially for the first alarm response. Except in heavily populated areas, one pumper and 2 or 3 men is all that can be expected initially. Considering the 250-500 gpm needed and the 50-100 ft range discussed earlier (*see "Available Water Supply"*) it could be very difficult for the first crew to accomplish much using only the equipment they have hauled to the scene.

Conclusion

It must be concluded that considerable study is needed to evaluate the capability of a fire department. In many instances the capability may be marginal at best. Nevertheless, the fire department remains the most flexible man-machine system possible to accomplish Objective 3 of the Emergency Control Objectives.

If the analysis initially reveals fire department control inadequacies, the next most frequent step is for the fire officer to recommend Special Protection. NFPA 58 is used to support such a recommendation. However, NFPA 58 stipulates that two conditions must be met—namely, that a serious hazard must exist and that the fire department is incapable—before Special Protection is required.

Even if the hazard can be shown to be serious (here an analysis of "Exposure to and from Other Properties" is a key), the next step should be to see if measures can be taken to upgrade the fire department.

Measures can include: provision of one or two fixed monitors arranged for the fire department to hook up a pumper to (as a substitute for having to get hose streams into service[1]); upgrading of the public water supply in the vicinity (often this is long overdue anyway as a result of construction of other facilities in the area); and special alarm systems to assure prompt response.

Often, partial Special Protection can change the picture drastically. For example, if a storage container is insulated and the fire department must only address a cargo vehicle, the initial response problem is greatly simplified. This can also be accomplished by providing more space between storage containers and cargo vehicles.

The most important single step is for the facility management and the fire department to work closely together in an atmosphere of cooperation. While vital to a sound analysis, it should be an ongoing practice. The personnel in both organizations are subject to frequent change and this relationship must be sustained for the good of the facility, its neighbors, the fire department, and the community at large.

Brief Outline for Conducting a Firesafety Analysis

The standard requires fire protection for an installation having storage containers with an aggregate water capacity of more than 4,000 gallons subject to exposure from a single fire, unless a firesafety analysis indicates a serious hazard does not exist. It also provides for alternate ways of protecting the installation.

[1]This could also reduce water demand by reducing the stream range needed.

I. A firesafety analysis considers the following:

1. Local conditions of hazards within the container site.
2. Exposure to and from other properties.
3. The available water supply.
4. The probable effectiveness of plant fire brigades.
5. The time of response and probable effectiveness of the fire department.

II. Prepare a plot plan to scale showing and locating:

1. The property lines.
2. Buildings, storage containers, loading and unloading sites, cylinder filling and storage areas, etc.
3. Fence, gates, rail sidings, vehicle parking areas and spacing, and roadways.
4. Buildings and other exposures on adjacent properties.
5. Fire hydrants or nearest water supply.
6. Existing protection required by building or other codes.

III. Determine water supply available, the flow quantity, the location and accessibility.

IV. Determine the effectiveness of the plant fire brigade or plant emergency procedures.

1. Emergency shut-off valves at loading and unloading locations.
2. Emergency shut-off valves for cylinder filling operations.
3. Product venting procedures and facilities.
4. Reliability and speed of communications within the plant area and with Public Emergency Services (fire, police, and medical, etc.)
5. Shut-offs for electric, gas pilots, and other ignition sources.

V. Determine the time of response and probable effectiveness of the local fire department.

VI. Take the information you have gathered, analyze it, and make a tentative judgement of your own. Then take it to the fire department and initiate dialogue.

Index

BUSINESS REPLY MAIL

FIRST CLASS PERMIT NO. 3376 BOSTON, MA

POSTAGE WILL BE PAID BY ADDRESSEE

NATIONAL FIRE PROTECTION ASSOCIATION
1 BATTERYMARCH PARK
PO BOX 9101
QUINCY MA 02269-9904

NO POSTAGE
NECESSARY
IF MAILED
IN THE
UNITED STATES

BUSINESS REPLY MAIL

FIRST CLASS PERMIT NO. 3376 BOSTON, MA

POSTAGE WILL BE PAID BY ADDRESSEE

NATIONAL FIRE PROTECTION ASSOCIATION
1 BATTERYMARCH PARK
PO BOX 9101
QUINCY MA 02269-9904